BK 628.53 B837I
INDUSTRIAL SOURCE SAMPLING
1973 18.00 FV
3000 329811 30019
St. Louis Community College
W9-COT-903

INDUSTRIAL SOURCE SAMPLING

INDUSTRIAL SOURCE SAMPLING

David L. Brenchley

Assistant Professor of Environmental Engineering, Purdue University, West Lafayette, Indiana

C. David Turley

Engineer for the Indiana State Board of Health, Indianapolis, Indiana

Raymond F. Yarmac

Consulting Engineer in Greenfield, Massachusetts

Second Printing, 1974

P.O. Box 1425, Ann Arbor, Michigan 48106

Library of Congress Catalog Card No. 72-96908
ISBN 0-250-40012-X

Manufactured in the United States of America

PREFACE

The control of *air pollution* has advanced to the stage where laws now govern the emissions from specific industrial sources. The enforcement of these *laws* often requires that field source tests be made. Recently the federal *Environmental Protection Agency* (EPA) has prescribed specific procedures; some local and state control agencies are also specifying the use of these procedures. This book covers the theory and practice of industrial source sampling. It focuses on the EPA methods.

Industrial Source Sampling is intended for use by industrial managers and engineers. It may also find use in the college classroom. *Chapters 1 through 5* provide a background and explain the complexities of source sampling. This will be of use to the administrator who must decide whether to hire a consultant or use his own personnel to conduct such tests. This information will be valuable to the engineer who must plan and perform source tests. *Chapters 6 through 8* cover the specific measurements to be made and identify the errors which can occur. These considerations are essential to improving the precision and accuracy of source tests. *Chapter 9* is a detailed consideration of "How to prepare for a source test." This compilation of practical experience can save you many headaches and dollars. *Chapters 10 through 14* present the step-by-step EPA methods. These include hints and comments on the selection and use of the materials and reagents. *Chapter 15* discusses the computations which must be made for the source tests. The example calculations are most helpful for preparing final source testing reports. *Chapters 16 and 17* present the specifications for commercially available source testing and source monitoring equipment. This information is essential for the selection and purchase of equipment. *Chapter 18* covers some of the more advanced monitoring methods now available. These methods are new to the air pollution measurements field and thus a detailed coverage of theory and operation is also given; some of these methods will find great use in the future. *Chapter 19* consists of some of the authors' constructive criticism and suggestions regarding the future for industrial source sampling. These comments are made with the thought of protecting both public and industrial interests.

D. L. Brenchley
C. D. Turley
R. F. Yarmac

TABLE OF CONTENTS

LIST OF TABLES. xiii
SYMBOLS xvii

CHAPTER 1. INTRODUCTION TO SOURCE SAMPLING . 1
Purpose of Sampling 1
The Problems Involved 3
Logistics 7

CHAPTER 2. SOURCE TESTING PROCEDURES 9
General Procedure 9
Measured Parameters 11
Sampling Strategy 11
Types of Sampling Procedures 18
Gaseous pollutants 18
Particulate matter 20

CHAPTER 3. INDUSTRIAL PROCESS INFORMATION . 27
Mass and Energy Balances 27
Emission Factors 29
Priority Sources 33
Fossil-fuel steam generators 34
Incinerators 45
Portland cement production 49
Nitric acid production 53
Sulfuric acid production 57

CHAPTER 4. SAMPLING TRAIN COMPONENTS 65
Temperature Measurement 65
Pressure Measurement 70
Velocity Measurement 73
Nozzles 79
Probes 81
Sample Collectors 82
Sample Flow Rate 84
Total Sample Volume 88
Gas Conditioning 93
Gas Movers 95
Flow Control 99

CHAPTER 5. CONDUCTING A SOURCE TEST 103
Identify Problem 103
Entry and Cooperation 106
Design Experiment 107
Pre-Test Survey 109
Final Preparation 112
Sampling and Analysis 114
The Report 115

CHAPTER 6. THE CALCULATION OF MOLECULAR WEIGHT 120
Ideal Gas Law 120
Dalton's Law 120
Computation Method 121
Nomographs 123

CHAPTER 7. CARRIER GAS MEASUREMENTS 127
Stack Gas Volume Flow Rate 127
Molecular Weight Determination 132
Sampling Trains for Carrier Gas Constituents 136
Moisture Content Determination 138
Sampling Train for Moisture 141

CHAPTER 8. ERRORS IN SOURCE TESTING 145
Types of Errors 146
Systematic errors 147
Random errors 150
Error Propagation 156
Random Errors in Source Sampling 157
Systematic Errors in Source Sampling . . . 172
Mistakes in Source Sampling 178

CHAPTER 9. PREPARATION FOR SOURCE SAMPLING . 183
Pre-Test Plant Survey 183
Preparation of Test Plan 190
Equipment Preparation 194

CHAPTER 10. DETERMINATION OF PARTICULATE MATTER EMISSIONS 199
Basic Method 200
Sampling Train 200
Maintenance and Calibration 207
Test Procedures 208
Analytical Procedures 215
Results 216

CHAPTER 11. DETERMINATION OF SULFUR DIOXIDE EMISSIONS 225
Basic Method 225
Sampling Train 225
Test Apparatus 227
Laboratory Procedures 230
Field Procedures 230
Analytical Procedures 232
Calculations 233

CHAPTER 12. DETERMINATION OF SULFURIC ACID MIST EMISSIONS 235
Basic Method 235
Sampling Train 235
Test Apparatus 236
Laboratory Procedures 238
Field Procedures 238
Analytical Procedures 240
Calculations 240

CHAPTER 13. DETERMINATION OF NITROGEN OXIDE EMISSIONS 243
Basic Method 243
Sampling Train 243
Test Apparatus 243
Laboratory Procedures 245
Field Procedures 245
Analytical Procedures 246
Calculations 249

CHAPTER 14. PROPERTIES OF AEROSOLS 251
Plume Opacity 251
Visual effects 253
Plume evaluation 255
Particle Sizing 262
Particle size statistics 262
Sample collection errors 264
Size analysis methods 265

CHAPTER 15. COMPUTATIONAL METHODS 273
Computational Techniques 273
Sample Calculations 278
Particulate emissions calculations . . 285
Sulfur dioxide emissions 292
Nitrogen oxide emissions 294
Sulfur trioxide and H_2SO_4 mist emissions 297
Isokinetic Sampling Field Calculations . . 298

CHAPTER 16. COMMERCIAL SOURCE TESTING EQUIPMENT 307
Aerosol Sampling 310
Method 5 trains 310
Other particulate trains 330
Gas Sampling 337
Method 6 trains 337
Method 7 trains 344

CHAPTER 17. COMMERCIAL CONTINUOUS MONITORING EQUIPMENT 347
Basic Monitoring Methods 348
Monitoring gases 348
Monitoring aerosols 357
Instrument Characteristics 360
Commercial Equipment 365
Calibration Methods 387
Selecting a System 388

CHAPTER 18. ADVANCED METHODS FOR SOURCE TESTING AND MONITORING 393
Beta Attenuation 394
Principle of operation 394
Application 396
Equipment 396
Piezoelectric Microbalance 400
Principle of operation 400
Thin film theory 401
Material requirements 403
Components 404
Crystal characteristics 404
Particle collection 404
Particle adhesion 405
Electronics system 407
Operating conditions 407
Aerosol concentration 408
Particle sizing 408
Calibration 408
Equipment 408
Applications 413
Chemiluminescence 416
Principle of operation 416
Ozone 417
Oxides of nitrogen 417
Carbon monoxide 420
Sulfur oxides 420
Hydrogen sulfide 420
Applications 421

Electrochemical Transducers 422
Principle of operation 422
Equipment 424
Flame Photometry 432
Principle of operation 432
Equipment 432
Applications 436

CHAPTER 19. AND LET THERE BE LIGHT 441
Present EPA Test Methods 441
So You Have a Source That Needs Testing? . 445
Your Day In Court 447

APPENDIX A. EPA STANDARDS OF PERFORMANCE FOR NEW STATIONARY SOURCES 449

APPENDIX B. SOURCE TEST GUIDELINES AND PROCEDURES: STATE OF CONNECTICUT . . . 473

SUBJECT INDEX 483

LIST OF TABLES

Table

2.1. Categorizing Sources According to Their Process Characteristics 12
2.2. Combinations of Source Conditions When Sampling for Gaseous Constituents . . 16
3.1. EPA Testing Methods for New Sources . . 28
3.2. Federal Standards for New Priority I Sources 31
3.3. Emission Factors for Bituminous Coal Combustion Without Control Equipment. 39
3.4. Emission Factors for Fuel Oil Combustion 40
3.5. Emission Factors for Natural-Gas Combustion 41
3.6. Emission Factors for Refuse Incinerators Without Controls 48
3.7. Particulate Emission Factors for Cement Manufacturing 52
3.8. Emission Factors for Nitric Acid Plants Without Control Equipment 56
3.9. Emission Factors for Sulfuric Acid Plants 60
5.1. Suggested Report Format 116
7.1. Velocity Traverse Data Sheet 131
7.2. Field Moisture Determination Data Sheet 143
8.1. Source Sampling Testing Results for Teams A and D (Example 8.2) 155
8.2. Estimated Error Terms Associated with Proportional Sampling Procedures . . 163
8.3. Operational Values for Source Test for Particulate Matter (Example 8.3) . . 165
8.4. Maximum Relative Errors for Isokinetic Source Sampling Operations 167
8.5. Estimated Error Terms Associated with Grab Sampling Procedure 170

Table

8.6. Operational Values for Source Test Oxides of Nitrogen (Example 8.4) . . 171
8.7. Common Mistakes in Source Sampling . . 179
9.1. Pre-test Survey Field Report 184
9.2. Location of Traverse Points in Circular Stacks 188
9.3. Test Plan Outline 191
9.4. Tool and Supplies Checklist 195
10.1. Routine Maintenance and Calibration of the Method 5 Particulate Train . . . 209
10.2. Particulate Test Data Sheet 212
11.1. Equipment and Chemical Check List Gas Sampling Tests 228
11.2. SO_2 Test Data Sheet 231
12.1. Field Data Sheet for Sulfuric Acid Mist Test 239
13.1. Data Sheet for EPA Method #7, NO_x Sampling 247
14.1. Field Data Sheet for Plume Opacity Tests 256
14.2. Opacity Requirements for New Priority I Sources 257
14.3. Smoke School Training Form 260
15.1. The Uncertainty of Source Sampling Instrument Measurements 274
15.2. Relative Uncertainty Allowed in Applying the Federal Performance Standards for Stationary Sources . . 277
15.3. Boiler Specification Sheet 279
15.4. Boiler Operation 280
15.5. Fuel Oil Analysis - Test #1 280
17.1. Continuous Monitoring Requirements for New or Modified Priority I Sources . 347
17.2. Status of Instrument Development for Continuous Monitoring of Gases . . . 349
17.3. Status of Instrument Development for Continuous Monitoring of Aerosols . . 349
17.4. Desirable Attributes for Ideal Source Monitoring Method 362
17.5. Process Information Required for Selecting Source Monitor 389
17.6. Check List for Rating Various Monitoring Systems 390
18.1. The Advantages and Disadvantages of Beta Gauging Method 397
18.2. The Advantages and Disadvantages of Piezoelectric Method 414
18.3. The Advantages and Disadvantages of Chemiluminescent Methods 421

Table

18.4. The Advantages and Disadvantages of the Electrochemical Transducer Method . . . 425

18.5. The Advantages and Disadvantages of Flame Photometric Methods 435

SYMBOLS USED IN THIS WORK

Symbol	Definition
A_s, A	cross-sectional area of stack
A,B,C	measurements
A_m	rotameter anulus area around float
A_n	internal cross-sectional area of sampling nozzle
b	light extinction coefficient
b_i	light extinction coefficient components
B_a	luminance caused by scattered light in a plume
B_p	luminance of the plume
B_b	luminance of the background
B_i	mole fraction of each component of a gas mixture
B_{H_2O}, B_{wo}	proportion by volume of water vapor in gas stream
(B)avg	average fuel flow rate
B_{wm}	mole fraction of moisture in gas at meter conditions
B_{ws}	mole fraction of moisture in gas at standard conditions
C_s	particulate concentration in stack gas (lb/dscf)
C_s'	particulate concentration in stack gas (gr/dscf)
$\overline{C}_s$	average particulate concentration in stack gas
C_m	measured concentration (Chapter 12)
C_m	meter coefficient (remaining chapters)
C_t	true concentration
C_i	concentration of specific pollutant
C_p, C_{pi}	pitot tube coefficient
CP	luminance contrast

Symbol	Definition
D_n, D	sampling nozzle diameter
emf	electromotive force
e_m	water vapor pressure for saturated condition at meter temperature
e_R	absolute error of calculated result
e_i	absolute errors associated with individual measurements
e_o	absolute error (bias) at observation conditions
E	normalized emission
%EA	percent excess combustion air
f( )	function of
F	frictional loss
g	local gravitation constant
g_c	universal gravitation constant
H_i	hypotheses
Δh	heating value of the coal
$\dot{H}$	heat input rate
I_{avg}	percent isokinetic variation
I_{pt}	percent isokinetic variation at traverse point
I	intensity of transmitted light
I_o	initial light intensity
K_m	meter proportionality factor
K_p, K	85.48 (ft/sec)(lb/lb-mole-OR)$^{1/2}$
L	light path length
m_f	mass of rotameter float
M_m	meter gas molecular weight
Δm	orifice meter pressure drop
M_s	molecular weight of stack gas
m	mass of gas (chapter 6)
M, M_{mix}	molecular weight of the gas
M_i	molecular weight of each component of mixture
M_{s_1}	molecular weight of sampled gas (ideal) at point 1
$M_{dry\ gas}$, M_d	dry gas molecular weight

Symbol	Definition
m_{wc}	mass of water collected in condensor
M_w, M_{H_2O}	molecular weight of water
m_{NO_x}, m, M	total mass of NO_x collected (Chapter 13)
m_s	mass of NO_x in sample measured by absorbance
m_a	mass of NO_x in blank measured by absorbance
m, m_n, M_n	mass of pollutant collected (Chapter 8)
m_c	coal usage rate
Δm_{pt}	orifice meter pressure drop at traverse point
Δm_{avg}	average orifice meter pressure drop
N	total number of observations (Chapter 8)
N	normality of barium perchlorate titrant (Chapter 11)
OR	overall rating of monitoring system
%OP	percent opacity
$\overline{PMR_s}$, $\overline{PMR_{s_c}}$, $\overline{PMR_c}$	average pollutant mass emission rate from a stack
P_{abs}	absolute pressure
P_{atm}, P_b, P_{bar}	atmospheric pressure
P_m	measured pressure (Chapter 4)
P_i	specific pressure
ΔP	velocity pressure in gas stream
P_s	absolute stack gas pressure
P_m	absolute meter gas pressure
P	absolute gas pressure
P_{mix}	absolute gas pressure of mixture
P_i	partial pressure of each component (Chapter 6)
$(\Delta P)^{1/2}_{avg}$, $(\sqrt{\Delta P})_{avg}$	average square root of velocity pressure
P( )	probability
P_{std}	29.92 inches Hg
P_i	initial absolute pressure of flask (Chapters 8, 13)
P_f	final absolute pressure of flask
$\overline{PMR_{Btu}}$	pollutant emission rate in terms of heat input

Symbol	Definition
$\overline{Q}_s$, Q_s	average volumetric stack gas flow rate
Q_m	meter flow rate
R	universal gas constant
R	calculated result (dependent variable) (Chapter 8)
r_i	rating assigned to factor i
S	sample standard deviation
%S	percent by weight sulfur in fossil fuel
T	absolute gas temperature
T_i	specific temperatures
T_s	absolute stack gas temperature
T_m, T_{mpt}	absolute meter gas temperature
t_d	average dry gas meter temperature
t_w	average wet test meter temperature
t_{do}	dry gas meter outlet temperature
T_{mix}	absolute temperature of gas mixture
$(T_s)_{avg}$, $T_{s_{avg}}$	average absolute stack gas temperature
t	students distribution
T_{std}	530°R
T_f	final absolute temperature of flask
T_i	initial absolute temperature of flask (Chapters 8, 13)
TR	fraction of background luminance transmitted through plume
%TR	percent light transmittance
$\overline{V}_s$, $(V_s)_{avg}$	average stack gas velocity
V_n	sampling nozzle velocity
V_s, $(V_s)_{pt}$	local stack gas velocity
V_w	gas volume passing through wet test meter
V_d	gas volume passing through dry gas meter
V	volume of gas
V_{mix}	volume of gas of mixture in a container
V_i	volume of gas of each component of mixture
V_{wo}	volume of extracted water from source at meter conditions

Symbol	Definition
V_{wc}	volume of water condensed referred to meter conditions
V_{wm}	volume of water passed meter at meter conditions
V_m, V	total volume of gas through meter at meter conditions (Chapter 8)
Var	sample variance
V_{sc}, $V_{m_{std}}$	total sample volume at standard conditions
V_f	flask volume
V_a	absorbing solution volume (Chapters 8, 13)
$V_{w_{std}}$	volume of water condensed referred to standard conditions
V_{l_c}	total volume of liquid water collected
$(V_n)_{pt}$	sample nozzle at traverse point
ΔV	sample volume at traverse point
V_t	volume of barium perchlorate titrant for sample
V_{tb}	volume of barium perchlorate titrant for blank
V_{soln}	total volume of SO_2 sample
V_a	volume of sample aliquot titrated (Chapters 11, 12)
$(V_n)_{avg}$	average sample nozzle velocity
w_i	filter weights
w_i	weighting factor assigned to factor i (Chapter 17)
x_c	corrected observation
x_o	observed value
x	independent variable (Chapter 8)
x_i	individual observations
$\overline{x}$	mean
x	plume thickness (Chapter 14)
y	relative frequency of occurrence
α	known degrees of freedom
γ	ratio of accuracy of wet test meter to dry test meter
η%	efficiency of a specific process
ρ_s	stack gas density

Symbol	Definition
ρ_{H_2O}	density of water
σ	standard deviation of the population
θ	total sample time
$\Delta\theta$	sample time at a traverse point
μ	population mean

CHAPTER 1

INTRODUCTION TO SOURCE SAMPLING

Source sampling is the experimental process of determining the characteristics of a waste gas stream evolving from combustion or other industrial operations. The parameters of most interest are gas temperature, flow rate and composition. This chapter outlines the basic reasons for performing source tests; in addition it indicates some of the problems associated with conducting these tests. Due to the complexity and expense involved, these tests should be attempted only when indicated by one of the reasons discussed in this chapter.

PURPOSE OF SAMPLING

There are several distinct reasons why a test may be necessary. Although the following reasons are each very different, the basic requirements for conducting each test are the same; this point will be emphasized in the following sections of this chapter.

Source Definition

A source test may be used to determine the nature and magnitude of air pollution emissions from a source. This information may determine if a problem exists, and if so the data can be used in the selection and design of air pollution control equipment. Many types of air pollution sources have already been characterized on the basis of source tests and their emission factors reported.[1] An emission factor is the pollutant mass emission rate per unit of some process characteristic. This process parameter may be the amount of raw material used or the amount of product produced. For example, if the burning of one ton of coal emits 85 pounds of particulate matter, then the emission factor is 85 pounds per ton. If 100 tons of similar coal were burned then the estimated amount of particulate emissions would be

$$100 \text{ tons of coal} \times 85 \frac{\text{lb. particulate}}{\text{ton of coal}} = 8500 \text{ lb. particulate}$$

For well-defined sources the emission factors may be sufficiently accurate and no source test need be performed. However, if any significant variation from source-to-source is expected, then a test on the specific source must be conducted.

Source test information may be used to improve emission inventory data for various air quality regions. An emission inventory is a compilation of all the air pollution sources by type, location and pollutant characteristics. Control agencies use the source test results as basic input data to various computerized air pollution dispersion programs which predict ambient air quality. Such prediction methods are often the basis for designating the boundaries of air quality regions and selecting air pollution control strategies to meet ambient air quality standards. The most widely-used such model is the Air Quality Display Model (AQDM) developed by the U.S. Environmental Protection Agency (EPA).[2]

Process Monitoring

Periodic source tests may be conducted to determine if a process is operating efficiently. If there are maladjustments in the plan, sometimes these can be diagnosed from a source test. In many cases these tests are an economic necessity; if the plant operated inefficiently for any length of time, resource materials and profits would be lost.

Control Efficiency

Source tests are a necessary step in determining the efficiency of pollution control equipment. An industry or control agency might require such a performance check prior to final acceptance of a newly purchased control device. The tests may be performed by the contractor or some previously agreed upon consultant. The test may be performed by making a source test before installation and then making another after the equipment is in operation; *before* and *after* emissions are then compared. A second possible method involves simultaneous source testing on the inlet and outlet side of the control device. Either method allows a removal efficiency to be calculated but the best method depends on the operational characteristics of the specific source.

Compliance

Compliance tests under specific operating conditions are required by law for many installations. For example, the 1970 Clean Air Act[3] indicates that the EPA Administrator may require the owner or operator to conduct performance emission tests and furnish a written report on the results. Such requirements for new sources are given in Appendix A. The states also have the authority to require source tests on any type of stationary source.[4] These tests are thought to be a necessary step in implementing air pollution control strategies and thereby meeting ambient air quality standards.

THE PROBLEMS INVOLVED

Conducting an accurate source test is a complex and often difficult task. Some measurements such as gas temperature and velocity must be made *in situ* while most of the measurements regarding as composition must occur external to the stack. Both the internal and external measurements are greatly affected by the process conditions.

Process Conditions

In most industrial operations the process conditions vary, thus causing a variation in the characteristics and quantities of gaseous effluents. Also there may be more than one in-plant source exhausting to the same stack and this will cause additional fluctuations. The source sampling process requires a substantial length of time; therefore consideration must be given to altering the sampling method and/or conditions to account for the changes in process conditions. The extreme case is that of the batch-type operation as shown in Figure 1.1. If the objective is to determine the "average" hourly emission rate, then any sample taken within a given hour will be biased. If the sample is taken during the first or third hours, the measured emission rate will be less than the true "average" value. If the

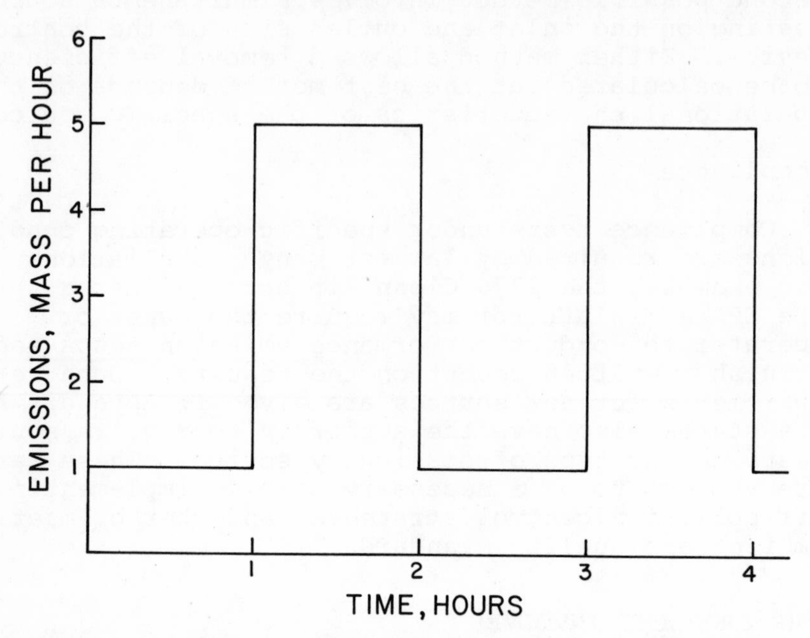

Figure 1.1. *Time variation of air pollutant emissions.*

sample is taken during the second or fourth hour the measured emission rate will be greater than the true "average" value.

In Situ Measurements

Source testing requires knowledge of the conditions within the stack, such as gas temperature and velocity. Special equipment must be used to withstand the corrosive environment. Due to poor mixing conditions these parameters often must be measured at many points within the stack. All this tends to lengthen the time required to perform a source test.

Gas Sampling

Since most waste gas streams have a high volume and mass flow rate, the total stream cannot be put through the sampling device. As shown in Figure 1.2 then, a part of the gas must be sampled and analyzed. The total pollutant mass emission rate from the source may be calculated after measuring (a) the pollutant mass concentration and (b) the gas volume

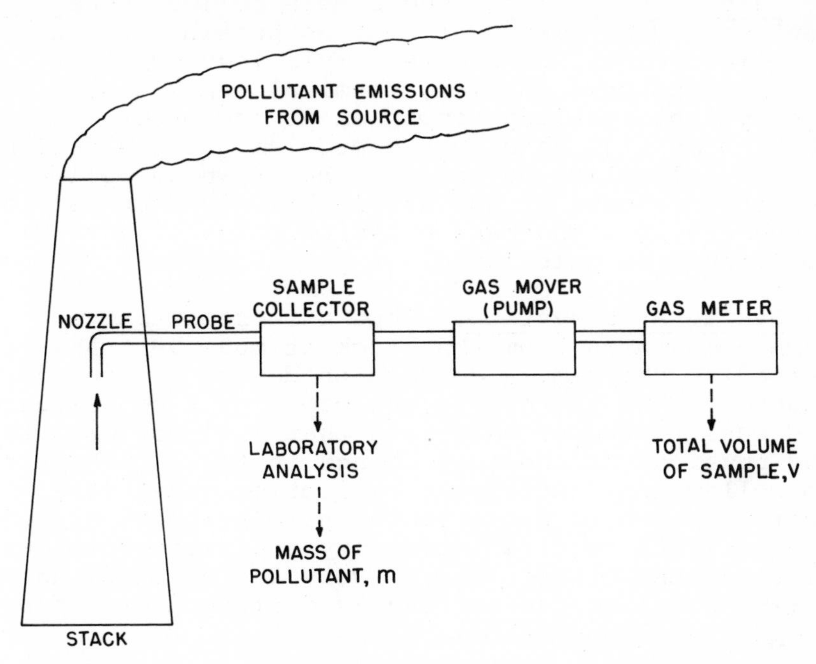

Figure 1.2. Simplified schematic diagram of the source sampling.

flow rate from the source. Basically the procedure entails obtaining a representative sample, efficiently collecting and preserving the sample, and then performing the desired analyses. Each of these operations requires many steps, each providing an opportunity for error.

Representative Samples. A representative sample is one which has the same characteristics as the parent gas stream at the point of sampling. For example, should a gas sample contain more or less pollutant than the parent gas stream, it is termed not representative or is said to be biased. If process conditions fluctuate widely then proportional sampling must be used. This simply means that the sampling rate varies in proportion to the stack gas velocity fluctuations. It should be noted here that this is quite difficult to achieve under actual field conditions because all flow rate adjustments generally must be made manually. The sampling strategies required to obtain representative samples are presented in Chapter 2.

Sample Collection. The sample collection equipment should be placed as near as possible to the sampling point. Otherwise sample loss due to absorption, condensation, chemical reactions, adsorption, settling, impingement and other mechanisms will be a severe problem. The efficiency of the collection device must be known as well as the total amount of gas sampled. Only then can the concentrations and mass emission rates of various pollutants be calculated.

Sample Preservation. Once a representative sample is taken from the stack it must be transported to the sample collection device without any change. Thus the nozzle and probe must be designed and constructed of materials that will not promote any physical or chemical change in the sample prior to collection. After the collection great care must be taken to preserve the sample. Most specific source tests require laboratory analysis of reagents, filters, and other items. Every degree of precaution during sampling can be readily destroyed by improper sample handling methods.

Analyses. An adequate amount of sample must be collected in order to perform the desired analyses. Thus often the consideration of sample size determines how the source test will be conducted. The analyses should be conducted immediately upon receipt of the sample at the laboratory because in most cases sample and/or reagent deterioration is a problem. If at all possible standard or prescribed analytical methods should be used. These methods have been developed for such sources and usually the interferences have been identified.

Sampling Location. Finding an acceptable location to withdraw a sample usually poses a problem. Most plants were not designed for such testing. The basic problem is two-fold. First of all there may not be any accessible or safe areas where sampling ports can be located and necessary equipment manipulated for the test. Only recently has this problem been stressed to those building new plants.[5] Secondly it is important that the sampling location be removed from any obstructions which will seriously disturb the gas flow in the duct or stack. If such obstructions as elbows or expansion unions exist, the test can be performed but it will be considerably more difficult. In this case a greater number of samples must be taken to satisfy the need to be representative.

Safety Considerations. Source sampling can be a hazardous task. Many of the sources to be sampled contain toxic constituents. The only available sampling location may be some distance from the ground where only temporary scaffolding is available. Hot, dusty, and noisy working environments are also commonly encountered. Also, because the source sampling equipment requires electricity, there is a danger of electrical shorts when sampling must be performed outside during inclement weather. Precautions must be taken under all sampling conditions and the test procedure and environment must comply with state and federal regulations. Stringent safety requirements have been promulgated under the federal 1970 Occupational Safety and Health Act.[6]

LOGISTICS

Performing a source test requires a considerable amount of manpower, equipment and money. Typically a field source testing team includes an engineer and a technician, and in addition someone with a knowledge of chemistry is needed for laboratory analyses. The capital cost of source sampling equipment to comply with source testing for gases and aerosols according to EPA procedures[3] may well exceed $6000-10,000. This does not include the cost of laboratory analytical equipment.

The cost of performing a source test varies greatly. It depends upon (a) the type and number of pollutants to be measured and (b) the physical conditions at the industrial plant. Consulting firms usually charge about $20-25 per man hour. At these rates it costs $2,000-3,000 to characterize a *single* source for one to three pollutants. Usually the industry bears the cost of preparing the source for making the test.

SUMMARY

Regardless of the eventual use of the data, every source test must meet the following basic requirements:

1. The gas stream being sampled from a source should represent either the total or a known portion of the emissions from the source.
2. Samples of the emissions collected for analysis must be representative of the gas stream being sampled.

3. The volume of gas sample withdrawn for analysis must be measured accurately in order to calculate the concentration of the analyzed constituents in the sampled gas stream.
4. The gas flow rate from the source must be determined in order to calculate emission rates for the various constituents.

REFERENCES

1. "Compilation of Air Pollution Emission Factors," Environmental Protection Agency, Office of Air Programs Publication No. AP-42, Washington, D.C. 1972.
2. "Air Quality Display Model," U.S. Public Health Service, Contract No. PH-22-68-60 TRW, Washington, D.C. 1969.
3. "Standards of Performance for New Stationary Sources," Federal Register, Vol 36 No 247, December 23, 1971.
4. "Air Programs: Approval and Promulgation of Implementation Plans," Federal Register, Vol 37 No 105, May 31, 1972.
5. "Designing Your Plant for Easier Emission Testing," Chem. Eng. June 26, 1972.
6. "Occupational Safety and Health Act--Rules and Regulations," Federal Register, Vol 36 No 105, May 29, 1971.

CHAPTER 2

SOURCE TESTING PROCEDURES

Obtaining a representative sample from a stack or duct is a complex task because usually all process parameters vary with both time and location within the stack. This chapter will identify and discuss the various types of source conditions which can occur and identify the proper sampling strategy to provide a representative sample. In short, it will outline the sampling strategies which must be used.

A number of different types of source testing procedures have been developed. This chapter will also identify the parameters which must be measured and the general configuration of various source sampling trains. The greatest emphasis in this and following chapters will be on those procedures required by state and federal law.[1]

GENERAL PROCEDURE

Three related terms describe the conditions in a stack: pollutant concentration, stack gas flow rate and pollutant mass emission rate. The latter term is the most important since many regulations specify allowable emissions on this basis. These three terms are defined and related as follows:

$$\overline{PMR}_s = \overline{C}_s \, \overline{Q}_s \qquad (2\text{-}1)$$

where

$\overline{PMR}_s$ is the average pollutant mass emission rate from the stack,

$\overline{C}_s$ is the average stack concentration of the pollutant, and

$\overline{Q}_s$ is the average volumetric stack gas flow rate.

Hence the overall objective is to determine $\overline{PMR}_s$ by measuring $\overline{C}_s$ and $\overline{Q}_s$. The average volumetric stack gas flow rate, $\overline{Q}_s$, is determined by measuring the average gas velocity, $\overline{V}_s$, and the area of the stack, A_s.

$$\overline{Q}_s = \overline{V}_s A_s \qquad (2\text{-}2)$$

Figure 2.1 illustrates the objectives of a source test. It should be noted that *average* values are indicated. This means that measurements must be taken at many points within the stack in order to arrive at average values.

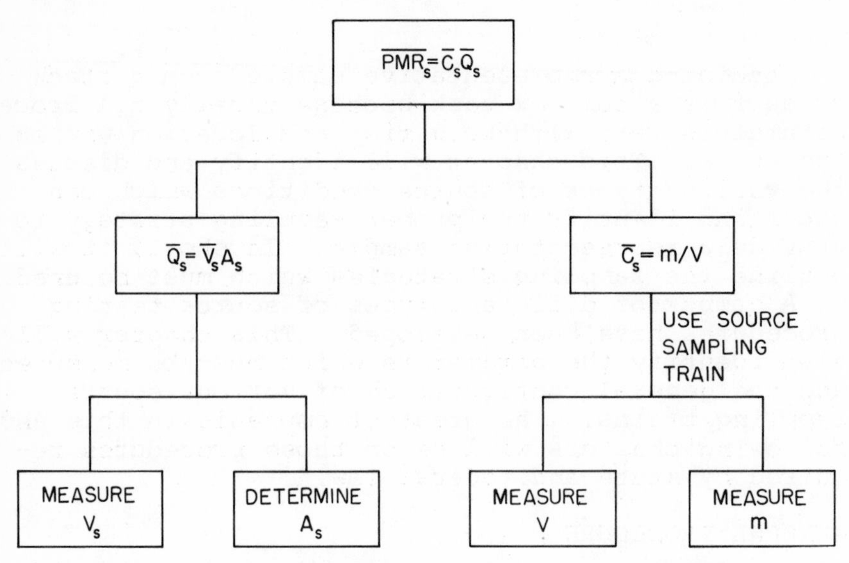

Figure 2.1. Overall objectives of source sampling.

Air pollutant emission standards are often expressed in terms of pollutant mass emitted per unit quantity of product produced (or unit quantity of material and/or energy input to the process). When this is the case, the average pollutant emission rate, $\overline{PMR}_s$, is determined and normalized with respect to the desired process variable. Dividing the average particulate emission rate, $\overline{PMR}_s$, of a coal-fired boiler by the average fuel heat input rate to the boiler during the emission test is an example. In this case, the normalized emission rate would be in terms of lb of particulate/million BTUs.

MEASURED PARAMETERS

For source sampling to be effective, a number of parameters must be measured. The theory and practice of making these measurements is the subject of Chapter 4. Basically the measurements may be divided into two categories: those internal and external to the stack.

The internal measurements are needed to define the state of the exhaust gas, namely temperature, pressure and flow rate. The latter parameter involves making velocity measurements at various points of the stack cross section. These measurements are essential for determining the sampling conditions to be used.

External measurements are made on the gas sample stream to establish the proper sampling rate and determine the total quantity of sample withdrawn. The important parameters are gas sample temperature, flow rate and total flow. For basic calculations the molecular weight of the stack gas also must be known. If the exhaust gas is from a combustion process, the carbon dioxide, carbon monoxide, oxygen and moisture content of the gas must be determined. In some cases the molecular weight of the exhaust gas can be estimated from other process parameters.

SAMPLING STRATEGY

A strategy for obtaining a representative sample must be determined for each source. If a test is performed according to a specific regulation, the agency should specify a procedure which provides a representative sample. However, if a source which has peculiar flow and concentration characteristics is to be sampled, the following discussion will be helpful in selecting an adequate sampling strategy.

Achinger and Shiegehara[2] characterized sources according to their cross-sectional and time variation. The resulting four possible source categories are shown in Table 2.1. This method is applied to the gas velocity parameter. It should be realized that pollutant concentration can also have time and cross-sectional variation. The following sampling approaches are recommended for these various categories. It must be remembered that $\overline{C}_s$ and $\overline{Q}_s$ must be measured in such a way that the resulting average pollutant mass emission rate, $\overline{PMR}_s$, is representative (unbiased).

Table 2.1

Categorizing Sources According to Their Process Characteristics

Category	*Variation of Conditions*	
	Time	*Cross-Sectional Velocity*
1	Steady	Uniform
2	Steady	Nonuniform
3	Unsteady	Uniform
4	Unsteady	Nonuniform

Category 1: Steady and Uniform

When source conditions are steady (no variation with time) and uniform (no variation from point to point within the cross section), only one measurement is needed for accurate results. This measurement may be taken from anywhere within the cross section and at any time. Sampling for a gaseous pollutant when the gas stream is turbulent (well mixed) usually falls into this category when the source operation does not vary with time.

Category 2: Steady and Nonuniform

When the source conditions are steady and non-uniform, a series of measurements must be made. These measurements can be made at any time and will still yield representative data. The number of measurements in the series is determined by the shape of the stack, the dimensions of the stack and the proximity of various obstructions such as elbows. As shown in Figure 2.2, the stack is divided into a number of equal areas and the measurements are taken at the centroid of each elemental area. The accuracy of this approach depends upon the degree of cross-sectional variation; the greater the number of measurements made, the more representative the source sampling data will be. This method assumes that the centroids are representative of the areas around them. The assumption becomes more valid as the size of the elemental areas decreases thus increasing the total number of sampling points.

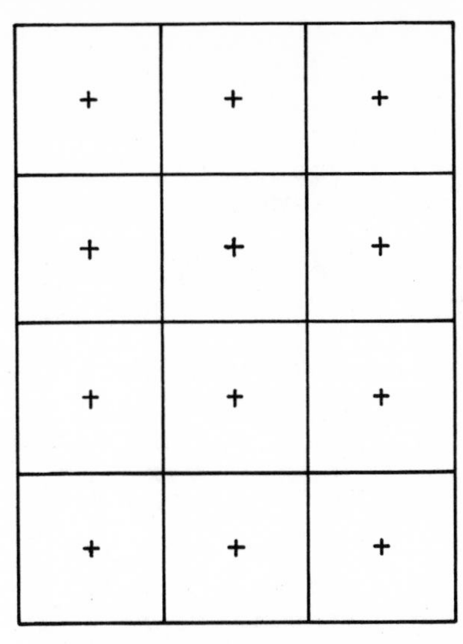

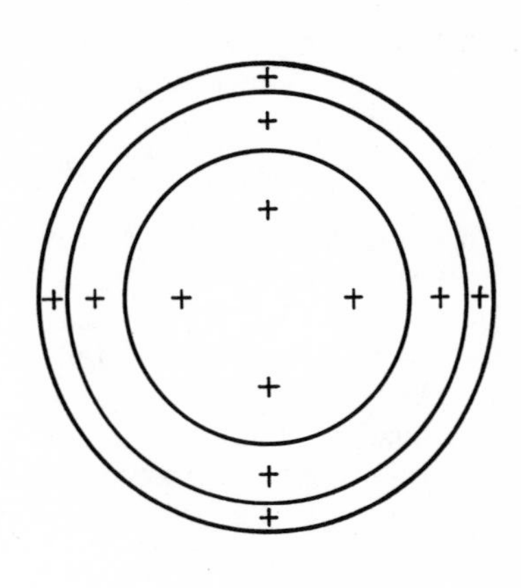

(a) Rectangular Duct *(b) Circular Duct*

Figure 2.2. Equal sampling areas for traversing ducts.

Sampling a well-controlled continuous process such as a coal-fired power plant for particulate matter is an example of this category. Usually a composite sample is proportionally extracted by traversing the cross section and sampling in proportion to the stack gas velocity at the centroid of each equal area. A composite sample is a sample made up of a series of pollutant measurements all collected in the same pollutant collection device as opposed to using a separate collector for each measurement.

Category 3: Unsteady and Uniform

When source conditions are unsteady and uniform, measurements are needed at only one location within the cross section. These measurements must be taken over the entire cycle of source operation for cyclic operations, or over as long a period as possible for noncyclic operations. The longer the sampling period for noncyclic operation, the more representative the measurement will be. An example of this category is the sampling of a cyclic (batch-type) operation for a gaseous pollutant at a point where the gas stream is turbulent.

Category 4: Unsteady and Nonuniform

When both the source and flow conditions are nonuniform, the most complicated sampling procedure is required. Practically speaking it is almost impossible to obtain a representative sample.

The source conditions can vary proportionally or nonproportionally. Source conditions vary proportionally with time when all measurements vary in the same ratio. For example, if one characteristic increases by 50% then all other measurements of that characteristic also increase by 50%. The nonproportional case may vary in a reproducible or nonreproducible (random) fashion.

Proportional Case. When source conditions vary proportionally, a series of measurements must be taken. The cross section of the duct is divided into equal areas as indicated in Figure 2.2. The series measurements are taken in simultaneous pairs of individual measurements. The composite sampling approach cannot be used. One of each pair of measurements is always taken at the same fixed reference location. The other measurement of each pair is taken at the centroid of the elemental area. The paired measurements are continued until every elemental area has been sampled. The measurements at the reference point provide a continuous record of the time variation. Since the values at the reference location vary in a known manner, and the values at any measurement location vary in the same manner as the values observed at the reference location, all measurements can be mathematically adjusted to a common point in time. An incinerator with an uneven charging rate might be an example of this category.

Nonproportional but Reproducible Case. Multiple measurements must be made when the source conditions are nonuniform. When the time variation is cyclic and all measurements are reproducible from cycle to cycle, pollutant measurements can be made at any time, providing the sampling period begins at the start of a cycle and continues to the end of the cycle or to the end of some succeeding cycle. The number of measurements required and the location within the cross section must be determined as mentioned for Category 2. Composite sampling is valid, providing the number of cycles sampled at each location is the same for every location sampled. Sampling an incinerator with a uniform charge frequency would be an example of this category.

Nonproportional and Nonreproducible Case. When the time variation is not cyclic but rather random, every location to be measured must be measured simultaneously. This procedure is extremely tedious and almost impossible to accomplish in field source tests. As shown in Figure 2.2 this situation would require 12 sampling trains. The requirements for equipment, personnel and space would prohibit making such a test.

Special Sampling Cases

There are some special sampling cases with respect to gases and particulate matter that are worthy of consideration.

Gaseous Pollutants. The cross-sectional concentration of gaseous pollutants is usually uniform because of the diffusion and turbulent mixing processes. Hence time variation is the major concern. The source is either steady or unsteady. Thus according to Table 2.1 categories 1 and 3 must be considered. Under these two categories there are four special cases in which gaseous pollutant concentration and velocity can vary with time as shown in Table 2.2. Since the concentration conditions are uniform it is only necessary to sample at one point within the stack to determine the average concentration, $\overline{C}_s$. Any constant sampling rate can be used provided that either the stack flow rate is steady or the pollutant concentration is steady; this covers cases 1.1, 1.2, and 3.1 in Table 2.2. For the unsteady concentration and velocity, case 3.2, proportional sampling must be used. In this last case there are two types of errors which can be made. First, if the sampling rate at any point during a test exceeds that prescribed by proportional sampling considerations, then the measured concentration, C_m, will be greater than the true concentration, C_t. And second, if the sampling rate at any point during a test is less than that prescribed by proportional sampling considerations, the measured concentration will be less than the true concentration.

Table 2.2

Combinations of Source Conditions When Sampling for Gaseous Constituents

Category & Case	*Time Variation of Conditions*	
	Concentration	*Velocity*
1-1	Steady	Steady
1-2	Steady	Unsteady
3-1	Unsteady	Steady
3-2	Unsteady	Unsteady

Particulate Matter. It was stated in the previous section that gaseous sampling biases occur if sampling rates above or below the proportional rate are used. This same situation is also true when sampling for particulate matter. A special case of the proportional sampling will serve as an example. It is known as *isokinetic sampling* which means that the gas sampling velocity entering the nozzle is equal to the stack gas velocity at that point. The sampling for particulate matter as shown in Figure 2.3 illustrates the different streamline conditions at the nozzle tip. If it is assumed that the particles are traveling at the same velocity as the parent gas stream and the gas sample is taken into the nozzle with a velocity, V_n, greater than the stack velocity, V_s, then the measured concentration, C_m, will be less than the true concentration, C_t. This error occurs because the particles have inertial characteristics; the larger and heavier particles are unable to follow the streamline flow into the nozzle. These particles pass the nozzle but the gas around them is drawn into it. Hence fewer particles are collected than should be and the measured concentration is less than the true concentration.

Using converse reasoning in Figure 2.3, it can be argued that if the nozzle sampling velocity, V_n, is less than the stack gas velocity, V_s, then a greater proportion of large particles are sampled than should be. Hence C_m is greater than

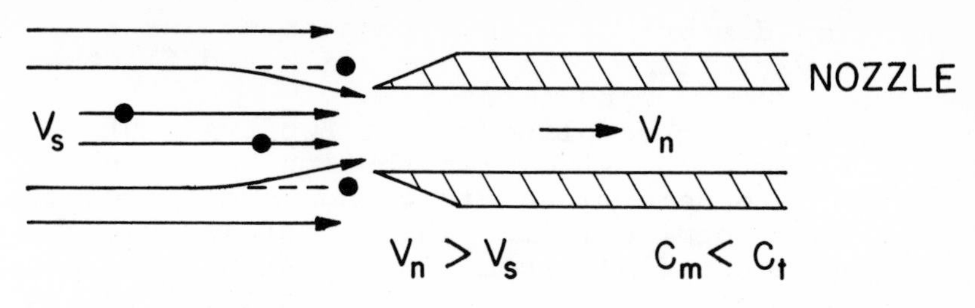

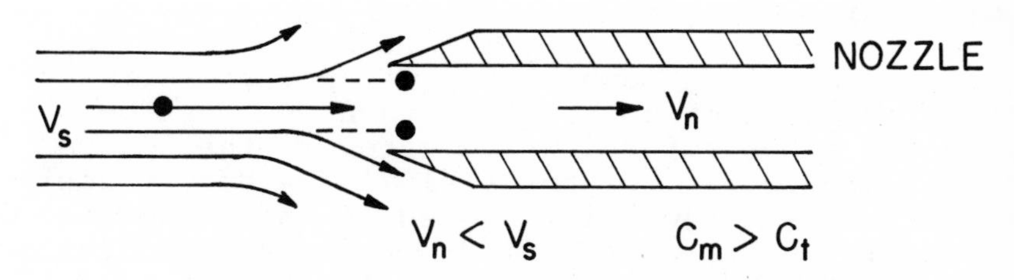

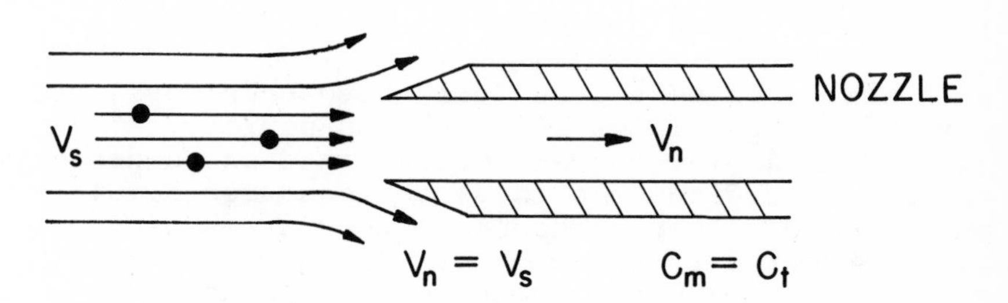

Figure 2.3. Representative and nonrepresentative aerosol sampling conditions.

C_t. In either of these cases the sample is biased or said to be nonrepresentative. If the gas sampling rate is adjusted so that V_n is equal to V_s, then a representative sample is obtained. Obtaining and maintaining isokinetic sampling conditions throughout a test can be difficult.

As mentioned previously, problems arise due to velocity variations across the stack and fluctuations of process conditions.

Nonisokinetic sampling causes two types of errors. Both the size distribution and mass emission rate determined are wrong. These collection errors occur only for larger particles. The very small ones (less than 3 microns in size) begin to act like gas molecules and thus are unaffected by inertial characteristics. These particles are effectively collected regardless of the sampling rate.

Determining Category

If an existing regulation does not prescribe the sampling strategy to be used, then one must be selected. Determining which category the source belongs in is part of the preliminary survey activities; this subject is covered in detail in Chapters 3 and 5. The information for making this decision can be obtained by reading the literature on similar processes, making a site visit to the plant, and discussing the situation with management and operational personnel.

TYPES OF SAMPLING PROCEDURES

A number of test procedures have been developed for obtaining samples from various sources.[1,3-6] These methods may be categorized into those designed for sampling gases and those designed for sampling particulate matter. The sampling systems used are often referred to as *sampling trains*, and are collections of special components such as gas meters, pumps, etc. assembled in a certain order so that representative samples are obtained.

Gaseous Pollutants

Gas samples may be taken on an *integrated* sample flow rate basis or a *grab* sample basis to meet the sampling strategy requirements discussed in the previous section. Examples of sampling trains to do this are shown in Figure 2.4. The basic components used in these trains are a filter, a probe, a gaseous pollutant collector, a gas meter, a flow regulator and a gas mover. Manometers and thermometers are also used. The filter serves to remove any particulate matter that may interfere in the

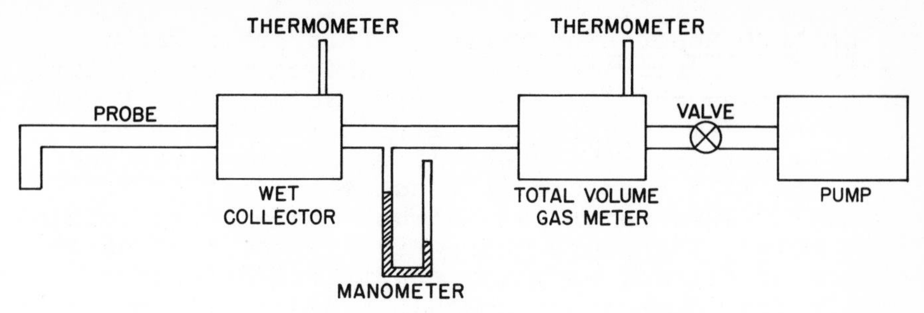

(a) LA APCD Absorption Train[3]

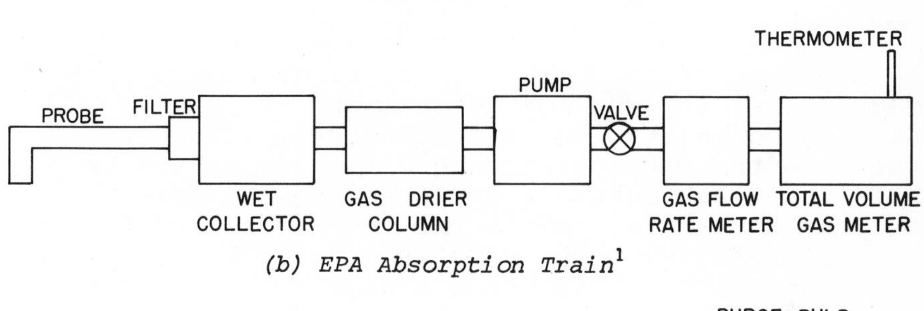

(b) EPA Absorption Train[1]

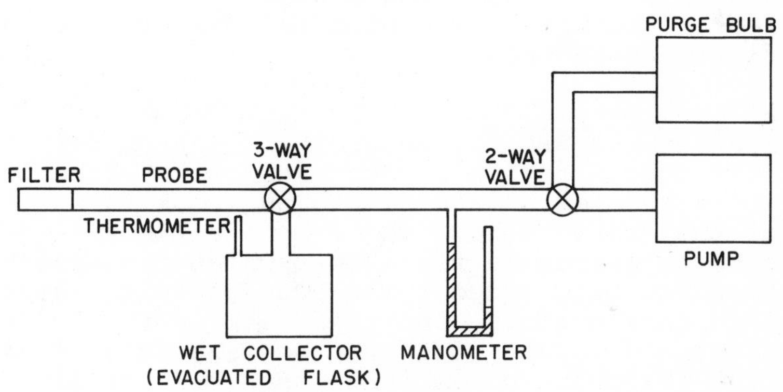

(c) EPA Grab Sampling Train[1]

Figure 2.4. Source sampling trains for gaseous pollutants.

analysis of the gaseous constituents. The probe, often made of glass and sometimes heated, transports the sample to the collector. The collector is usually some sort of wet chemical device which absorbs the gaseous pollutant. The gas meter measures the sampling flow rate and/or the total gas volume sampled. The gas mover is a pump or other vacuum source which provides the suction to move the gas sample through the other sampling components.

LA APCD Absorption Train.[3] This train was designed for integrated sampling and collects pollutants by absorbing them in a scrubbing solution in a wet collector (impinger). A filter (glass wool) is usually put into the probe line to remove the particulate matter before it reaches the collector. A total volume meter is used but there is no volume rate meter, therefore making it difficult to maintain proportional sampling conditions with this unit. The flow rate through the system is controlled by a hose clamp between the gas meter and the vacuum pump.

EPA Absorption Train.[1] This is the Method 6 train required by the EPA for determining the sulfur dioxide concentration in stack gases. The glass wool and the wet collector are similar to the LA APCD sampling train. A dry chemical gas drier column is used to protect the gas meters and the pump from damage. The pump is located before the two gas meters. This means that the pump must be leakless; otherwise serious errors will result in the flow measurements.

EPA Grab Sample Train.[1] This is the Method 7 train required by the EPA for determining the oxides of nitrogen concentration in stack gases. Particulate matter is removed by stuffing glass wool into the end of the glass probe. The sample collector consists of an evacuated flask into which the absorbing solution has been added. The NO_x absorbing solution is added to the flask, and the flask is then evacuated and valved off. The squeeze bulb is used to purge the probe of ambient gases. The flask valve is then opened and the flask is filled with sample. The pollutant concentration is then determined in the laboratory by spectrophotometric methods.

Particulate Matter

The particulate sampling trains basically are composed of a nozzle, a probe, a device for collecting the particulate matter, one or more gas meters and a gas mover. Generally a pitot tube is used to measure the stack gas velocity. As was the case in gas sampling for gaseous constituents, manometers and thermometers are also used. Figure 2.5 illustrates several different arrangements of these components. The nozzle provides the opening through

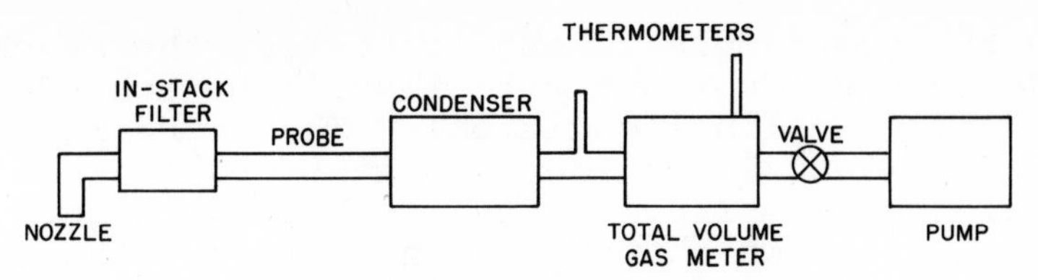

(a) Joy WP-50 Particulate Train[10]

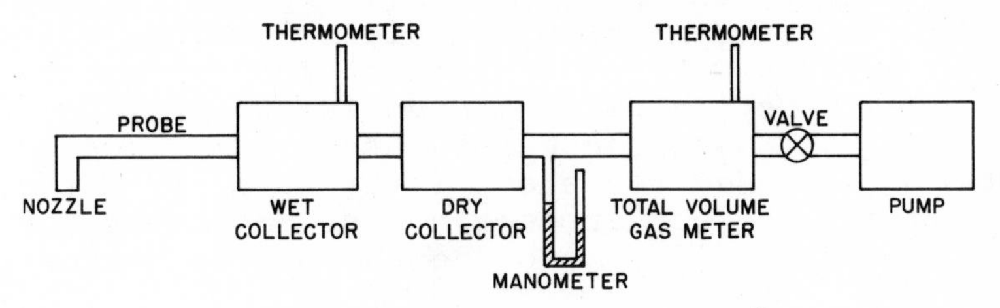

(b) LA APCD Particulate Train[3]

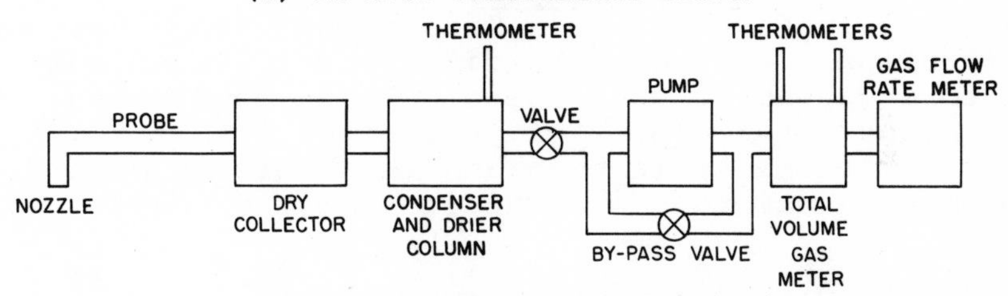

(c) EPA Particulate Train[1]

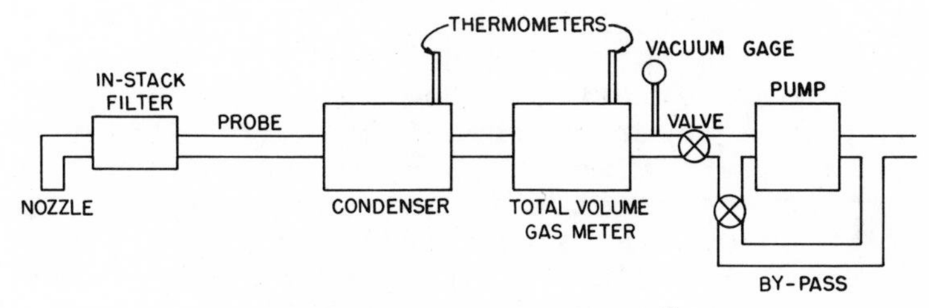

(d) ASTM Particulate Train[7]

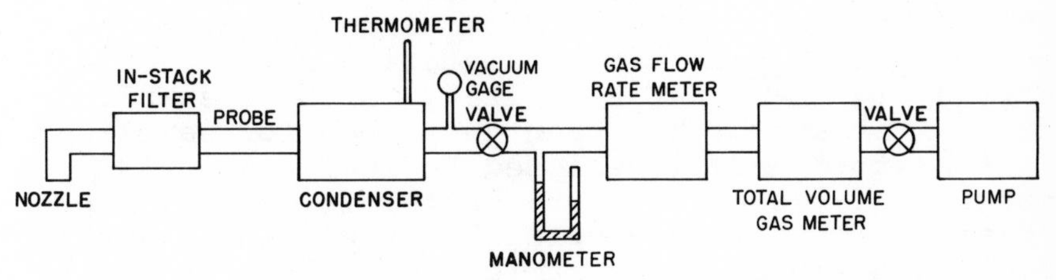

(e) SF APCD Particulate Train[11]

Figure 2.5. Source sampling trains for particulate matter.

which the gas sample is extracted from the stack; its design is important with respect to achieving isokinetic sampling conditions. The probe is rigid so that the nozzle can be accurately positioned at the selected sampling points as shown in Figure 2.2. The collector, which removes the particulate matter from the sample and retains it for later analyses, may operate on any number of principles: filtration, electrical precipitation, liquid scrubbing, condensation or combinations of these. The gas meter and gas mover function as mentioned in the section on sampling for gaseous constituents. Sometimes a condenser is used to remove water vapor from the gas stream, thereby protecting the gas meter and pump from damage due to condensation and corrosion.

Joy WP-50 Particulate Train.[10] The WP-50 train is shown in Figure 2.5a. This method uses in-stack filtration and recommends different types of collectors depending upon the stack gas conditions. The Alundum thimble is often used with this train. No gas rate meter is used and therefore it cannot adequately maintain isokinetic sampling conditions.

LA APCD Particulate Sampling Trains.[3] Three particulate sampling trains have been developed by the Los Angeles Air Pollution Control District for various applications. The first one, shown in Figure 2.5b, is the one most commonly used to collect particulate matter. The wet collector is followed by a paper thimble filter to collect any particulate matter which escapes the wet collector. A gas volume rate meter is not used; hence it is difficult to obtain and maintain isokinetic sampling conditions. A second sampling train (not shown) uses ceramic thimble filters to collect particulate matter from the high temperature gas streams. This one is similar to the ASTM sampling train.[7] The wet collector then catches material not collected by the high temperature filter. A third sampling train (not shown) has been developed for sampling incinerators and is designed to indicate what portion of the particulate matter is above 5 microns in size. This is accomplished by placing a small cyclone ahead of the ceramic type filter and a wet collector.

EPA Particulate Train.[1] This train is the Method 5 federal sampling train required for testing new process facilities such as power plants, cement plants and incinerators. It is also the method prescribed by many states and used to test all types of sources. The details of this method are presented in Chapter 10. A number of companies manufacture equipment similar to this sampling train and these are discussed in Chapter 16.

The gas sample passes through a heated glass probe into a filter-type dry particulate collector, then into an ice bath condenser and drier column. The pump is located ahead of the two gas meters, making it imperative that the pump be leakless in order to eliminate errors. Since both a total volume gas meter and a gas flowrate meter are used, it is possible to operate the system under isokinetic conditions and make the necessary manual adjustments with the valves to maintain such conditions.

ASTM Particulate Train.[7] The general sampling train recommended by ASTM is shown in Figure 2.5d. Although not shown here, a gas flow meter should be used also. The type of in-stack filter used is determined by stack gas conditions. The parameters considered in selecting the type of particulate collector are the expected sample weight, flue temperatures and the presence of liquid droplets. For example, Alundum thimbles are recommended only if the expected sample weight exceeds 100 milligrams and no liquid droplets are present. This method is basically quite flexible, enabling alterations to be made which allow applicability to a wide range of source types.

ASME Particulate Train.[4] The American Society of Mechanical Engineers has described in its Power Test Codes the use of a sampling train to determine particulate emission from fossil fuel combustion processes. To meet the ASME specifications a sampling train must have the following parts:

1. a tube or nozzle for insertion into the gas stream and through which the sample is withdrawn; thin edged nozzles with diameters at least 0.25 inches are preferred.
2. a filter (thimble, flat disk, or bag form) for removing the particulate from the sampled gas. For the purpose of the Power Test Code, 99.0% by weight efficiency is considered satisfactory.
3. a means of checking the equality of the velocity of the gas entering the nozzle and

the velocity of the gas in the flue at the point of sample withdrawal.

4. a means for measuring the quantity of gas sampled.
5. an exhausting device for drawing the gas through the sampling nozzle, filter, and metering device together with the necessary connecting tubing. It is important that the temperature of the gas be above the dew point until after it has passed the filter.

This sampling procedure is not very restrictive. In fact most of the sampling methods previously mentioned actually comply with ASME requirements.

SF APCD Particulate Train.[11] The San Francisco Bay Air Pollution Control District uses a sampling train with an in-stack filter. The nozzle and filter holder are made of Pyrex. The filter is a glass wool packing. This arrangement can be used for gas temperatures up to 1000°F. In this system both the gas flow rate meter and the total volume gas meter operate under vacuum conditions, making it essential that no leaks exist in the system. If this occurs, ambient air will leak into the system and thereby cause meter reading errors.

Comparison of Procedures

Source sampling procedures may be compared mainly on the basis of equipment advantages and analytical advantages. Often however these considerations succumb to "what is required by law?"

In sampling for gaseous constituents the grab sampling method requires little supporting equipment. The sampling flasks are light in weight but care must be taken not to break them. Probably the best feature of the method is that the amount of time the sampling team must spend in an adverse sampling environment is small. As presently used neither the grab method (Figure 2.4c) or the integrated sampling methods (Figures 2.4a and 2.4b) provide any analytical advantages; both use wet chemical analyses.

In sampling for particulate matter the WP-50 and ASTM trains (Figures 2.5a and 2.5d) are definitely more portable and rugged. They do not use as much fragile glassware as the EPA train (Figure 2.5c). However the in-stack Alundum filter used is quite heavy and fragile. An analytical disadvantage occurs here because the weight of particulate matter collected may be very small compared to the tare weight of the filter. Also the collection efficiency of such a filter is open to some question. The filters are cleaned and reused, allowing for the possibility of a changing collection efficiency with use.

Comparison of Results

The results obtained by various types of sampling trains usually are not the same because often a slight variation in the design or configuration of the sampling train greatly alters the type of material collected. This situation is best illustrated by considering the controversy regarding the proper sampling method for particulate matter. Originally the EPA proposed[8] that any material collected in the wet collector (condensor) portion of the sampling train shown in Figure 2.5c must be added to that collected in the dry collector (filter) portion. For such sources as incinerators this meant that more material often was collected in the wet collector than in the dry collector. After numerous objections from people in the field, the proposed method was altered[1] so that compliance now is based only upon material collected in the filter and in the probe preceding the filter. Hemeon and Black[9] contend, however, that even this modification does not render the EPA method valid. They maintain that considerable condensation and chemical reaction occurs in the probe prior to the filtering of the sample; hence, the SO_2 in the gas has ample opportunity to form sulfates which are later collected on filter. Hemeon and Black call these sulfate compounds pseudo-particulate matter and insist that they should not be included in the test results. Others might argue that such reactions, if they occur, will also occur in the atmosphere and therefore the material should be included. An important point to be noted here is that the definition of particulate matter depends strictly upon the sampling method. Thus, if compliance tests or performance tests are to be compared from source-to-source, the same method must be used.

SUMMARY

The problems of source sampling are many, with the procedures being various and the considerations highly technical. The necessary precision and accuracy for conducting source tests can occur only when the equipment is capable of achieving the desired sampling conditions and the test personnel are adequately trained. Before any source tests are performed, the method and equipment must comply with the law and/or the specific needs of the individual testor.

REFERENCES

1. "Standards of Performance for New Stationary Sources," Federal Register, Vol 36 No 247, December 23, 1971.
2. Achinger, W. C., and Shiegehara, R. T. "A Guide for Selected Sampling Methods for Different Source Conditions," J. Air Pollut. Control Assoc. Vol 18 p. 605 (1968).
3. Dejorkin, H. et al. "Source Testing Manual," Los Angeles, Country Air Pollution Control District, Los Angeles, Calif. 1965.
4. "Determining Dust Concentration in a Gas Stream," Performance Test Code 27-1957 American Society of Mechanical Engineers, New York, New York. 1957.
5. Cooper, H. D. H. and Rossano, A. T. "Source Testing for Air Pollution Control Environmental Science Services, Wilton, Conn. 1971.
6. Haaland, H. H. "Methods for Determination of Velocity, Volume, Dust and Mist Content of Gases," Bulletin WP-50 Western Precipitation Corporate, Los Angeles, California, 1968.
7. "Standard Method for Sampling Stacks for Particulate Matter," 1971 Annual Book of ASTM Standards, Part 23 Water and Atmospheric Analysis D2928-71, p. 832-858.
8. "Proposed Standards of Performance for New Stationary Sources," Federal Register, Vol 36 No. 159, August 17, 1971.
9. Hemeon, W. C. L. and Black, A. W. "Stack Dust Sampling: In-Stack Filter or EPA Train," J. Air Pollut. Control Assoc. Vol 22 No. 7, p. 516. July 1972.
10. "Methods for Determination of Velocity, Volume, Dust and Mist Content of Gases," Bulletin WP-50, Joy Manufacturing Company, Los Angeles, California.
11. Karels, G. G. "Improved Sampling Method Reduces Isokinetic Sampling Errors," Presented at the 12th Methods Conference in Air Pollution and Industrial Hygiene Studies, University of Southern California, Los Angeles, California, April 6-8, 1971.

CHAPTER 3

INDUSTRIAL PROCESS INFORMATION

Industrial source sampling as described in this text entails the quantifying of industrial process air pollutant emissions. As stressed in the preceding chapters, calculated emissions are of little use when they are not representative of the true emission rates. The emissions of any process are always functions of the operating conditions of the process. Most federal, state and local air pollutant emission standards take this into consideration. Emission regulations usually stipulate how the process is to be operated during the source test or how to normalize emission data to a set of standard operating conditions, given the actual operating conditions during the source test.

The purpose of this chapter is to familiarize the reader with the process information needed to conduct a successful source test. Mass and energy balances, emission factors and specific processes are discussed. Due to the scope of this text, only the air pollution sources to which present federal performance standards[1] apply are presented. Table 3.1 identifies the various testing methods set forth in these standards.

MASS AND ENERGY BALANCES

The preservation of mass and energy are basic scientific principles which can be used as an accounting system for studying the operation of industrial processes. Given any process, a simplified block diagram accounting for all material and energy inputs and outputs can be constructed as shown in Figure 3.1. The process inputs are always materials and/or energy. The process outputs can usually be described as useful and waste products,

Table 3.1

EPA Testing Methods for New Sources

Method	*Objective*
1	Determination of sampling location and traverse points for stationary source sampling.
2	Determination of stack gas velocity and volumetric flow rate (S-type pitot tube).
3	Gas analysis for carbon dioxide, excess air, and dry molecular weight (Orsat analysis).
4	Determination of moisture in stack gases (condensation technique).
5	Determination of particulate emissions from stationary sources (fiber glass filter and isokinetic conditions).
6	Determination of sulfur dioxide emissions from stationary sources (proportional sampling technique).
7	Determination of nitrogen oxide emissions from stationary sources (grab sample method).
8	Determination of sulfuric acid mist and sulfur dioxide emissions from stationary sources (proportional sampling technique).
9	Visual determination of the opacity of emissions from stationary sources.

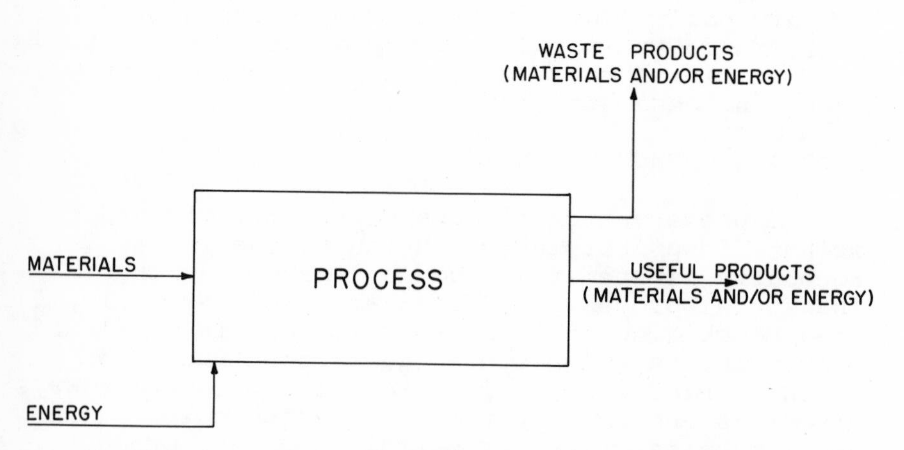

Figure 3.1. Process balance of materials and energy.

both of which can be materials and/or energy. The efficiency of the process is usually described as the ratio

$$\eta\% = \left(\frac{\text{Useful products}}{\text{Total input}}\right) 100$$

on a mass or energy basis.

Obviously most processes are too complex to allow the calculation of complete and accurate energy and mass balances. However, many processes can be simplified to the point where a rudimentary material or energy balance will yield a great deal of useful information about the process and its waste products. Examples would be solvent degreasing, paint spraying, and fossil fuel combustion. The appropriate material balances are shown in Figure 3.2. Given the input data and the quantity of useful output, reasonably accurate estimates of the amounts of the air pollutants emitted by these processes can be determined. In some cases, where highly accurate estimates of wastes emitted are not important, mass and energy balance estimates will preclude the need for costly source sampling.

Currently most air pollutant emission standards are written in normalized terms. That is, most emission standards are not written in terms of maximum allowable pollutant concentrations, or mass emission rates, but in terms of pollutant mass per unit of process input or output. The present federal performance standards in Table 3.2 are an example. When emission standards are in terms of lb of particulate/10^6 BTU input, lb of SO_2/lb H_2SO_4 produced, etc., process material balances must be conducted in conjunction with the source tests. The federal performance standards for new sources[1] require that the process input (or output as required by the particular process) be determined along with pollutant concentration and mass emission rates to establish compliance. The standards stipulate that the process input (or output) be determined using the available process monitors in service and substantiated with a material balance.

EMISSION FACTORS

When sufficient information is not available to perform a material balance and the expense of a source test is not justified, emission factors may be used to estimate air pollution emissions. Emission factors are developed by conducting a sufficient number of tests on a given process to

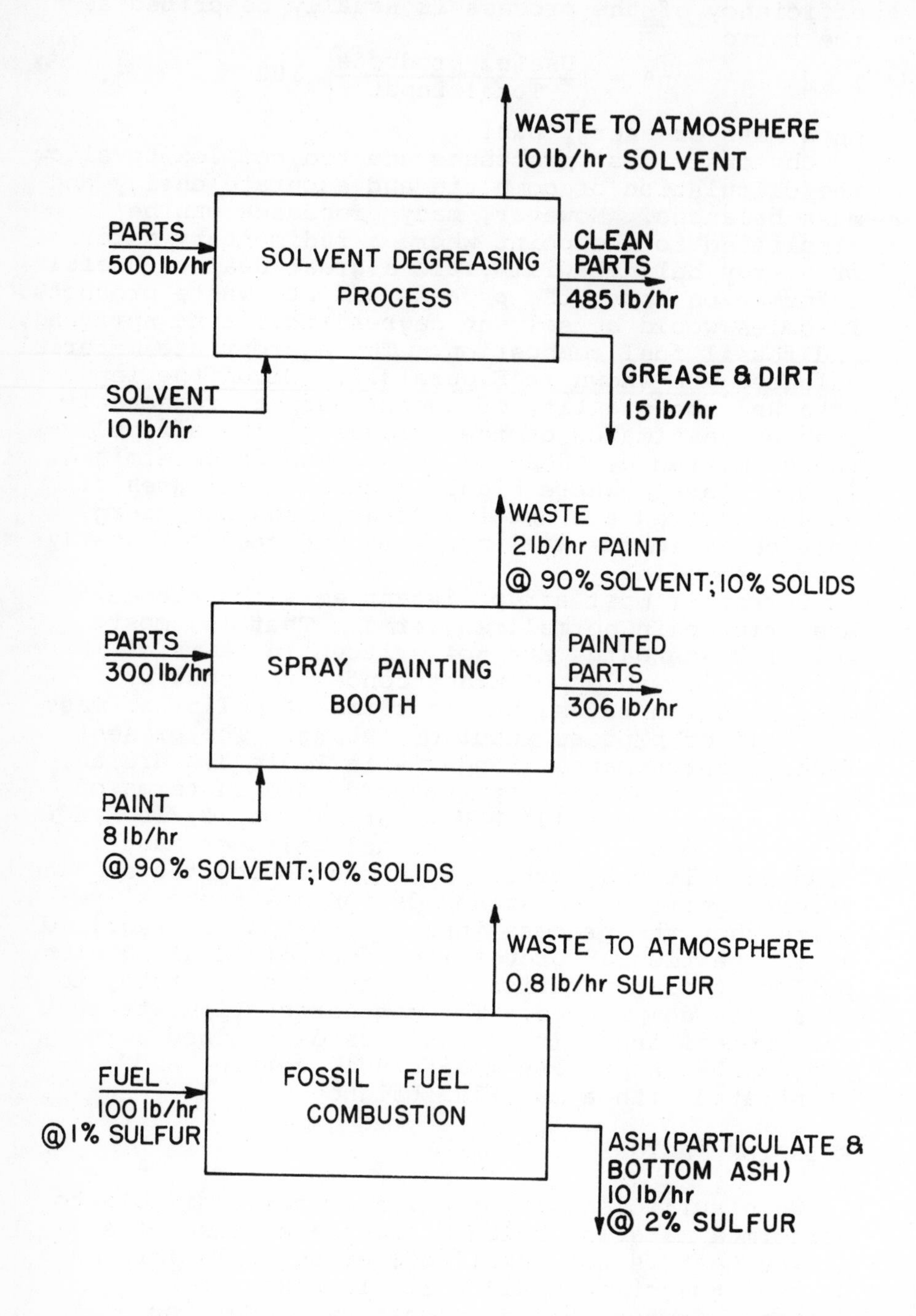

Figure 3.2. Examples of process material balances.

Table 3.2
Federal Standards for New Priority I Sources

Source (Constructed or modified after 8/17/71)	*Air Pollutant Emission Standards*	*Federal EPA Test Procedures*
Fossil fuel steam generators	Particulate	Particulate
(250×10^6 BTU/hr input)	1. 0.10 lb/10^6 BTU	1. EPA method 5
	2. 20% opacity (40% for 2 min/hr)	2. EPA method 9
	Sulfur dioxide	Sulfur dioxide
	1. 0.80 lb/10^6 BTU (liquid fuel)	1. EPA method 6
	2. 1.20 lb/10^6 BTU (solid fuel)	2. EPA method 6
	Nitrogen oxides	Nitrogen oxides
	1. 0.20 lb/10^6 BTU (gaseous fuel)	1. EPA method 7
	2. 0.30 lb/10^6 BTU (liquid fuel)	2. EPA method 7
	3. 0.70 lb/10^6 BTU (solid fuel)	3. EPA method 7
Incinerators (burn rate > 50 ton/day and content > 50% municipal)	Particulate	Particulate
	1. 0.08 gr/scf (corrected to 12% CO)	1. EPA methods 5 & 3
Portland cement plants	Particulate	Particulate
	1. Rotary kiln: 0.30 lb/ton of feed	1. EPA method 5
	2. Clinker cooler: 0.10 lb/ton of feed	2. EPA method 5
	3. All facilities: 10% opacity	3. EPA method 9
Nitric acid plants (producing 30-70% acid by pressure of atmospheric process)	Nitrogen oxides	Nitrogen oxides
	1. 3 lb NO_x/ton HNO_3 (100%)	1. EPA method 7
	2. 10% opacity	
Sulfuric acid plants (contact process)	Sulfur oxides	Sulfur oxides
	1. 4 lb SO_2/ton H_2SO_4 (100%)	1. EPA method 8
	2. 0.15 lb H_2SO_4 mist/ton H_2SO_4	2. EPA method 8
	3. 10% opacity	3. EPA method 9

accurately characterize the emissions of the process. The measured emissions are then normalized with respect to process material inputs and/or operating conditions and used to estimate the emissions of identical or similar process equipment.

Two compilations of air pollutant emission factors have been published by the federal EPA.[2,3] The emission factors in these documents are intended for use in conducting source inventories. The emission factors reported were determined mostly by source tests which were conducted by many different individuals using various types of source sampling procedures. Therefore, the factors presented are not usually precise enough to predict accurately the emissions of individual sources. As an example, the data used to develop the emission factor for particulate emissions from pulverized coal boilers is shown in Figure 3.3.[4] In this case the median value, 80%, the fraction of total ash content that is emitted as particulate, was chosen as representative. Therefore, the emission factor for

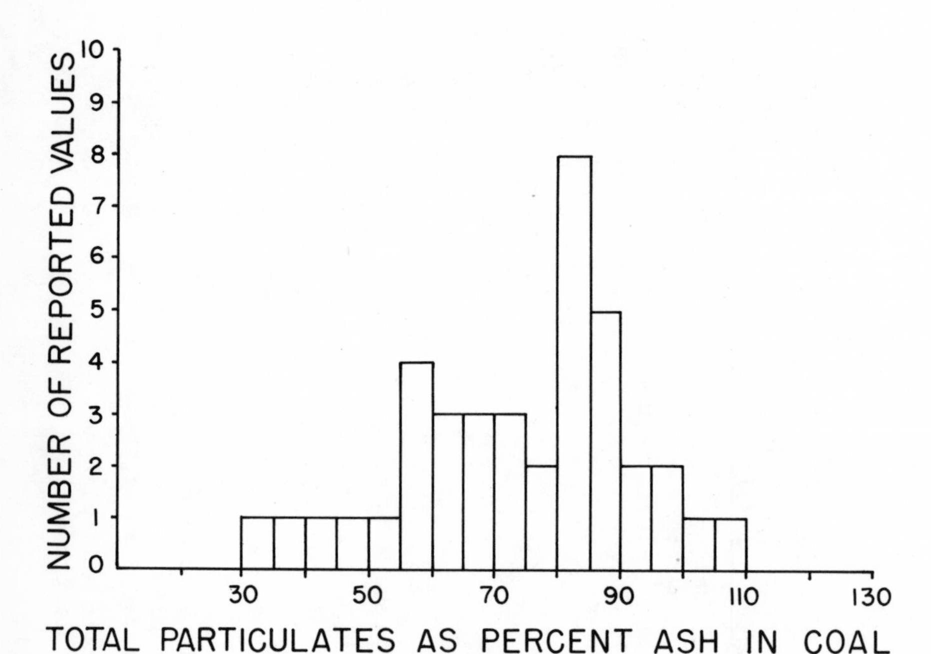

Figure 3.3. Particulate emission from pulverized coal-fired units.

pulverized coal boilers is 80% x 2000 lb/ton x % ash = 16 lb particulate/ton of coal. However, the range of data reported is from 30% to 110% of the ash content and the relative frequency of the interval of 75-85% is only 0.28 (11/39). The limitations of using the 80% of ash content figure to estimate the particulate emissions from pulverized coal boilers are obvious.

The authors of AP-42[3] recognize the limitations of the reported emission factors. The emission factors have been ranked alphabetically, A (excellent for emission source inventory work) through E (poor). The emission factor for particulate emissions from pulverized coal boilers, 80% x 2000 lb/ton x % ash = 16 lb particulate/ton of coal, has an applicability rating of A. AP-42 also includes supplementary data such as typical particle size distributions from various sources and average control efficiencies for particulate control equipment. These emission factors are often very useful in estimating the concentrations to expect when performing source emission tests.

PRIORITY SOURCES

The federal EPA has recently promulgated standards of performance for new stationary sources.[1] These regulations, given in Appendix A, establish performance standards for all steam generators (with heat inputs greater than 250 x 10^6 BTU/hr), incinerators (with burn rates greater than 50 ton/day), cement plants, nitric acid plants, and sulfuric acid plants, the construction and modification of which commenced after August 17, 1971. Table 3.2 shows a summary of the Priority I sources presently covered by the federal standards. These federal standards include and make provisions for:

1. construction plan reviews
2. notification of start up
3. public availability of emission data
4. more stringent state standards
5. emission standards
6. performance test methods and procedures
7. continual emission and fuel monitoring.

Thus, the processes to which the federal performance standards pertain will be discussed here. A number of good general references are given at the end of the chapter.

Fossil-Fuel Steam Generators

The combustion of fossil fuels, coal, oil, and natural gas, supplies approximately 95% of the total heat energy used in the United States. Most of this energy is used to produce steam for electrical power generation, process heating, and space heating. The major air pollutants associated with fossil fuel combustion are particulate, sulfur oxides, and nitrogen oxides. A simplified block diagram of a steam generation process is shown in Figure 3.4. The fuel--coal, oil, or gas--is combusted in the boiler. Combustion air is provided by forced draft fans. The heat of combustion is used to vaporize water in a heat transfer system on the interior of the boiler. The steam can then be used for heating purposes or to do work. To increase the overall efficiency of the system the combustion gases are usually passed through heat transfer devices such as economizers (to preheat boiler feed water) and air preheaters (to preheat the forced combustion air). Induced draft fans are used to keep fossil fuel combustion processes under negative pressure with respect to ambient conditions. This precludes the leaking of noxious combustion gases into the boiler house. Air pollution control equipment is usually added to the system between the boiler and the induced draft fans.

There are many different types of steam boilers. Boilers are classified in accordance with their rated energy output, the fuel combusted, and how the fuel is burned. Steam boilers range from the small domestic type for home heating to large electric utility boilers, which generate millions of pounds of steam per hour. Gas-fired boilers use either atmospheric or power burners to mix fuel and combustion air for combustion. Combustion air is added by the venturi principal in atmospheric burners and is provided by forced draft fans for the power gas burners. Oil-fired boilers combust oil droplets in an air suspension. The oil is vaporized with heat or atomized mechanically. Solid fuel, usually bituminous coal, is commonly burned in four different ways:

1. in stationary fuel beds
2. in moving fuel beds
3. in air suspensions
4. in combinations of air suspension and moving beds.

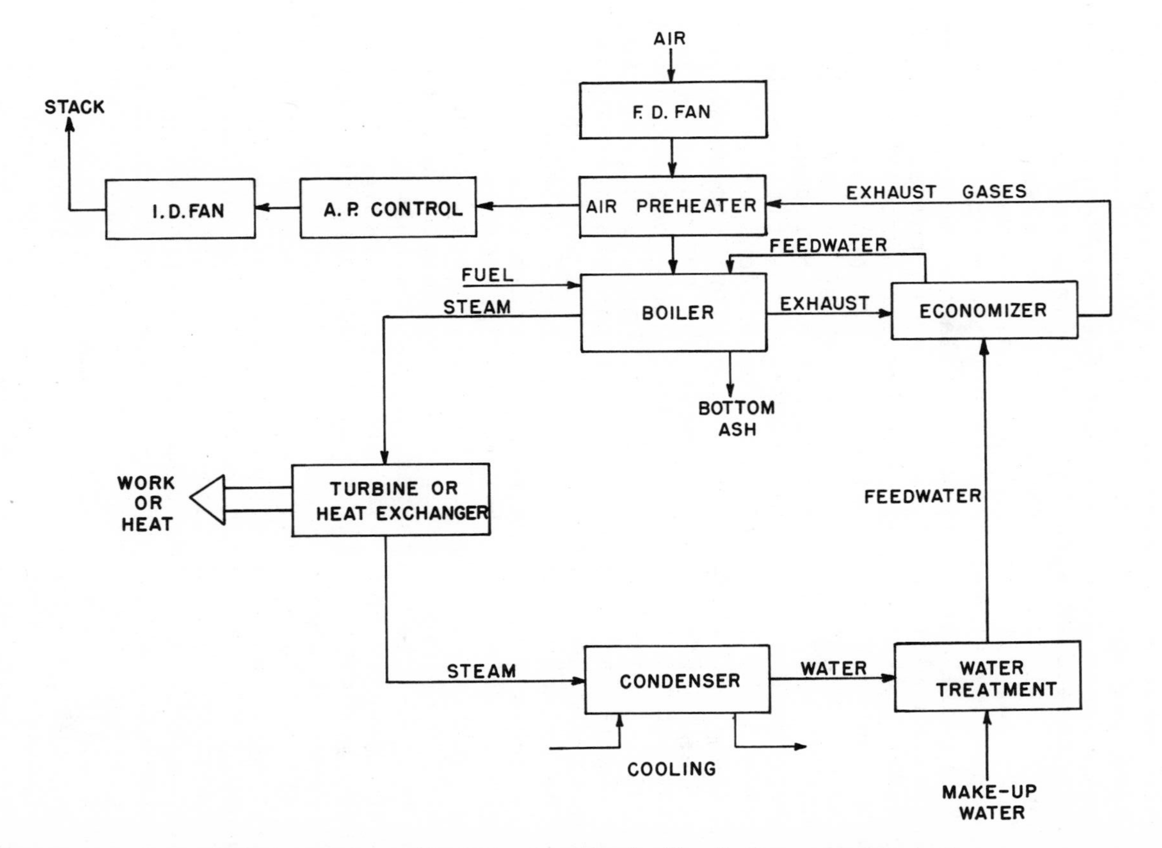

Figure 3.4. *Simplified process diagram for fuel-fired steam generator.*

Stationary fuel bed combustion units are hand fired and very inefficient. Thus, the method is only used in outdated domestic and small commercial boilers applications. Moving fuel beds are still routinely used, generally in industrial, commercial, and domestic applications. In these stoker-fired boilers, as they are called, the coal is fed onto moving grates from above (overfeed stokers) or below (underfeed stokers) the moving fuel bed. In stokers the primary combustion air is always added from beneath the moving fuel bed. Pulverized and and spreader stoker boilers are the most efficient and thus are generally used in large industrial and electric utility applications. Both pulverized and spreader stoker boilers utilize air suspension techniques for coal combustion. In pulverized boilers the fuel is finely ground and is totally combusted while in air suspension. Spreader stokers burn a larger size of coal particle, which undergoes the initial oxidation process while in a suspension and the final oxidation process while on a moving grate. Several of the general references at the end of this chapter deal with specific boiler designs, their operation and air pollutant emissions. All significant air pollutant emissions are exhausted from the boiler stack.

The principal pollutants emitted from fossil fuel steam generation processes are particulate, SO_2, and NO_x. References 5 through 7 present air pollution control methods for the pollutants. When the combustion process is inefficient, carbon monoxide and hydrocarbon emissions can be a problem. This is not usually the case, however, since steam cannot be economically produced without efficient fuel combustion.

Particulate emissions from steam boilers are directly related to the ash content of the fuel burned. Coal ash content up to 20% is not uncommon. Compared to this the ash content of oil and gas is much smaller and almost insignificant by comparison. Because of the type of fuel burned, pulverized coal boilers have the highest particulate emission rates. Uncontrolled particulate emission rates from pulverized coal boilers of up to 2000 lb/hr are not uncommon. Mechanical collectors and electrostatic precipitators are usually used to control particulate emissions from coal-fired boilers. Precipitators can be designed for control efficiencies of over 99%. Generally, particulate control is not needed on oil- and gas-fired boilers.

Sulfur dioxide emissions from fossil-fueled steam generators are a function of the sulfur content of the fuel burned. Typically bituminous coal has a sulfur content of 1-6%, residual fuel oils from 0.5-4%, and distillate oils less than 1%. The sulfur content of natural gas is usually insignificant. Some natural gases, however, contain appreciable amounts of reduced sulfur compounds and are therefore a problem. Sulfur dioxide emissions of hundreds of pounds per hour are typical when bituminous coal is burned. Currently there are numerous methods of sulfur dioxide control under development but as yet no method is in widespread use.

Nitrogen oxide emissions from fossil-fueled steam generators depend on combustion conditions such as high flame temperatures, long residence times, and excess oxygen available for the oxidation of N_2 to NO_x. Oxidation of nitrogenous compounds in the fuel burned also contributes to NO_x emissions. To date, the only method of controlling NO_x emissions has been the improving of furnace designs to reduce flame temperatures and oxygen needed for combustion.

The federal emission standards[1] for new (after August 17, 1971) fossil-fired steam generators with heat inputs greater than 250 x 10^6 BTU/hr are:

A. Particulate matter
   1. 0.10 lb/10^6 BTU input
   2. 20% opacity (40% opacity allowed for 2 minutes per hour)

B. Sulfur dioxide
   1. 0.80 lb/10^6 BTU input (liquid fuel)
   2. 1.20 lb/10^6 BTU input (solid fuel)

C. Nitrogen oxides
   1. 0.20 lb/10^6 BTU input (gaseous fuel)
   2. 0.30 lb/10^6 BTU input (liquid fuel)
   3. 0.70 lb/10^6 BTU input (solid fuel)

All sampling and analytical methods for emissions and fuels are specified by the federal EPA.[2] The federal test methods and procedures for fossil-fueled steam generators require that EPA methods 1, 2, 3, 5, 6, 7, and 9 be used to determine compliance. The particulate test consists of three repetitions of method 5, the sulfur dioxide test of three repetitions of two samples each according to method 6, and the nitrogen oxide test of three repetitions of four samples each according to EPA method 7. All performance tests should be conducted while the process is at or above the maximum steam rate at which the unit will be normally operated. All fuels and relevant operational conditions should also be representative of the normal process operation. Emission factors for typical steam generators are illustrated in Tables 3.3, 3.4, and 3.5.

To determine compliance with the federal performance standards boiler heat inputs have to be determined for each test. The BTU per hour input should be determined by suitable fuel flow measurement and confirmed by a material balance of the steam generation system. Generally the material balance can be done in two ways. The heat input is calculated either theoretically from fuel analysis data and the determined exhaust gas flow rate or from an estimated steam efficiency and steam output. Examples 3.1 and 3.2 illustrate such heat input calculations. As made obvious by these examples the steam efficiency method is quite a bit easier than the fuel analysis method. However, the steam efficiency method relies on the accuracy of the boiler house steam measurement instruments and estimated efficiency. In practice it is hard to convince management that the boiler operation instruments should be recalibrated before a source test. Typically the instruments are reliable enough for routine boiler operation only. Furthermore, boiler steam efficiency is usually an involved and questionable value.

An example of the detailed heat balance needed to determine boiler steam efficiency is shown in Figure 3.5. Boiler efficiency can only be determined by conducting well-defined extensive tests. Since these tests are costly, accurate boiler efficiency tests usually are not done routinely. Therefore, at the time of most source tests the exact boiler efficiency is unknown and can only be approximated by a previous efficiency test or manufacturer's rating. Calculating boiler heat input from fuel analysis and stack flow rate data is also a rather crude approximation. Representative fuel samples, nonstochiometric combustion, and measurement errors in determining stack flow rate are obvious shortcomings. Therefore, it is good practice to substantiate boiler heat inputs determined by fuel flow measurement by both methods.

Table 3.3
*Emission Factors for Bituminous Coal Combustion without Control Equipment**

Furnace size, 10^6 BTU/hr heat input	Particulates[a] lb/ton coal burned	Sulfur Oxides[b] lb/ton coal burned	Carbon Monoxide lb/ton coal burned	Hydro-Carbons[c] lb/ton coal burned	Nitrogen Oxides lb/ton coal burned	Aldehydes lb/ton coal burned
Greater than 100 (Utility and large industrial boilers)						
Pulverized						
General	16A	38S	1	0.3	18	0.005
Wet bottom	13A[d]	38S	1	0.3	30	0.005
Dry bottom	17A	38S	1	0.3	18	0.005
Cyclone	2A	38S	1	0.3	55	0.005
10 to 100 (large commercial and general industrial boilers)						
Spreader stoker[e]	13A[d]	38S	2	1	15	0.005
Less than 10 (commercial and domestic furnaces)						
Spreader stoker	2A	38S	10	3	6	0.005
Hand-fired units	20	38S	90	20	3	0.005

*Emission Factor Rating: A (after Reference 3)

[a]The letter A on all units other than hand-fired equipment indicates that the weight percentage of ash in the coal should be multiplied by the value given. Example: If the factor is 16 and the ash content is 10%, the particulate emissions before the control equipment would be 10 times 16, or 160 pounds of particulate per ton of coal.

[b]S equals the sulfur content (see footnote a above).

[c]Expressed as methane.

[d]Without fly-ash reinjection.

[e]For all other stokers, use 5A for paeticulate emission factor.

Table 3.4

*Emission Factors for Fuel Oil Combustion**

	Type of Unit			
		Industrial and Commercial		
Pollutant	*Power Plant* $lb/10^3$ *gal*	*Residual* $lb/10^3$ *gal*	*Distillate* $lb/10^3$ *gal*	*Domestic* $lb/10^3$ *gal*
Particulate	8	23	15	10
Sulfur dioxide[a]	157S	157S	142S	142S
Sulfur trioxide[a]	2S	2S	2S	2S
Carbon monoxide	0.04	0.2	0.2	5
Hydrocarbons	2	3	3	3
Nitrogen oxides (NO_2)	105	(40 to 80)[b]	(40 to 80)[b]	12
Aldehydes (HCHO)	1	1	2	2

*Emission Factor Rating: A (after Reference 3).

[a]S equals percentage by weight of sulfur in the oil.

[b]Use 40 for tangentially fired units and 80 for horizontally fired units.

Table 3.5

*Emission Factors for Natural-Gas Combustion**

	Type of Unit				
	Power Plants $lb/10^6\ ft^3$	*Industrial Process Boilers* $lb/10^6\ ft^3$	*Domestic and Commercial Heating Units* $lb/10^6\ ft^3$	*Gas Turbines* $lb/10^6\ ft^3$	*Gas Engines* $lb/10^6\ ft^3$
Particulates	15	18	19	-	-
Oxides of sulfur[a] (SO_2)	0.6	0.6	0.6	-	-
Carbon monoxide	0.4	0.4	20	-	-
Hydrocarbons	40	40	8	-	-
Oxides of nitrogen (NO_2)	390	(120 to 230)[b]	(50 to 100)[c]	200	(770 to 7,300)[d]
Aldehydes (HCHO)	3	3	10	-	-
Organics	4	7	1	-	-

*Emission Factor Rating: B (after Reference 3)

[a]Based on average sulfur content of natural gas of 2,000 grains/10^6 ft^3.

[b]Use 120 for smaller industrial boilers <500 boiler horsepower and 230 for larger industrial boilers >7,500 boiler horsepower.

[c]Use 50 for domestic heating units and 100 for commercial units.

[d]Use 770 for oil and gas production; 4,300 for gas plants; 4,400 for refineries; and 7,300 for pipelines.

Example 3.1. *Heat Input as Derived from Fuel Analysis*

	Ultimate Analysis of Fuel (bituminous coal)	
	% by weight	*Molecular weight*
Carbon	65.25	12
Hydrogen	4.23	2
Oxygen	9.09	32
Nitrogen	1.10	28
Sulfur	3.11	32
Ash	17.22	-
BTU Value	11,780 BTU/lb	

Assumptions:

1. Boiler bottom and fly ash is 3% carbon.
2. The combustion process is stoichiometric and follows the relationships:

$$C + O_2 \rightarrow CO_2$$
$$H_2 + O \rightarrow H_2O$$
$$S + O_2 \rightarrow SO_2$$

Combustion Air (per 100 lb of Coal)

O_2 for carbon	65.25/12 =	5.438 lb mole
O_2 for hydrogen	4.23/2x2 =	1.058 lb mole
O_2 for sulfur	3.11/32 =	.097 lb mole
O_2 in coal	-9.09/32 =	-.284 lb mole
O_2 for carbon in ash	-17.22x.03/12 =	.043 lb mole
O_2 required Total	=	6.266 lb mole O_2/100 lb coal

Oxygen from Excess Combustion Air (as determined by EPA method 3 at sampling location)

82.9% XA: 6.266 x .829 = 5.194 lb mole

Total O_2 required:

6.266 + 5.194 = 11.460 lb mole O_2/100 lb coal

Total air required:

$$11.460 \frac{\text{lb mole } O_2}{\text{100 lb coal}} \times 4.76 \frac{\text{lb mole air}}{\text{lb mole } O_2} = 54.55 \frac{\text{lb mole air}}{\text{100 lb coal}}$$

Volume calculation (at 32°F):

$$54.55 \frac{\text{lb mole air}}{\text{100 lb coal}} \times 359 \frac{\text{ft}^3}{\text{lb mole}} = 19{,}583 \text{ ft}^3 \text{ air/100 lb coal}$$

Coal feed rate calculation (using stack flowrate as determined by EPA method 2; 45,400 SCFM):

$$\frac{\text{45,400 SCFM x 60}}{\text{19,583 SCF air/100 lb coal}} = 13{,}900 \text{ lb coal/hr}$$

Heat input calculation:

$$13{,}900 \text{ lb coal/hr} \times 11{,}780 \text{ BTU/lb coal} = 164 \times 10^6 \text{ BTU/hr}$$

Example 3.2. Heat Input Calculation as Derived from Steam Flow.

Assumptions:

1. There are no steam losses from the boiler during the test.
2. All boiler house instruments for monitoring temperatures, pressure, and flow rate of steam and feed water are accurate.
3. The steam efficiency of the boiler is 0.78 as estimated from a previous test.

Heat Input Calculations:

Average flow rate; steam & feed water 119×10^3 lb/hr

Steam conditions 414 PSIG; 650°F; h = 1333.2 BTU/lb

Feed water conditions 225°F; h = 1140 BTU/lb steam

$$\text{Heat Input} = \frac{(1140 \text{ BTU/lb})\ (119 \times 10^3 \text{ lb/hr})}{0.78} = 174 \times 10^6 \text{ BTU/hr}$$

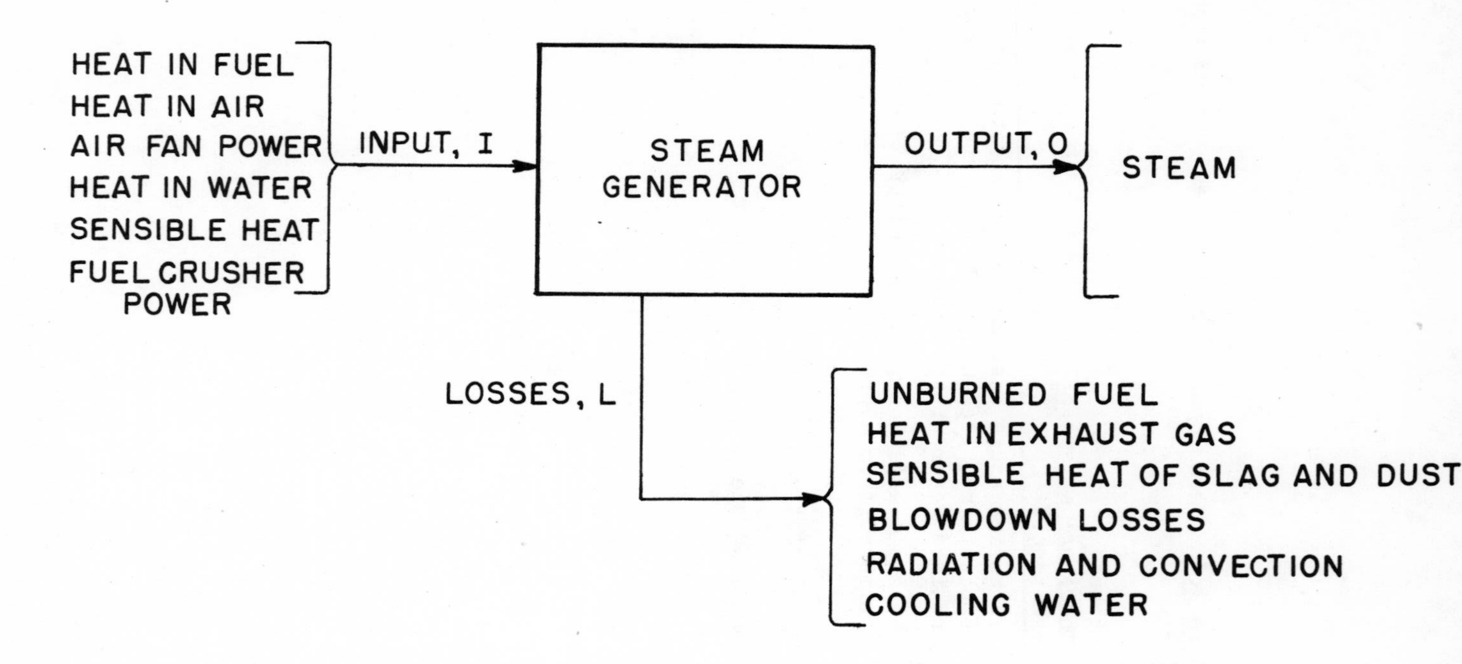

$$\text{EFFICIENCY \%} = \eta = O/I \times 100 = (I-L)/I \times 100$$

Figure 3.5. Heat balance of steam generator.

The federal EPA new source performance standards for fossil fuel fired steam generating units also require

1. continuous monitoring of emission opacity
2. continuous monitoring of SO_2 emissions, both directly and by daily fuel analyses
3. continuous monitoring of NO_x emissions
4. daily determination of fuel flow rate
5. weekly determination of BTU and ash content of fuel
6. daily recording of the average electrical output and the minimum and maximum hourly generation rate (where applicable)
7. maintaining records of the above measurements for two years following the date of measurement.

Incinerators

An incinerator is any furnace used to reduce the volume and/or mass of solid or liquid waste by combustion. Incinerators range in size from large municipal refuse incinerators with capacities of hundreds of tons per day to small pathological waste incinerators with burn rates of less than 100 lb/hr.

The present federal performance standards for new sources apply only to incinerators with burn rates exceeding 50 ton/day of solid waste. In the context of the standards solid waste is defined as refuse that is greater than 50% municipal waste, which consists of paper, wood, yard wastes, food wastes, plastics, leather, rubber, and other combustible and noncombustible materials such as glass and rock.

There are two basic types of municipal incinerators, batch and continuous feed. For complete and efficient combustion of solid waste on a large scale, continuous operation is a necessity. Batch feed operations usually are employed only when cost is a dominant factor. In general, batch feed incinerators have smaller capacities than continuous feed units due to waste handling and charging problems. Continuous feed incinerators are fed with stoker mechanisms, with traveling grate stokers being the most common. Other drivers such as reciprocating and rocking grate stokers, rotary kiln, and circular grate stokers are also widely used. A schematic of a typical traveling grate municipal incinerator is shown in Figure 3.6. In this traveling grate incinerator the refuse is charged through a hopper

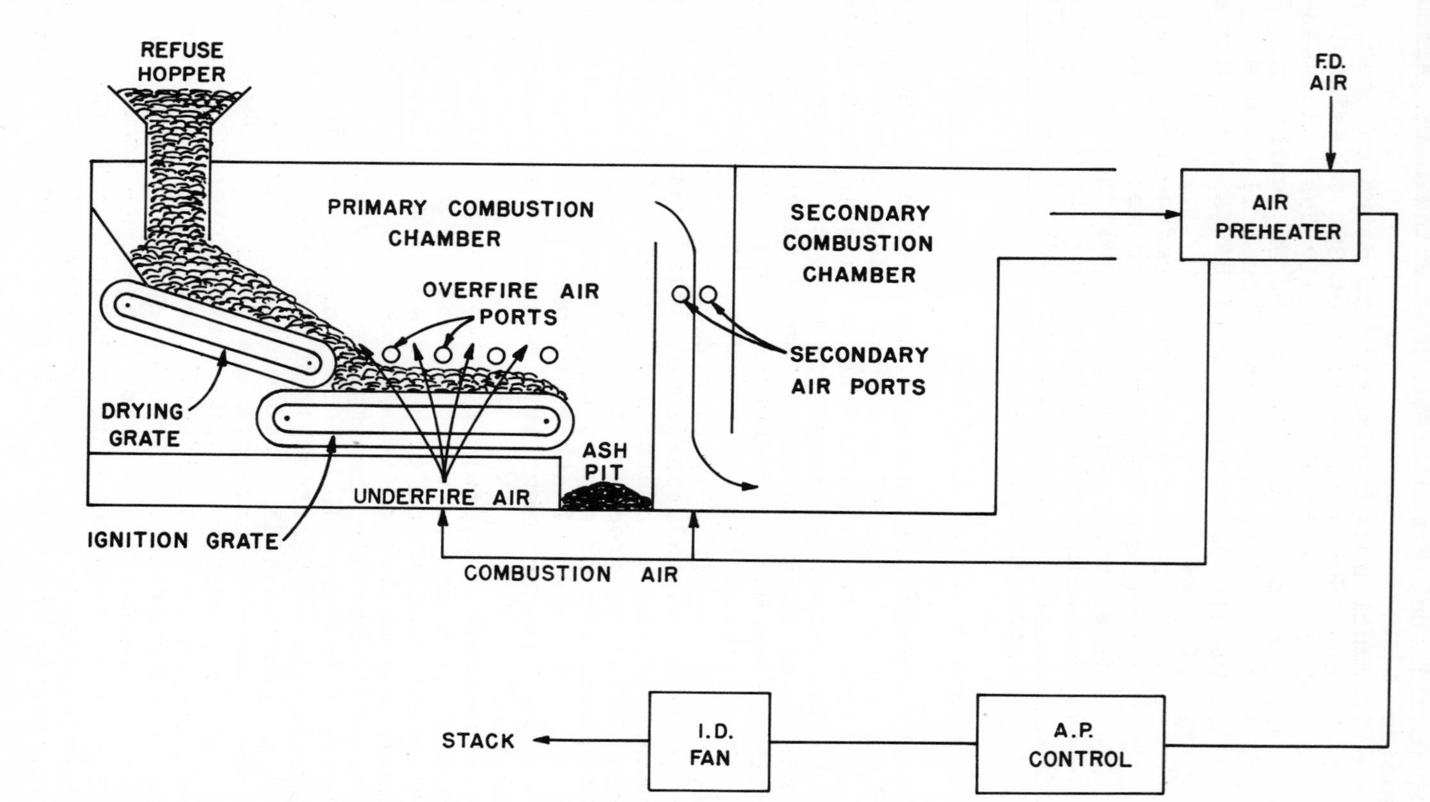

Figure 3.6. *Schematic diagram of a traveling grate incinerator.*

above the furnace. An inclined traveling grate partially dries and then delivers the refuse to the ignition grate. Combustion takes place on the traveling fuel bed, with underfire air promoting primary combustion of the refuse solids. The combustible gases driven off of the fuel are partially combusted in the primary chamber when mixed with overfire air. In multiple chamber incinerators secondary air is added and turbulently mixed with the remaining combustible gases to facilitate further combustion. Thus the combustion process is usually more complete in multiple chamber incinerators.

The principle air pollutants emitted from incinerators are particulate, SO_x, NO_x, CO, and hydrocarbons. Particulate emissions are a function of the ash content of the waste burned and the entrainment of fly ash by primary combustion air. Sulfur oxide emissions are dependent only on the sulfur content of the waste. Nitrogen oxide, carbon monoxide, and hydrocarbon emissions are functions of the combustion process. High temperatures, long residence times, and high levels of excess combustion air increase NO_x emissions. Insufficient temperature, mixing, combustion air, and residence time lead to higher CO and HC emissions.

Emission factors have been developed for most types of incinerators and are shown in Table 3.6. The current federal EPA emission standards for new municipal incinerators are for particulate only. They are 0.08 grams/SCF, corrected to 12% CO_2 as determined by EPA methods 1, 2, and 5 (3 repetitions). These performance standards further require that the source tests conducted for compliance be conducted using a fuel representative of normal operation, charged at or above the maximum burn rate. In order to normalize the measured particulate concentration to 12% CO_2, the standards require that integrated stack gas analyses be performed (EPA method 3) concurrently with the particulate emission tests. Operators of such incinerators must also record daily burn rates and all particulate emission tests performed. These records must be maintained for at least two years.

Typical control equipment used on municipal incinerators include wetted baffles, multiclones, and electrostatic precipitators (listed in order of increasing control efficiency). Scrubbers are also used, but their use usually causes liquid waste problems.

Table 3.6

Emission Factors for Refuse Incinerators Without Controls[a]*

Incinerator type	*Particulates lb/ton*	*Sulfur oxides*[b] *lb/ton*	*Carbon Monoxide lb/ton*	*Hydrocarbons*[c] *lb/ton*	*Nitrogen Oxides*[d] *lb/ton*
Municipal					
multiple chamber, uncontrolled	30 (8 to 70)	1.5	35 (0 to 233)	1.5	2
with settling chamber[e] and water spray system	14 (3 to 35)	1.5	35 (0 to 233)	1.5	2
Industrial/commercial					
multiple chamber	7 (4 to 8)	1.5[f]	10 (1 to 25)	3 (0.3 to 20)	3
single chamber	15 (4 to 31)	1.5[f]	20 (4 to 200)	15 (0.5 to 50)	2
controlled air	1.4 (0.7 to 2)	1.5	Neg	Neg	10
Flue-fed	30 (7 to 70)	0.5	20	15 (2 to 40)	3
Flue-fed (modified)[g]	6 (1 to 10)	0.5	10	3 (0.3 to 20)	10
Domestic single chamber					
without primary burner	35	0.5	300	100	1
with primary burner	7	0.5	Neg	2	2
Pathological	8 (2 to 10)	Neg	Neg	Neg	3

*Emission Factor Rating: A (after Reference 3)

[a]Average factors given based on EPA procedures for incinerator stack testing. Use high side of particulate, HC, and CO emission ranges when operation is intermittent and combustion conditions are poor.

[b]Expressed as SO_2.

[c]Expressed as methane.

[d]Expressed as NO_2.

[e]Most municipal incinerators are equipped with at least this much control.

[f]Based on municipal incinerator data.

[g]With afterburners and draft controls.

Municipal incinerator fireboxes are either refractory lined or of the water wall design. Water-walled incinerators are equipped with steam boilers and have the advantage of recovering most of the heat of combustion. Space heating and steam or power generation are possibilities. There is less heat recovery from refractory lined incinerators. Without a waste heat boiler only the preheating of combustion air is possible. In fact, combustion gases must usually be cooled by the addition of air and/or water to protect the refractory lining and emission control equipment. Diluting the particulate concentration in this manner leads to higher control costs per unit mass of particulate. In the past the simplicity of refractory lined incinerators has limited the use of the more costly water-walled units in the United States. Possibly the trend will reverse in the future due to the increasing importance of control equipment and the depletion of fuel reserves.

Portland Cement Production

The raw materials needed to produce portland cement are lime, silica, alumina, and iron; they are usually supplied by the mixing of limestone and clay, shale, or iron ore. There are four major steps in the portland cement process: quarrying and crushing, grinding and blending, cement clinker production, and finish grinding and packaging. Figure 3.7 is a schematic of the portland cement process. The raw materials are quarried and then crushed to approximately 1 inch in diameter. These are then blended and finely ground to the desired mixture using one of two common methods: the wet or dry process. In the wet process water is added to the blended crushed materials, which are subsequently ground in wet tube or ball mills. The dry process simply involves the drying and grinding of the crushed raw materials after they are blended to the desired mixture. Rotary grinders or hammer mills followed by tube mills are usually used in the dry process. Simultaneous drying and grinding is also used in combination mill-dryers.

Though the wet process is the original method of material preparation, it is gradually being replaced by the dry blending and grinding process. In newer plants the dry process results in heat savings and better process control. Once the material is blended and finely ground, the mixture

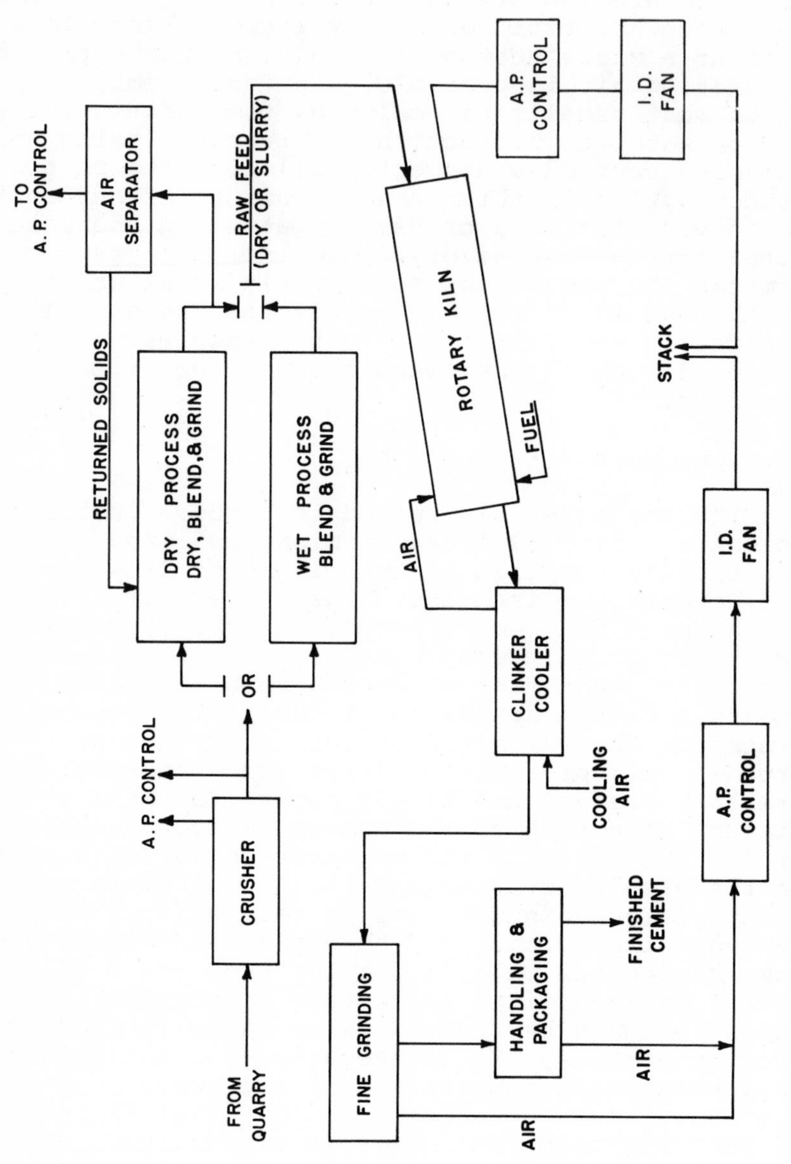

Figure 3.7. Simplified process diagram for cement plant.

is introduced as a slurry or raw dry meal into an inclined rotary cement kiln, which is 6-20 ft in diameter and up to 600-700 ft long. Here the raw materials are heated as they pass counterflow to hot combustion gases. While in the kiln (1-4 hours) the mixture is heated to 2900°F, the fusion point at which the cement clinker is formed. The cement formed can be approximated by the five component $CaO-SiO_2-Al_2O_3-Fe_2O_3-MgO$ system. The chief strength-producing component of cement is tricalcium silicate, $3CaO-SiO_2$. Heat is provided for the clinker formation by the combustion of fossil fuel (pulverized coal, oil, or natural gas) at the outlet end of the inclined rotary kiln. Waste heat boilers are sometimes added to recover some of the heat from dry process kiln exhaust gases, which may be on the order of 1200-1500°F.

Following formation the clinkers are air cooled. The cooling air from these clinker coolers is sometimes reused as preheated combustion air for the kilns. Once cooled the cement clinkers can be finely ground and 4-6% gypsum ($CaSO_4$) is added to control the setting time of the cement as it is used in mortar and concrete. Typical portland cement plants produce 1-2 million barrels (376 lbs per barrel) of cement per year.

Since the portland cement process involves repeated handling of both crushed and finely ground dry material, particulate emissions are a problem in virtually all parts of the process. However, the rotary kiln, aggregate driers, and clinker coolers used are the major particulate sources in the portland cement process. Uncontrolled emission dust concentrations from these parts of the process are on the order of 1-12 gr/SCF. At these concentrations, uncontrolled emissions of 20-100 ton/day from typical plants are common. With the exception of the CO_2 produced as the limestone is calcined in the kiln, the portland cement process produces no significant gaseous air pollutant emissions. Sulfur oxide emissions from fossil fuel combustion within the kiln are usually controlled to a significant extent by the alkalai content of the raw materials used. Emission factors for the portland cement process are shown in Table 3.7.

The magnitude of particulate emissions from rotary cement kilns and driers is so great that efficient control and collection is required because of the air pollution potential and the economics of the process. Typically the particulate size

Table 3.7

Particulate Emission Factors for Cement Manufacturing[a]*

Type of Process	*Uncontrolled Emissions*[b] *lb/bbl*
Dry process	
kilns	46 (35 to 75)
dryers, grinder, etc.	18 (10 to 30)
Wet process	
kilns	38 (15 to 55)
dryers, grinders, etc.	6 (2 to 10)

*Emission Factor Rating: B (after Reference 3)

[a]One barrel of cement weighs 376 pounds (171 kg).

[b]Typical collection efficiencies are: multicyclones, 80%; old electrostatic precipitators, 90%; multicyclones plus old electrostatic precipitators, 95%; multicyclones plus new electrostatic precipitators, 99%; and fabric filter units, 99.5%.

distribution requires that combinations of inertial separators (such as multiclones) and more efficient collectors such as electrostatic precipitators and baghouses be used for particulate control. The particulate collected is of sufficient quantity and quality to be recycled to the process. A portion of the collected particulate emissions, however, is always wasted due to quality restrictions.

The federal performance standards apply to all new portland cement plants using either the wet or the dry process, and the emission standards apply to particulate emissions only. They are:

1. Rotary kiln emissions
   a. 0.30 lb/ton of feed to kiln
   b. 10% opacity
2. Clinker cooler emissions
   a. 0.10 lb/ton of feed to kiln
   b. 10% opacity
3. Any other facility* other than the kiln or clinker cooler
   a. 10% opacity

*These facilities include the raw mill system, finish mill system, raw mill dryer, raw material storage, clinker storage, product storage, conveyor transfer points, and the bagging and bulk loading and unloading systems.

The performance standards also require that source emission tests be conducted according to EPA methods 1, 2, 3, 5, and 9 at or above the maximum production rate of the process. To determine emissions in the form of lb of particulate/ton of kiln feed, total kiln feed (except fuels) in tons/hour must be determined concurrently with each repetition (three in all) of the source test. Operators of new cement plants are also required to maintain records of daily production and kiln feed rates and any particulate emission measurements for two years following the date of each measurement.

Nitric Acid Production

Nitric acid is most economically produced by the oxidation of anhydrous ammonia. The chief use of nitric acid is in the production of nitrates, with most of the nitric acid produced today being used to produce ammonium nitrate for fertilizers and explosives. The significant materials used are simply anhydrous ammonia, air and water. A platinum-rhodium catalyst is also used.

Currently most of the nitric acid produced in the United States is produced by the oxidation of ammonia ($NH_3 \rightarrow NO_2$) and subsequent absorption of NO_2 in water ($NO_2 + H_2O \rightarrow HNO_3$). The basic chemical equations describing the process are

$$4NH_{3\,(g)} + SO_{(g)} \xrightarrow[1650°F]{cat.} 4NO_{(g)} + 6H_2O_{(g)} \quad (3\text{-}1)$$

$$2NO_{(g)} + O_{2\,(g)} \rightarrow 2NO_{2\,(g)} \quad (3\text{-}2)$$

$$3NO_{2\,(g)} + H_2O_{(g)} \rightarrow 2HNO_{3\,(aq)} + NO_{(g)} \quad (3\text{-}3)$$

The oxidation of NH_3 and consequent absorption of NO_2 in water to form nitric acid can be carried out at atmospheric or intermediate pressures. However, the economics involved favor the pressure process. The oxidation rate of $NO \rightarrow NO_2$ is a function of the square of the pressure under which it is carried out, and absorption of NO_2 in water is also more efficient under pressure. Thus, lower pressure processes need process equipment with larger volumes and retention times to produce equivalent amounts of nitric acid. In most cases the capital investment needed for nitric acid production by the pressure process ($\simeq$ 100 psig) is less than that of lower pressure processes.

In the pressure process, compressed anhydrous ammonia is vaporized and mixed with preheated compressed air (100 psig; 400°F). The pressurized mixture (approximately 10% NH_3 and 90% air) is passed over a hot (1650°F) catalyst of platinum and rhodium (90-10%). In the presence of the hot catalyst the ammonia is oxidized to NO and water vapor. The process gas (a mixture of NO, O_2, N_2 and H_2O vapor) is then cooled to promote the secondary oxidation of the nitric oxide by the remaining oxygen. Air heaters and boilers are generally used to recover some of the wasted heat. The process gas (now a mixture of NO, NO_2, N_2, and H_2O vapor) is then further cooled to approximately 100°F using a cooler condenser where part of the water condensed reacts with the NO_2 to form nitric acid, HNO_3. A liquid separator is then used to direct the liquid (HNO_3 and H_2O mist) and gaseous (NO, NO_2, O_2, and H_2O vapor) process streams to an absorption tower. In the absorption tower more air and water are added to facilitate further oxidation of NO $\rightarrow$ NO_2 and the formation of nitric acid, HNO_3. Internal cooling within the absorption tower is used to enhance these reactions. Tail gas from the absorption tower is then vented to a mist separator to recover HNO_3 mist followed by subsequent exhausting to a stack or air pollution control equipment.

Usually the pressurized tail gas is expanded doing compressor work, thus recovering some of the energy necessary to run the pressure process. The product, 55-65% HNO_3, is emitted from the bottom of the absorption tower. When higher strength acid is desired, a concentration process must be used. Since the HNO_3-H_2O system has its maximum boiling point at approximately 68% HNO_3, a dehydrating agent must be used to concentrate the acid. Concentrated sulfuric acid which has a boiling point over 500°F, is commonly used. The sulfuric and nitric acids are added to the top of a dehydrating column. The mixture flows downward and the heat generated causes the HNO_3 to vaporize. The concentrated HNO_3 vapor is drawn off from top of the column and condensed. Dilute H_2SO_4 is recovered at the bottom of the tower. The nitrogen oxides and HNO_3 vapor in the tail gas from the acid condenser can usually be recovered by using a water absorption tower.

The major air pollution problem in the nitric acid production process is the tail gas from the absorption column. The tail gases are predominantly N_2 and O_2 with less than 1% unconverted nitrogen oxides, chiefly NO and NO_2. HNO_3 mist entrainment is also possible. Nitrogen oxides are emitted

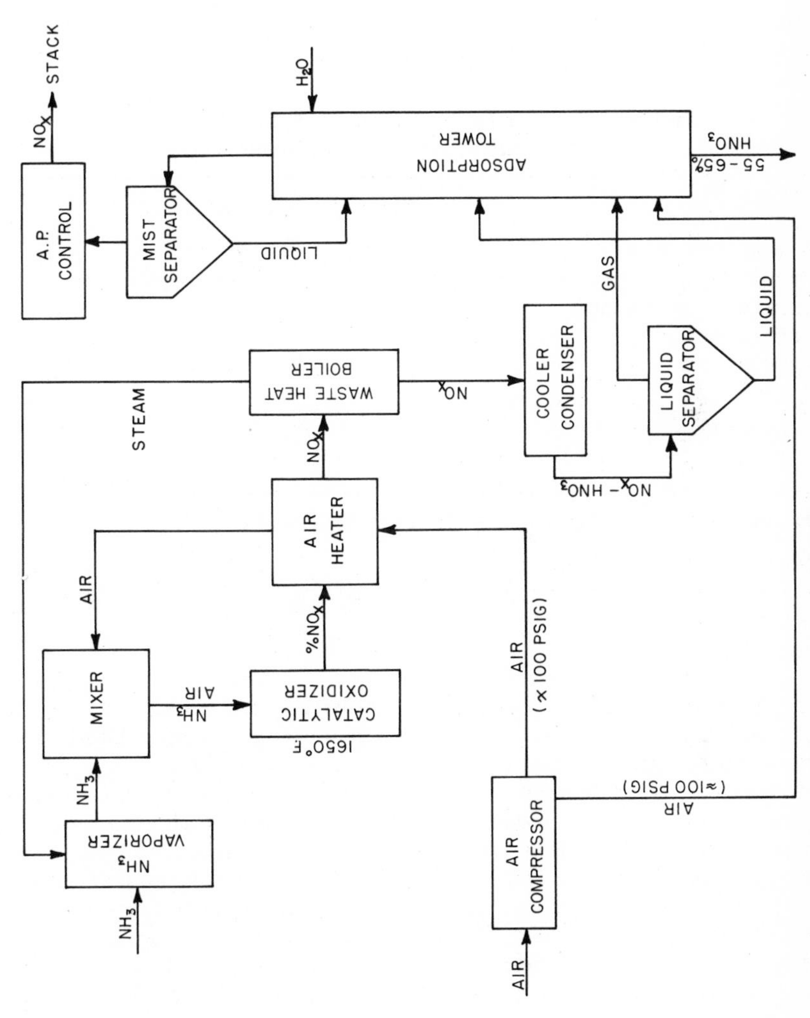

Figure 3.8. *Simplified Process Diagram for Nitric Acid Production*

because the absorbing efficiency of the tower decreases as HNO_3 content in the tower increases. Increasing the oxygen concentration and absorption pressure while the absorption pressure is increased will lower the NO_x emissions from absorbing towers. Emission factors for nitric acid production are given in Table 3.8.

Table 3.8

*Emission Factors for Nitric Acid Plants Without Control Equipment**

Type of Process	*Nitrogen Oxides (NO_x)*[a] *lb/ton*
Ammonia - oxidation	
old plant	57
new plant	2 to 7
Nitric acid concentrators	
old plant	5
new plant	0.2

*Emission Factor Rating: B (after Reference 3)

[a]Catalytic combustors can reduce emissions by 36-99.8%, with 80% the average control. Alkaline scrubbers can reduce emissions by 90%.

The principal control techniques used to reduce NO_x emissions from absorption towers are alkalai scrubbers and catalytic reduction of the NO_x. Typically scrubbers cause liquid waste problems. However, in this case, if NO_x concentrations are high enough, usable nitrite and nitrate salts can be produced. In plants using the pressure process to produce nitric acid, catalytic reduction is usually the most economical and efficient (up to 90%) control method. The heat generated by NH_3 oxidation and the catalytic reduction of NO_x tail gases to nitrogen can be used to do compressor work and/or provide process steam. In the catalytic reduction control process a hydrocarbon fuel such as natural gas is added to the reheated absorbing tower tail gases. As the oxygen in the tail gas is depleted, the nitric oxides are reduced to nitrogen according to the following reactions:

$$CH_4 + 2O_2 \rightarrow CO_2 + 2H_2O \quad (3\text{-}4)$$

$$CH_4 + 4NO_2 \rightarrow 4NO + CO_2 + 2H_2O \quad (3\text{-}5)$$

$$CH_4 + 4NO \rightarrow 2N_2 + CO_2 + 2H_2O \quad (3\text{-}6)$$

The federal performance standards for new nitric acid plants producing acid which is 30-70% in strength specify the following NO_x emission standards:

1. 3 lb NO_x/ton HNO_3 (as 100% HNO_3)
2. 10% opacity

Performance tests are to be run (three repetitions) using EPA methods 1, 2, and 7 and at or above the maximum acid production rate of the facility. The acid produced expressed as tons/hr of 100% HNO_3 also has to be determined and substantiated with material balances during the performance tests. The standards also require the continuous monitoring of NO_x emissions, and the recording of daily production data, which must be filed and retained for two years following each measurement.

Sulfuric Acid Production

Sulfuric acid is a very strong and economically produced inorganic acid. For this reason it is one of the most widely used chemicals. Phosphate fertilizer production, the chemical production industry, and the petroleum refining industry use most of the sulfuric acid produced in the United States.

There are two basic production processes for sulfuric acid, the chamber process and the contact process, the former being the original commercial production method. This process utilizes the reduction of nitrogen dioxide to nitric oxide in a lead chamber as the oxidation mechanism for converting sulfur dioxide to sulfur trioxide. The SO_3 is then subsequently hydrated to form sulfuric acid, H_2SO_4. The chamber process characteristically produces acid that is less than 80% H_2SO_4. For this reason the chamber process has been virtually replaced by the more efficient and economic contact process.

As in the chamber process, elemental sulfur, metal sulfide ores, refinery sludges, and by product SO_2 or hydrogen sulfide can be used to produce H_2SO_4 by the contact process. Elemental sulfur is usually used. Elemental sulfur is of high purity and relatively easy to handle. A schematic of a contact process sulfuric acid plant is shown in Figure 3.9. Molten sulfur is mixed with dried and preheated air

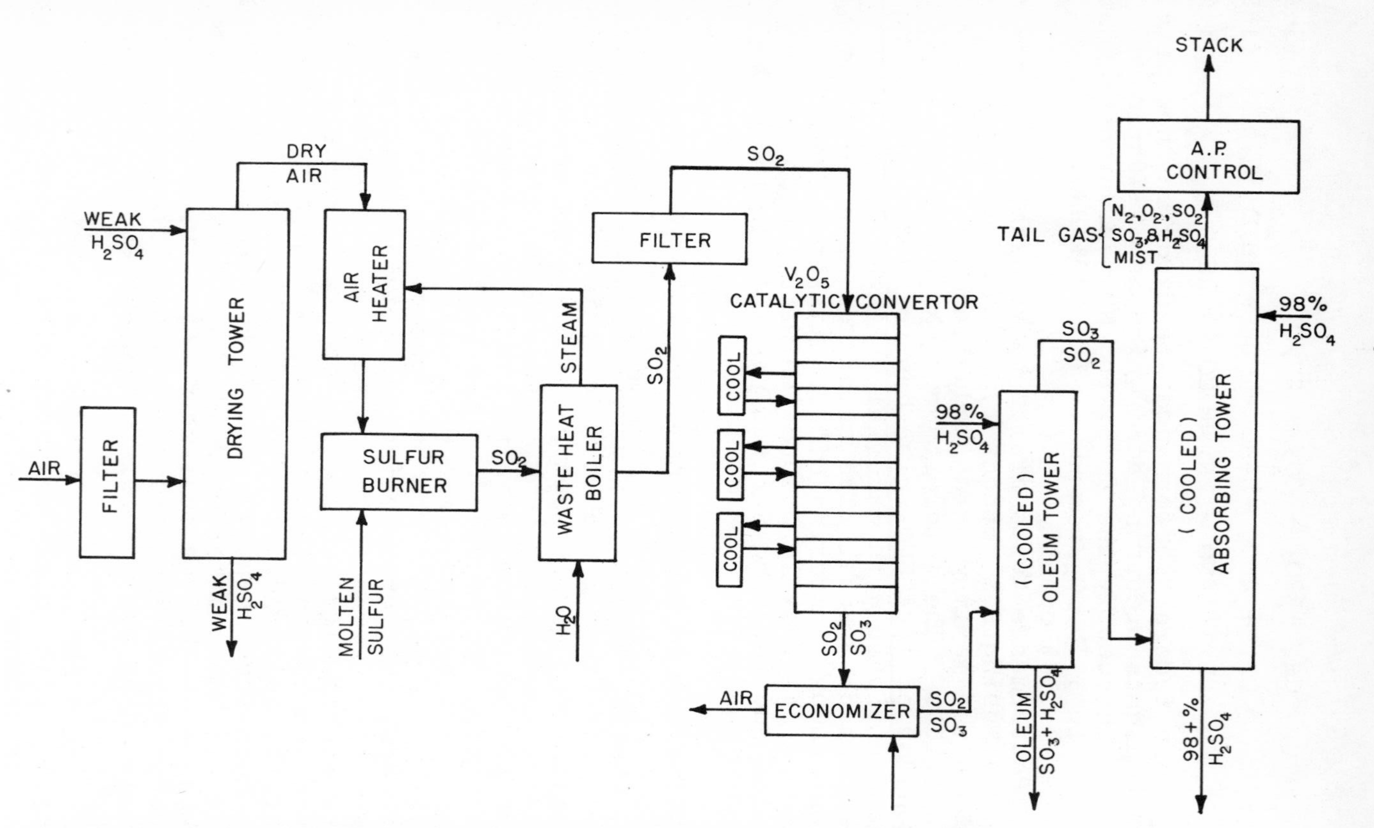

Figure 3.9. *Simplified process diagram for sulfuric acid production.*

and burned to form sulfur dioxide. Then the process gas (7-11% SO_2) is cooled by means of a waste heat boiler and filtered to remove all particulate impurities. The process gas is then introduced to a catalytic converter. In the converter the oxidation of sulfur dioxide to sulfur trioxide by oxygen is catalyzed by vanadium pentoxide, V_2O_5. The conversion is accomplished using 3-4 catalyst beds in series with intermittent cooling between the beds to increase conversion efficiency. Conversion efficiencies of 98% are common.

The converted gas, mostly N_2, O_2, and SO_3 with smaller amounts of SO_2, is then cooled in an economizer and delivered to an oleum tower. In the oleum tower the process gases are passed countercurrent to a flow of 98-99% sulfuric acid. Here, most of the SO_3 is absorbed to form what is called oleum or fuming sulfuric acid, a mixture of SO_3 dissolved in H_2SO_4. Oleum can be diluted with water to form 99-100% H_2SO_4. Thus it is an economic way to produce sulfuric acid for shipment. The absorbing efficiency of sulfuric acid for SO_3 varies inversely with concentration. Therefore, the oleum tower is followed by an absorbing tower which collects the remaining SO_3. The absorbing towers use 98% H_2SO_4 as the absorbing fluid, and 98+% H_2SO_4 is produced. Because of reaction heat, the oleum and absorbing towers are usually cooled to enhance their efficiency.

The major air pollutant emissions from sulfuric acid production are the tail gases from the final absorbing tower. These tail gases are principally nitrogen and oxygen but they also contain unconverted SO_2, unabsorbed SO_3 and entrained H_2SO_4 acid mist. The unconverted SO_2, which may be on the order of 0.5% of the tail gas volume, is the most serious problem. The SO_3 and H_2SO_4 mist emissions are not as much of a problem. Minor air pollutant emissions (SO_2, SO_3 and H_2SO_4 mist) are encountered in acid handling processes, such as the filling of storage tanks and tank cars, and in acid concentration processes. Process gas leaks prior to the absorbing towers also pose air pollution problems. The emission factors developed by the federal EPA for sulfuric acid plants are given in Table 3.9. As might be expected, the emission factors indicate that SO_2 emissions from sulfuric acid plants are inversely related to the conversion efficiency of SO_2 to SO_3 in the process.

Table 3.9

Emission Factors for Sulfuric Acid Plants[a]*

Conversion of SO to SO_3, *%*	SO_2 *Emissions* *lb/ton of 100%* H_2SO_4[b]
93	97
94	84
95	70
96	55
97	40
98	26
99	15
99.5	7

*Emission Factor Rating: B (after Reference 3)

[a]Acid-mist emissions range from 0.3 to 7.5 pounds per ton of acid produced for plants without acid mist eliminators, to 0.02 to 0.2 pound per ton of acid produced for plants with acid-mist eliminators.

[b]Use 40 (20) as an average factor if per cent conversion of SO_2 to SO_3 is not known.

When SO_2 control is required, ammonia or alkalai scrubbers are usually employed. Efficiencies on the order of 90% are attainable. However, scrubbing tail gases from the final absorber is a costly process. Markets are not usually available for the salts formed in the scrubbing process and the liquid effluent from the scrubbers can be a liquid waste problem. However, sulfuric acid mist is usually effectively controlled. Mist eliminators (wire mesh or glass fiber) and electrostatic precipitators are commonly used. When preceded by humidification, mist eliminators and electrostatic precipitators will also remove most of the SO_3, as it is converted to H_2SO_4 mist by the addition of water vapor. Mist eliminators will also partially control SO_2 emissions. For this reason SO_2 scrubbers are generally preceded by mist eliminators.

The present federal performance standards apply to all contact process sulfuric acid plants constructed after August 17, 1971. The standards, however, do not apply to facilities where the primary purpose is to control atmospheric emissions of SO_2

or other sulfur compounds. The emission standards are:

1. 4 lb SO_2/ton H_2SO_4 (as 100% H_2SO_4)
2. 0.15 lb H_2SO_4 mist/ton H_2SO_4 (as 100% H_2SO_4)
3. 10% opacity.

Performance tests must be conducted according to EPA methods 1, 2, 8 and 9. The tests (three repetitions) must be conducted while the process is operated at or above its maximum acid production rate. All test results must be normalized and expressed in terms of emissions per ton of 100% H_2SO_4. Therefore, the process output has to be measured and substantiated by a material balance over the production system. The operators of the process are also required to install continuous monitoring equipment for SO_2 and to keep records of all pertinent production data. Records of such data must be filed for two years.

REFERENCES

1. "Performance Standards for New Sources," Federal Register, Vol 36 No. 247, December 23, 1971.
2. Compilation of Air Pollution Emission Factors, Publ. No. AP-42, U.S. Public Health Service, Durham, North Carolina. 1968.
3. Compilation of Air Pollution Emission Factors, Publ. No. AP-42. U.S. Environmental Protection Agency, Research Triangle Park, North Carolina. 1972.
4. Atmospheric Emissions from Coal Combustion - An Inventory Guide, Publ. No. 999-AP-24. U.S. Public Health Service, Division of Air Pollution, Cincinnati, Ohio. 1966.
5. Control Techniques for Particulate Air Pollutants, Publ. No. AP-51. U.S. Public Health Service, National Air Pollution Control Administration, Washington, D.C. 1969.
6. Control Techniques for Sulfur Oxide Air Pollutants, Publ. No. AP-52. U.S. Public Health Service, National Center for Air Pollution Control, Washington, D.C. 1969.
7. Control Techniques for Nitrogen Oxide Emissions from Stationary Sources, Publ. No. AP-67. U.S. Public Health Service, National Air Pollution Control Administration, Washington, D.C. 1969.

GENERAL REFERENCES

1. Air Pollution Aspects of Emission Sources: Nitric Acid Manufacturing--A Bibliography with Abstracts, Publ. No. AP-93. U.S. Environmental Protection Agency, Research Triangle Park, North Carolina. 1971.

2. Air Pollution Aspects of Emission Sources: Municipal Incineration - A Bibliography with Abstracts, Publ. No. AP-92. U.S. Environmental Protection Agency, Research Triangle Park, North Carolina. 1971.
3. Air Pollution Aspects of Emission Sources: Sulfuric Acid Manufacturing - A Bibliography with Abstracts, Publ. No. AP-94. U.S. Environmental Protection Agency, Research Triangle Park, North Carolina. 1971.
4. Air Pollution Aspects of Emission Sources: Cement Manufacturing - A Bibliography with Abstracts, Publ. No. AP-95. U.S. Environmental Protection Agency, Research Triangle Park, North Carolina. 1971.
5. Air Pollution Aspects of Emission Sources: Electric Power Production - A Bibliography with Abstracts, Publ. No. AP-96. U.S. Environmental Protection Agency, Research Triangle Park, North Carolina. 1971.
6. Air Pollution Aspects of Emission Sources: Boilers - A Bibliography with Abstracts, Publ. No. AP-105. U.S. Environmental Protection Agency, Research Triangle Park, North Carolina. 1972.
7. Atmospheric Emissions from Fuel Oil Combustion - An Inventory Guide, Publ. No. 999-AP-2. U.S. Public Health Service, Division of Air Pollution, Cincinnati, Ohio. 1962.
8. Messersmith, C. W., Warner, C. F., and Olsen, R. A. Mechanical Engineering Laboratory, John Wiley & Sons, Inc., New York, New York. 1958.
9. Danielson, J. A. editor. Air Pollution Engineering Manual, Publ. No. 999-AP-40. U.S. Public Health Service, National Center for Air Pollution Control, Cincinnati, Ohio. 1967.
10. Fryling, G. R. editor. Combustion Engineering, Combustion Engineering Inc., New York, New York. 1966.
11. Steam, Its Generation and Use, The Babcock and Wilcox Company, New York, New York. 1963.
12. Steam Generating Units, ASME Power Test Code 4.1. American Society of Mechanical Engineers, New York, New York. 1964.
13. Stern, A. C. editor. Air Pollution, Vol III, Academic Press, New York. 1968.
14. Shreve, R. N. Chemical Process Industries, McGraw-Hill Book Company, New York, New York. 1967.
15. Atmospheric Emissions from Nitric Acid Manufacturing Processes, Publ. No. 999-AP-27. U.S. Public Health Service, National Center for Air Pollution Control, Cincinnati, Ohio. 1966.
16. Atmospheric Emissions from Sulfuric Acid Manufacturing Processes, Publ. No. 999-AP-13. U.S. Public Health Service, Division of Air Pollution, Cincinnati, Ohio. 1965.

17. Emissions from Coal-Fired Power Plants: A Comprehensive Summary, Publ. No. 999-AP-35. U.S. Public Health Service, National Air Pollution Control Administration, Durham, North Carolina. 1967.
18. DeMareo, J., Keller, D. J., Leckman, J. and Newton, J. L. Incinerator Guidelines - 1969, PHS Publ. No. 2012. U.S. Public Health Service, Bureau of Solid Waste Management, Washington, D.C. 1969.
19. Atmospheric Emissions from the Manufacturing of Portland Cement, Publ. No. 999-AP-17. U.S. Public Health Service, Bureau of Disease Prevention and Environmental Control, Cincinnati, Ohio. 1967.

CHAPTER 4

SAMPLING TRAIN COMPONENTS

Many different components are required to conduct source tests. First mentioned in Chapter 2, these basic components are discussed further in this chapter, which presents their theory of operation and provides practical comments regarding their calibration and use. Many references provide detailed coverage of the components presented in this chapter, with references 1 through 8 providing the most extensive coverage. The intent of this chapter is to discuss the components and identify their relative merits for use in source sampling applications.

TEMPERATURE MEASUREMENT

Several temperature measurements are required in conducting a source test, with the particulate train requiring the most. These measurements include the stack gas, filter, and cooled sample stream temperatures. Since absolute temperatures are used in all gas law calculations, the relative errors of temperature measurements are usually small. In source testing, dial thermometers and thermocouples are commonly used.

Temperature Scales

The two scales used in temperature measurement are degrees Fahrenheit and degrees Centigrade. The two absolute scales which will be used in calculation are degrees Rankine, °R, and degrees Kelvin, °K. Conversion for these is shown in the following equations and in Figure 4.1.

$$°R = °F + 460° \quad (4\text{-}1)$$

$$°K = °C + 273° \quad (4\text{-}2)$$

$$°F = (9/5)°C + 32° \quad (4\text{-}3)$$

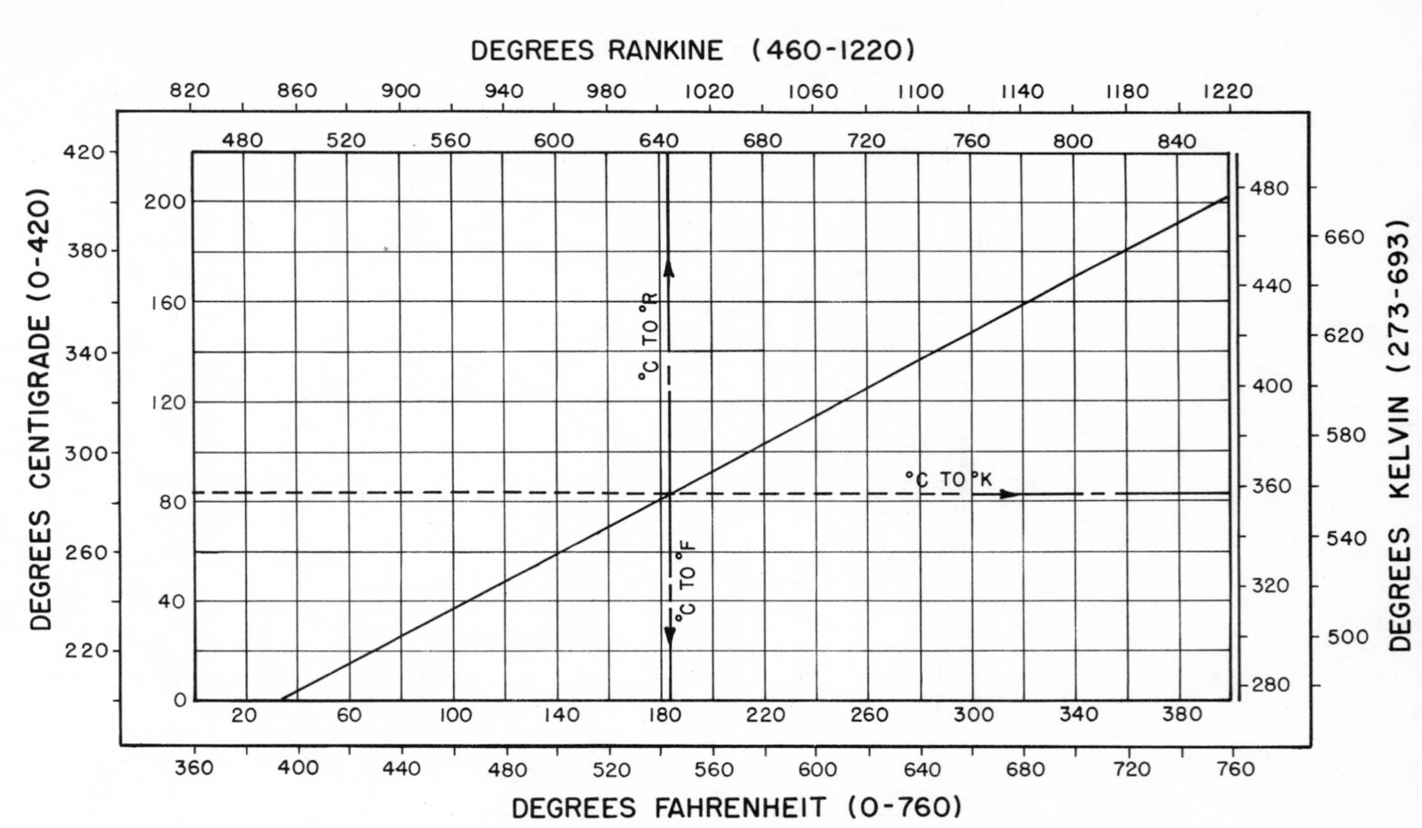

Figure 4.1. Temperature equivalent graph, Centigrade, Kelvin, Rankine, Fahrenheit.

Mercury Bulb Thermometers

The mercury bulb thermometer is simply a measurement of mercury expansion with temperature. This expansion is linear over the range of the thermometer. The thermometers are constructed of a glass capillary tube with a mercury reservoir at its base and the temperature calibration on the capillary tube. Glass thermometers do break easily and this is definitely a risk when they are used for source sampling. However, there is one mercury bulb thermometer rugged enough for field use, but this unit is quite heavy, inhibiting its practical use for measuring stack gas temperatures. To measure temperature, a large Hg reservoir is placed in the stack. This reservoir is connected to a gauge with a flexible tubular lead. The expansion of Hg in the bulb and lead is converted to mechanical motion on a dial face containing a temperature scale.

Dial Thermometers

In addition to the Hg dial thermometer, two basic types of dial thermometers are used in source testing. One is a bimetallic and the second is the gas bulb thermometer.

The bimetallic thermometer is based on a strip of two metals bonded together.[3] Because of different thermal coefficients of the two metals, the bonded strip will deform with temperature. Depending on the configuration of the strip, this deformation is transformed into a dial movement which contains a temperature scale. These thermometers should not be used above scale limit because the strip may not recover from the excess deformation. The accuracy of these thermometers depends on strip configuration and manufacturing care.

The gas bulb thermometer relies on the expansion of an inert gas with temperature.[3] The expansion is sensed as an increase in pressure. The dial temperature scale, then, is actually a pressure scale. Gas bulb thermometers are used for lower temperature ranges. Hence, this type of thermometer would be used mostly to measure gas temperatures within the sampling train.

Thermocouples

The thermocouple is the most popular device for measuring higher temperatures. As shown in Figure 4.2,

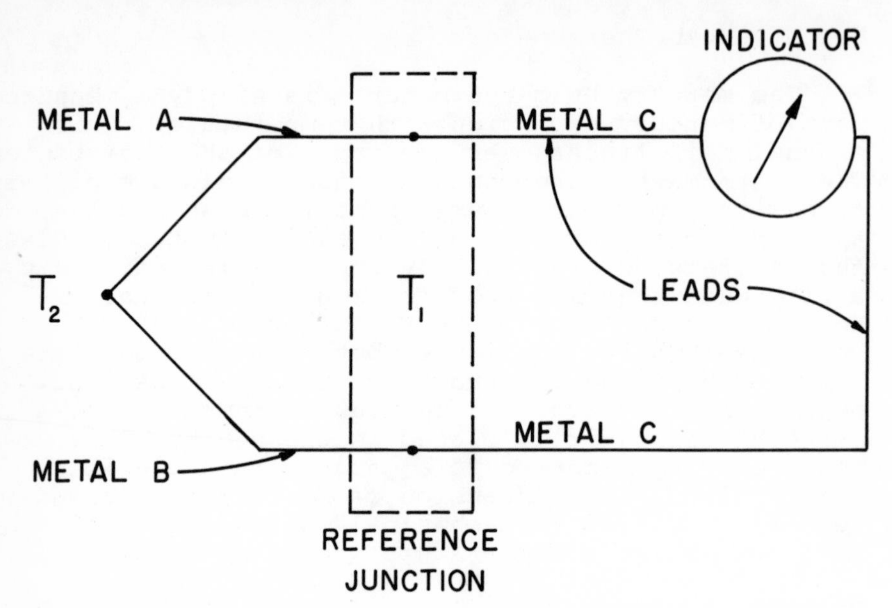

Figure 4.2. Thermocouple junction.

it consists of two dissimilar metal wires, A and B, connected together at one junction, with their other ends connected to a third wire, C. With the last two junctions held at a constant reference temperature, T_1, an electrical potential, emf, is developed due to the temperature, T_2. The magnitude of the potential depends upon the types of metals, A and B, and the magnitude of the gas temperature, T_1.

Several metal pairs can be used. Generally the Chromel/Alumel will be the best choice because of its resistance to oxidation. This pair is useful for a range of -300°F to 2300°F.[3] Other commonly used metal pairs are:

Copper/Constantan	-300 to 660°F
Iron/Constantan	-240 to 1850°F
Platinum/Platinum 10% Rhodium	32 to 2800°F

Several errors can occur in the use of a thermocouple. Local exothermic chemical reactions on the surface of the thermocouple may cause high readings. This can be minimized by applying special coatings to the surface. In some cases, the thermocouple may be contaminated by absorption of a surrounding third metal. It then is a different junction of unknown emf characteristics. At high temperature, the measured temperature may be lowered by radiation heat losses from the thermocouple surface. It is very difficult to estimate the magnitude of such an error. Such errors are more significant at higher temperatures because the amount of heat loss is proportional to the fourth power of the absolute temperature. This problem can be somewhat alleviated by using shielded thermocouples. The shield will tend to increase the response time but this is not critical for most source testing applications.

Depending on how the measurement is made, the two metals may be soldered. The safest approach however is to weld all junctions. Bare thermocouples will not have the life expectancy of those protected from oxidation and corrosion. However, thermocouples are relatively inexpensive.

The most commonly used method of measuring thermocouple output is the potentiometer.[1] The potentiometer operates by matching the thermocouple emf with a bucking voltage, thereby eliminating current flow in the system. This eliminates the problem of error introduced by wire resistance and allows the use of any length of wire. The emf may then be converted to temperature using tables such as those found in References 2 and 9. The thermocouple output may also be applied directly to a voltage measuring device such as a galvanometer. These may have two configurations. The first uses a reference junction where either the thermocouple wire or a third wire may be used as an extension. An electrical temperature compensation may replace the reference junction with the use of the voltmeter. This configuration is generally termed a pyrometer. The use of any voltage measuring device requires calibration before use. The Hg bulb thermometer is an acceptable calibration standard for temperature measurement in source sampling.

Thermistor

The thermistor belongs to the large class of resistance thermometers,[3] which in turn has two general classes, the conductive and semiconductive sensors. The thermistor, a semiconductor, measures temperature by the measurement of resistance change of the thermistor with temperature in a balanced bridge circuit. Thermistor electronics are inexpensive compared with that of the thermocouple potentiometer. The thermistor response is linear for the limited temperature range over which it is designed. Outside this range its response is very nonlinear.

PRESSURE MEASUREMENT

Pressure is defined as force per unit area. Most pressure measurements are taken with atmospheric pressure as a reference. The pressure above atmospheric is considered positive while pressure below is negative (Figure 4.3). Absolute pressure, then, is always the algebraic sum of the atmospheric pressure and the measured pressure above or below atmospheric. This is expressed

$$P_{abs} = P_{atm} + P_m \tag{4-4}$$

where,

P_{abs} is the absolute pressure,
P_{atm} is the atmospheric pressure, and
P_m is the measured pressure.

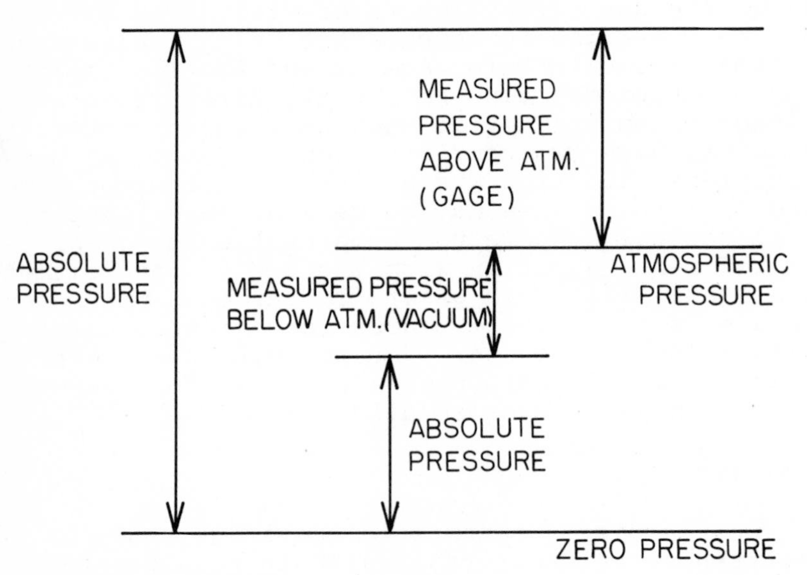

Figure 4.3. Pressure relationships.

Pressure Gauges

The most common method of measuring high pressures is the Bourdon type of gauge.[4] Vacuum gauges use the same principle. This device is shown in Figure 4.4. The unit consists of a bent tube which is open at one end. Its free end is connected to a pointer

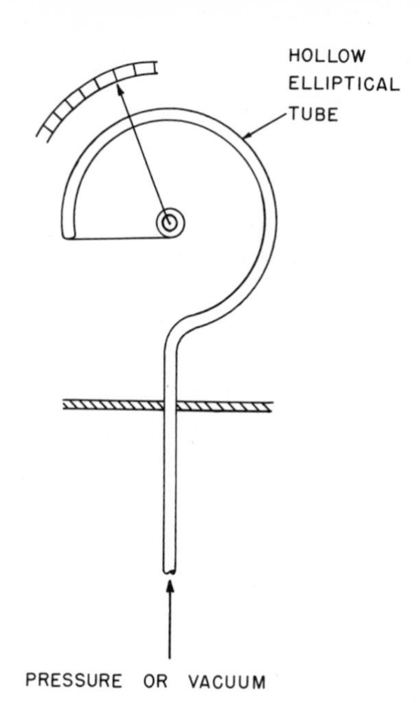

Figure 4.4. Bourdon pressure gauge.

by means of a link. When a positive pressure is exposed to the inside of the tube, it tends to straighten, causing the pointer to move along the scale. This type of device is not highly precise but when carefully calibrated it can be used in source sampling. Its accuracy will usually be a function of quality (or cost).

Manometers

An easy way to measure a low pressure is to balance a column of liquid against the pressure.[8] By observing the height of liquid supported by the pressure, calculations can be made to determine the magnitude of the pressure. Devices making this measurement are called manometers. Figure 4.5 shows a U-tube manometer filled with one fluid #2 (water) being exposed to another #1 fluid (air) with pressure P_1. Fluid #2 could also be oil or mercury while fluid #1 could be the stack gas. The operation of the manometer is based upon the simple fact that at point X the pressure exerted by the left column is equal to that exerted by the right column. Applying this principle we obtain

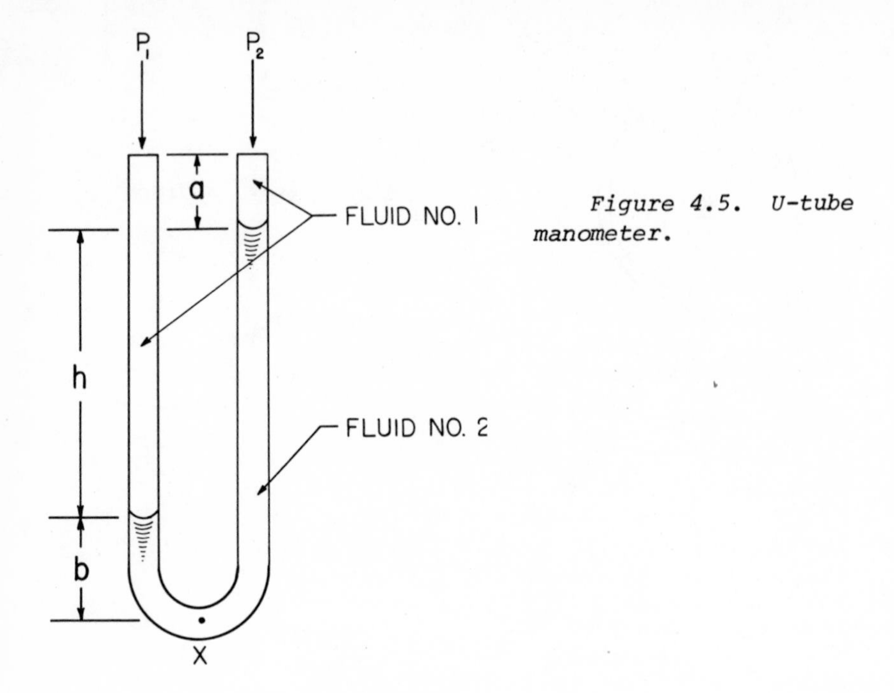

Figure 4.5. U-tube manometer.

$$P_1 + (a + h)\ \gamma_1 + b\gamma_2 = P_2 + a\gamma_1 + (h + b)\gamma_2 \qquad (4\text{-}5)$$

where,

P_1 is the higher gas pressure
γ_1 is the specific weight of fluid #1
γ_2 is the specific weight of fluid #2
P_2 is the lower gas pressure (atmospheric) and
a, b, h are dimensions as shown in Figure 4.5.

This equation allows manometers to be used in measuring differential pressures. In source sampling trains, manometers are often used in conjunction with determining the stack gas velocity and the sample flow rate through the sampling train. For small values of $P_1 - P_2$, the manometer is often inclined to increase the sensitivity.

The inclined manometer is used to measure the stack velocity pressure and sample stream orifice head. The range encountered is usually less than one inch H_2O for velocity pressure and less than one inch H_2O for orifice head.

The inclined manometer is sensitive to level and to some expansion of fluid with temperature. One operational problem with these manometers is that the fluid can be completely or partially blown out if the connections are made improperly. A trained operator can read an inclined manometer to the nearest 0.005 inch if the velocity pressure is constant. However, if fluctuating velocity conditions occur, then this is not possible.

Mechanical pressure gauges are also available to measure the low differential pressures commonly encountered as velocity pressure and orifice meter heads in source sampling. The Magnehelic gauge manufactured by F. W. Dwyer, Mfg. in Michigan City, Indiana, is an example of such an instrument. The Magnehelic gauge is a flexible diaphragm pressure gauge.using a magnetic linkage to amplify and display the movement of the diaphragm in such a way that the instrument can be calibrated for pressure differentials. Applicable gauge ranges that are available are 0-0.5 in. H_2O, 0-1.0 in. H_2O, and 0-5 in. H_2O. Maximum sensitivity of 0.01 in. H_2O per division is available only with the 0-0.5 in. H_2O gauge. The gauges are susceptible to damage from dusty and corrosive atmospheres. When a high degree of accuracy is needed as in source sampling, frequent calibration is necessary. Inclined manometers are more reliable and simpler; they are easier to repair in the field than the diaphragm pressure gauges. Furthermore, inclined manometer malfunctions are usually more obvious and can be quickly corrected in the field. Manometers also never need to be calibrated. The major positive attributes of the diaphragm pressure gauges are (a) the relatively low cost and (b) the good portability.

VELOCITY MEASUREMENT

The measurement of velocity in a duct utilizes one of the simplest devices in most sampling trains, the pitot tube. To understand why the pitot tube is the primary means of velocity measurement in stack sampling, one must first look at the description of flow through a duct. This is describedcby the following equation;

$$\overline{Q}_s = A_s \overline{V}_s \tag{4-6}$$

where,

$\overline{Q}_s$ is the average volumetric flow through the duct, cfm

$\overline{V}_s$ is the average duct velocity, fpm

A_s is the cross-sectional duct area, sf.

If the total flow was measured through the duct for a known period of time, then it could be divided by the cross-sectional area to obtain the average velocity. But, there are no volumetric gas meters or bags large enough to handle the total flor for most stacks. An instantaneous meter, such as the orifice to be discussed in a following section, would not be sufficient either; it presents the problem of additional frictional loads to system fans and must be calibrated. The alternative is to use a direct measurement of velocity. The pitot tube has this capability. However, it will not measure directly the average duct velocity but measures only the instantaneous velocity at the point at which it is located.

Principle

The total pressure in a duct is due to the static pressure and the velocity pressure. To understand what the pitot tube actually measures, one must first understand what happens in the gas stream. If a smooth round object is placed in a duct, the gas will flow around it. This is represented by the streamlines in Figure 4.6. At point p the gas stagnates against the object and thus total pressure exists. The difference between the total and static pressures is called the velocity pressure, ΔP.

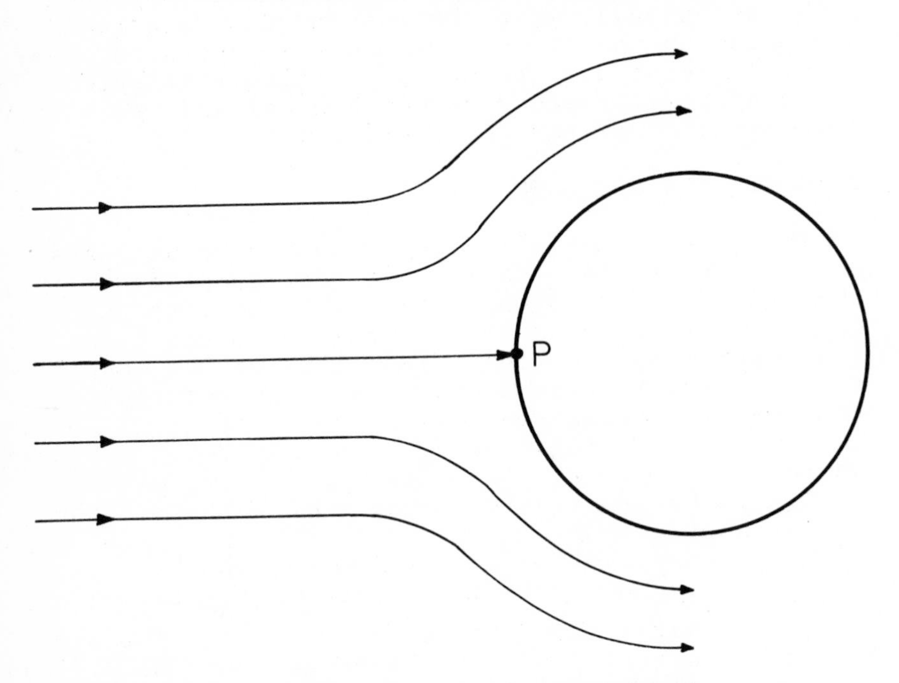

Figure 4.6. Gas stagnation against an object.

$$\Delta P = P_p - P_s \qquad (4\text{-}7)$$

where,

ΔP is the velocity pressure at point P
P_p is the total pressure at point P
P_s is the static pressure at point P.

These relationships are demonstrated in Figure 4.7 for a duct which has the *same velocity at every point*. The velocity can be calculated by performing an energy balance at points P and S in the gas stream.

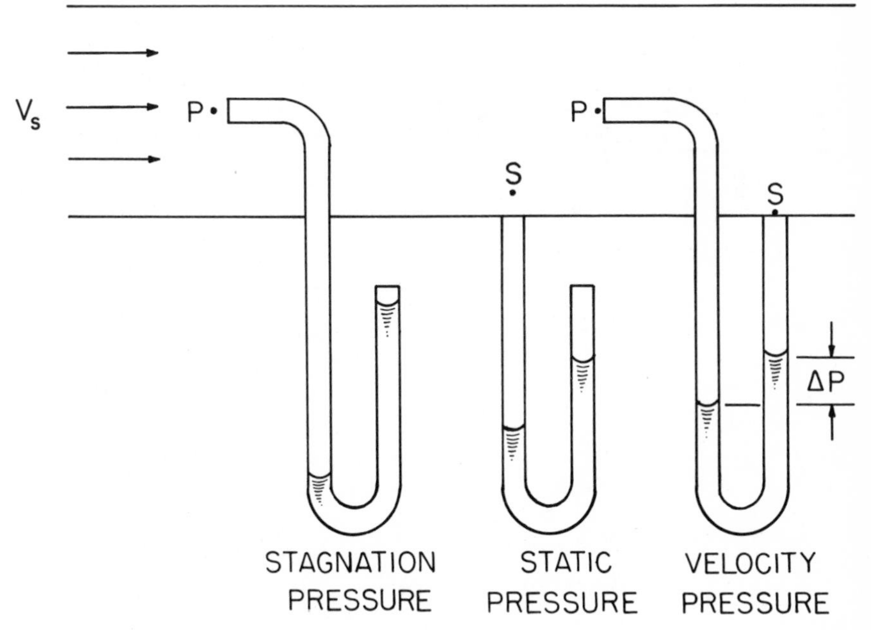

Figure 4.7. Components of total pressure.

$$\frac{P_p g_c}{\rho_s\, g} + \frac{V_p^{\,2}}{2g} = \frac{P_s g_c}{\rho_s\, g} + \frac{V_s^{\,2}}{2g} \qquad (4\text{-}8)$$

where,

P_p is the absolute total stack pressure at point P
P_s is the absolute stack pressure at point S
V_p is the velocity at point P
V_s is the stack velocity
g_c is the universal gravitation constant

g is the local gravitation constant and

ρ_s is the density of the stack gas.

By definition the stagnation velocity, V_p, at point P must be zero. Therefore Equation 4-8 reduces to

$$V_s = \left(\frac{2g_c(P_p - P_s)}{\rho_s} \right)^{1/2} \tag{4-9}$$

Configurations

Several configurations are possible for pitot tubes. One is shown in Figure 4.8. The measured

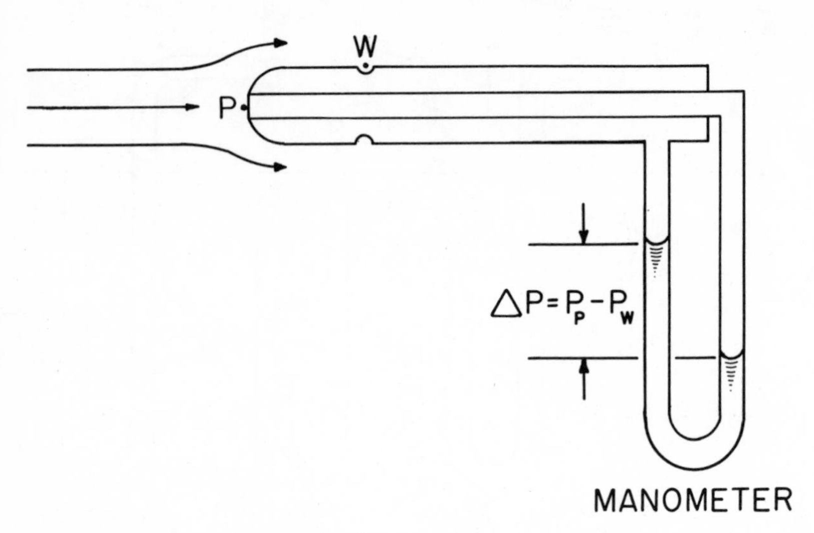

Figure 4.8. Standard pitot tube.

static pressure is done at point W. Pressure at W approximates P_s. Equation 4-8 becomes

$$V_s = C_p \left(\frac{2g_c(P_p - P_w)}{\rho_s} \right)^{1/2} \tag{4-10}$$

where,

P_w is the measured static pressure and

C_p is the pitot coefficient.

The pitot tube shown in Figure 4.8 is usually termed a standard pitot tube. The C_p value for this configuration is approximately 1.00. There are several variations on this type of tube which have the same C_p value.

One configuration which does not resemble the standard pitot is the reversed or S-type. The S-type pitot is used primarily for stack testing because of one major advantage: it will not clog in dust ladden gas streams. This tube is shown in Figure 4.9. The S-type pitot does not give the same velocity pressure as the standard pitot tube.

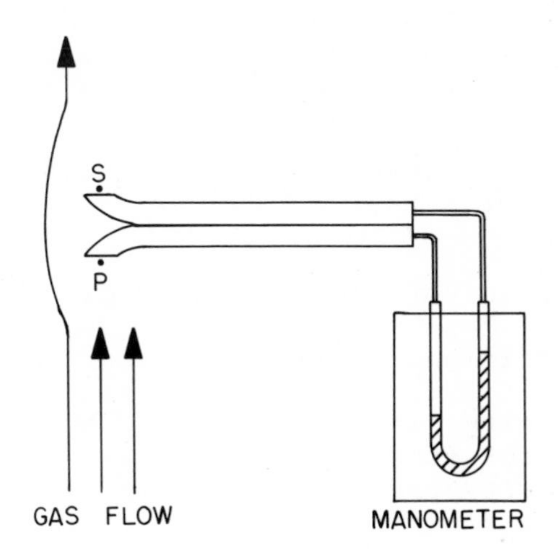

Figure 4.9. S-type pitot tube.

The observed ΔP is actually larger for a given velocity because the rear face of the tube points away from the gas stream. Thus P_w is a wake pressure which is lower than the true static gas pressure. This is the second advantage of the S-type pitot. The measured Δp will be a larger readable value than that of a standard pitot.

When the S-type pitot is used with a water manometer, the pitot Equation 4-10 takes the following form:

$$V_s = 85.48\ C_p \left(\frac{T_s \Delta P}{P_s M_s} \right)^{1/2} \qquad (4\text{-}11)$$

where,

ΔP is the velocity head, in H_2O

T_s is the stack gas temperature, °R

P_s is the absolute stack gas pressure, in Hg and

M_s is the molecular weight of the stack gas.

Calibration

For the limits of 0.01-10 inches of water velocity pressure of a Standard pitot, C_p usually takes on values at 0.98-1.00. The C_p of S-type usually ranges from 0.83-0.87. However, each must be calibrated individually prior to each test.

The calibration of a pitot tube requires a gas stream of known velocity, which means that a wind tunnel facility of known characteristics should be available for use. However, the S-type pitot tube can be calibrated against the standard-type pitot tube. In fact this is the procedure specified by EPA Method 2.[10] The following equation applies:

$$C_{P_{test}} = C_{P_{std}} \left(\frac{\Delta P_{std}}{\Delta P_{test}} \right)^{1/2} \qquad (4\text{-}12)$$

where,

$C_{P_{test}}$ is the coefficient of the S-type pitot tube

$C_{P_{std}}$ is the coefficient of the standard pitot tube

ΔP_{std} is the velocity pressure measured by the standard pitot tube and

ΔP_{test} is the velocity pressure measured by the S-type pitot tube.

To calibrate the S-type pitot tube, the velocity pressuresis measured at the *same* point in a flowing gas stream with both the S-type and standard type pitot tube. Both pitots must be properly aligned to the gas stream.

The observed velocity pressures are applied to Equation 4-12 and the coefficient for the S-type pitot tube is determined. If $C_{P_{std}}$ is not known then a valud of 0.99 should be used.

The coefficients for the S-type pitot should be determined with first one leg and then the other pointed downstream. If the computed coefficients differ by more than 0.01, the pitot tube should not be used without properly labeling the two tubes. Then care must be used to insure that the desired tube is facing downstream during source tests. EPA actually recommends[10] that S-type pitot tubes not meeting this 0.01 constraint should be discarded.

The determination of the average stack gas velocity, $\overline{V}_s$, is one of the greatest sources of error in source testing (see Chapter 8). Therefore it is recommended that the pitot be recalibrated between each field test. Remember, the physical abuse which occurs during field testing can and does alter the coefficient.

Other Measurement Methods

At extremely low or high velocities the pitot method is inaccurate and unreliable. There are several other mechanical and electronic methods available: hot wire anemometers, rotating vane anemometers, and fluidic devices. For further details on these and other methods see References 1, 3, 5 and 6.

NOZZLES

The nozzle is considered the initial sampling train boundary. It removes a portion of the effluent from the duct and delivers it to the sampling probe. The nozzle has the following restrictions:

1. It must not disturb the duct stream flow.
2. It must not alter the pollutant being sampled.
3. It must not add to the sample being collected.
4. It must be of a size which allows entry to the duct.

In the case of particulate sampling the nozzle must disturb the flow in the duct as little as possible or the sample will not be representative. In the case of gases this is of no importance. Any nozzle in a duct will disturb the flow, but a thin-walled, sharp-edged nozzle disturbs it the least.

The nozzle must not alter the constituents of the sample and must be chemically inert. That is, it must not react with any of the pollutants or act as a catalyst for reactions of pollutants. In the case of particulates, any bends in the nozzle will cause the impingment of the larger particulates. The nozzle must then be cleaned carefully and the material added to the total particulate sample collected.

The nozzle must add nothing to the sample. This simply means it must have structural integrity, not breaking or chipping and being able to withstand the environs of the duct. It must fit into the sampling port, which is usually 3 or 4 inches in diameter.

A null balance type nozzle is also now available.[11] The null balance works on the principle of matching the stack static pressure with the nozzle static pressure to obtain isokinetic conditions. This is shown in Figure 4.10. The one problem with

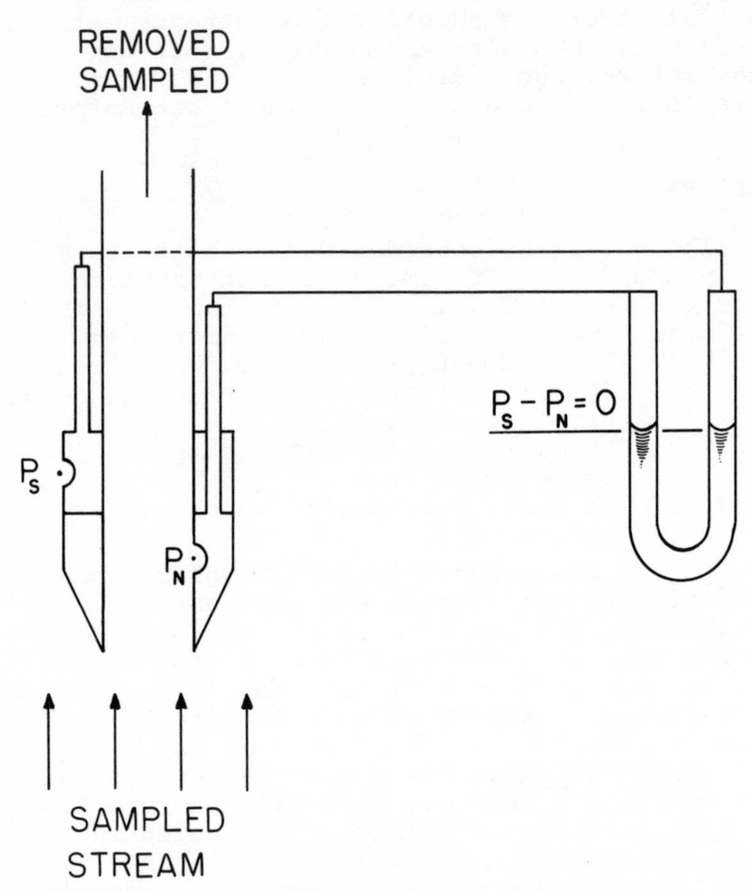

Figure 4.10. Null balance nozzle.

this type of probe is the friction developed in the nozzle. Ideally, $P_s = P_n$ for isokinetic conditions, but actually $P_s = P_n + F$ for isokinetic conditions where F is the frictional loss.[12,13] Therefore the null balance can be used only when it is calibrated for a specific duct or stack and is accompanied by a S-type pitot tube and total sample volume measurement to verify isokinetic conditions.

PROBES

The probe is the sampling interface between the gas stream in the duct and the external sampling train. It is exposed to high temperatures on the nozzle end and ambient temperatures on the train end. The probe must meet several requirements:

1. It must not alter the sample.
2. It must be structurally strong enough to support itself.
3. It must be easily cleaned.
4. It must not add to the sample.

Ideally the sample should be delivered to the sampling train at the stack temperature. Basically this requires that the probe be heated to maintain sample conditions and prevent condensation. The practicality of achieving this is discussed in a following section on gas conditioning.

Structurally the probe must support itself, the nozzle, the pitot tube, and possibly a thermocouple. This is usually accomplished by the use of an outer sheath around the probe and heating element. Figure 4.11 shows a nozzle and probe assembly.

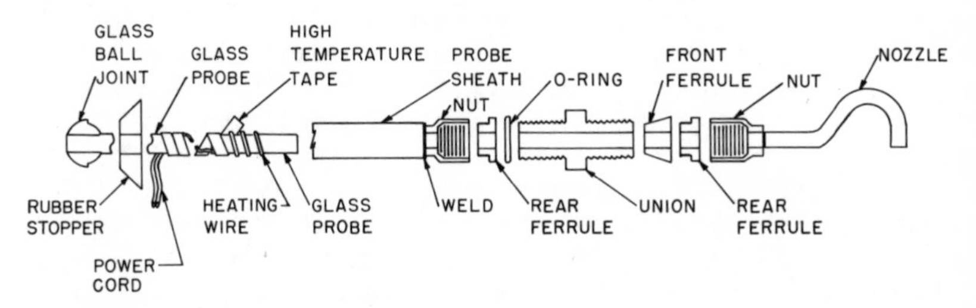

Figure 4.11. Probe assembly.

The requirement of probe cleaning conflicts with the structural requirement. The surface of glass tube is much more easily cleaned than a metal tube, but it is also more fragile. In most cases glass probes over 6-7 feet long are impractical. If a metal tube is used, some material will become trapped in the roughness of the surface of the tube. In cases of extremely high temperature the only choice possible is to use a water-cooled high quality stainless steel probe.

Glass has another advantage which should be considered. It is for all practical purposes chemically inert--it will not promote any chemical alterations of the sample. This is not true of stainless steel, especially if there are acid gases present in the gas stream being drawn to the sampling train.

SAMPLE COLLECTORS

This section will consider gaseous and particulate collectors. The choice of a specific collector will depend on the pollutant, the characteristic of the gas stream, and the purpose of the test.

Gaseous Collectors

The collection of gas samples can be divided into four main types of collection devices. One type is the cold trap, an approach which condenses pollutant vapors in the sample flow stream. The second type of device is that containing a solid adsorbent. This removes the pollutant gas from the stream by its adsorption on a solid surface. An example of this device is the indicating detector tubes which are commercially available.[6] The third type of device is the grab sample container, which may either be a rigid container or a flexible bag. The rigid containers are evacuated prior to the sampling. The bag collection technique draws a sample by expanding the bag in a second rigid container. If the sampling rate is low, integrated samples may be taken using the devices used for grab sampling. The fourth type of gaseous collectors is that which utilizes the principle of gas absorptions by a liquid media.

Gas absorption devices are generally called impingers. There are several different types of these devices available and they are shown in Figure 4.12. Their efficiency depends on the diffusivity of the gases, the retention time in the devices, the bubble size, and the gas solubility. The first of these devices is a midget impinger. It contains small amounts of absorbing liquid, and therefore requires low flow rates. Impingment takes place on the bottom surface due to close placement of impinger tip. This accomplishes a reduction of bubble size. The second type of device is the Greenburg-Smith Impinger. It has a total volume of 500 ml and allows a flow rate of approximately one cfm. The tip of the device locates the impingment surface 5 mm from the impinger tip. This type of device is usually used when higher flow rates are required. The third type of device uses fritted

a. Midget impinger

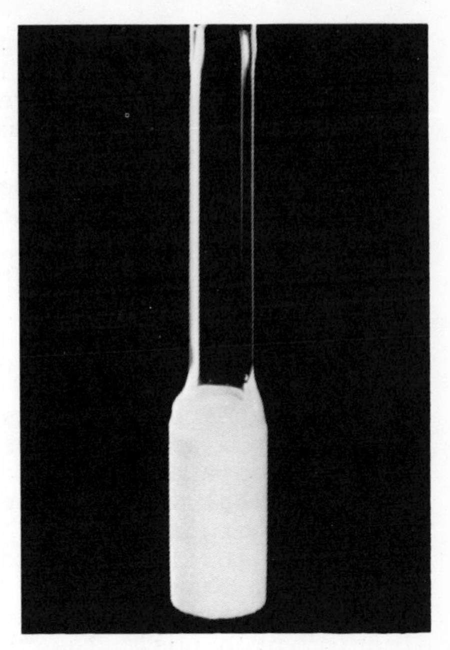

b. Fritted bubbler

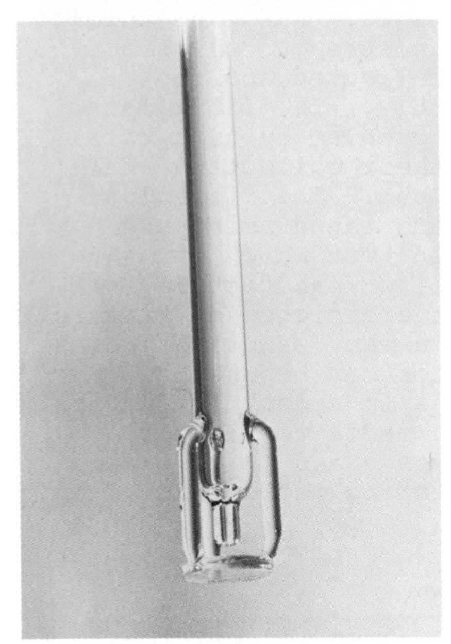

c. Greenburg-Smith impinger tip

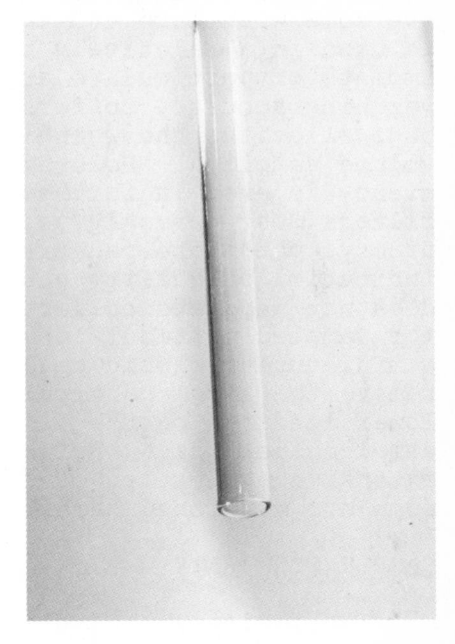

d. Modified impinger tip

Figure 4.12. Source sampling impingers (courtesy of Glass Innovations, Inc., Addison, New York).

glass as the means of dispersing the gas into fine bubbles. This is generally the most efficient means of gas absorption. However it has the problem of clogging if the gas stream is not completely free of particulate. The last classification of device is the bubbler. This type has no special tip which means that scrubbing efficiency will be low because of large bubble size. It is quite effective, however, for water condensation from the sample stream.

Dry Collectors

This type collector is used primarily for the collection of particulates matter. Filtration is the basic method and there are three major types: the fiberglass filter, the ceramic Alundum thimble and the fiberglass bag. A paper thimble is sometimes used for low temperature conditions.

Alundum thimbles are reused and therefore their collection efficiency is not constant because of cleaning problems. The thimble is made of fragile ceramic material but any breakage problems can be overcome with care and technique. When the thimble is used in a relatively clean stream, only small amounts of particulate are collected and thus weighing accuracy suffers badly. The fiberglass bags alleviate the weighing problem by high total volume sampling. However, their collection efficiency is even in further doubt. The fiberglass filters most commonly used are rated using a dioctyl phthalate penetration test which gives an indication of quality control on manufacture. It does not indicate collection efficiency of filters for sampled particulates. In many sampling trains a miniature cyclonic collector can be used to remove the large particulates in order to maintain lower loadings on the filter media for longer sampling periods.[7] References 6 and 14 give an extensive evaluation of filtering medias including some not mentioned here.

SAMPLE FLOW RATE

Two devices, the rotameter and the orifice, are used in sampling trains.

Rotameter

The rotameter is a variable area meter which consists of a vertically tapered transparent tube containing a float.[15] This is shown in Figure 4.13. Flow passes through the tube and supports the float.

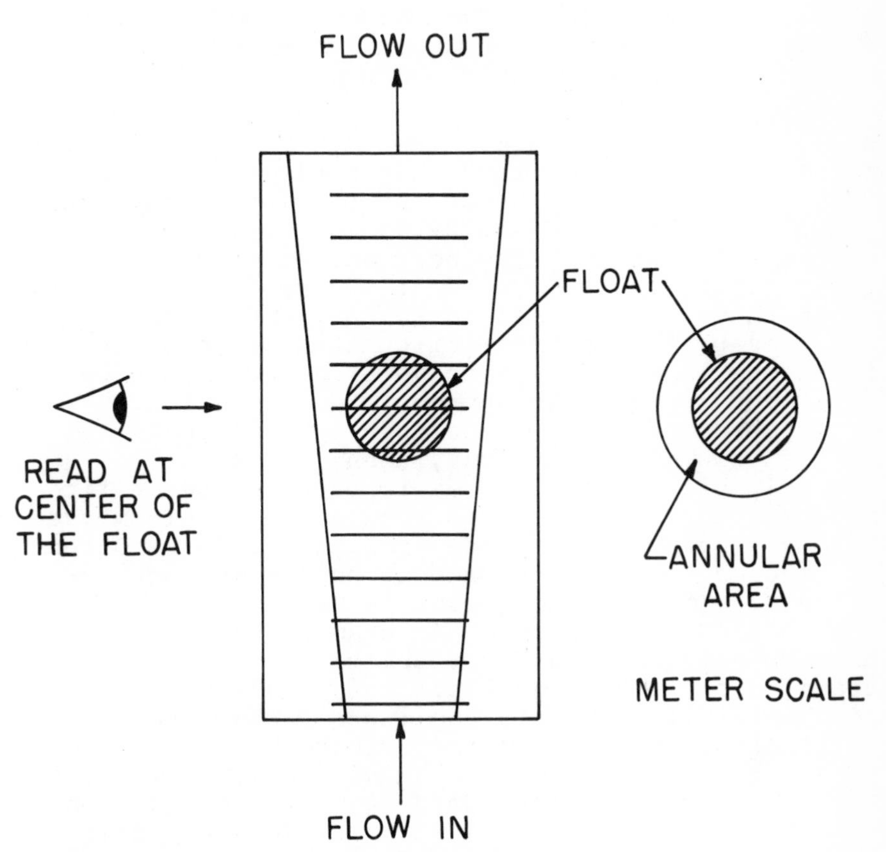

Figure 4.13. Schematic of a rotameter.

When the gravitational, buoyant, and drag forces acting on the float are balanced, the float will stop. The height at which the float stops will be proportional to the flow rate through the tube. This relationship is given as

$$Q_m = \frac{C_m A_m}{D_f} \left(\frac{g m_f R T_m}{P_m M_m} \right)^{1/2} \quad (4\text{-}13)$$

where,

Q_m is the meter flow rate
C_m is the meter coefficient

A_m is the area of the anulus around the float which is dependent on height
g is the local gravitational constant
m_f is the mass of the float
R is the Universal Gas Constant
T_m is the absolute meter gas temperature
P_m is the absolute meter gas pressure
M_m is the meter gas molecular weight
D_f is the diameter of the float.

Rotameters are usually equipped with a flow rate scale on the meter face. The flow rate values printed on the meter face only apply to the conditions at which the meter was originally calibrated (usually air at 70°F and 29.92 in. Hg). Therefore, before using rotameters under any other conditions, they have to be calibrated at the conditions at which they will be used (gas, gas temperature, and gas pressure). In those circumstances when the ideal gas laws apply, a conversion factor can be developed to make the appropriate correction.

If a rotameter is to be used to make total sample volume calculations, then it must be properly calibrated. However, because the orifice meter is more precise, the rotameter is seldom used for this purpose in source sampling trains. Often it is used to maintain proportional sampling conditions, and a dry gas meter is used to determine the total sample volume. An example of this is the EPA sampling train for sulfur dioxide. This train is shown in Figure 2.4 and is also the subject of Chapter 11. In this application absolute calibration is not necessary since the rotameter is being used to adjust relative sample flow rates. Any error due to nonstandard gas stream conditions will be included in all readings. Thus the device can still be used to make relative flow rate changes even though the absolute flow rate is now known.

Orifice Meters

The sharp-edge orifice meter (Figure 4.14) is a simple and accurate method to measure volume rate, providing an instantaneous flow measurement. In source sampling it is often used on conjunction with a total sample volume type meter. As the gas passes through the restriction (orifice) a pressure drop occurs. Using the energy equation and the physical characteristics of the orifice the following expression can be developed. For the details of this derivation see References 1 and 4.

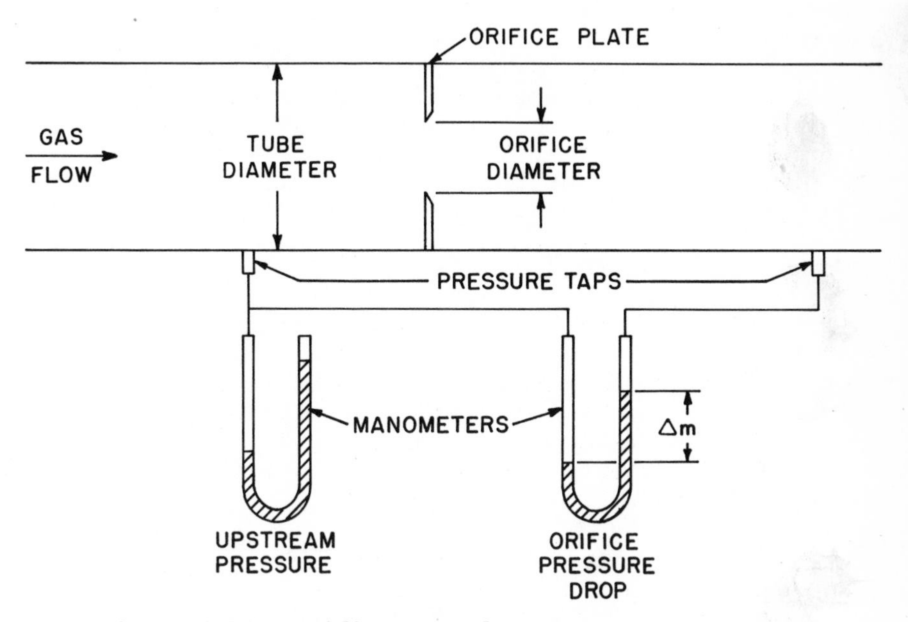

Figure 4.14. Orifice type flow meter.

$$Q_m = K_m \left(\frac{T_m \Delta m}{P_m M_m} \right)^{1/2} \quad (4\text{-}14)$$

where,

Q_m is the gas flow rate, cfm

K_m is a proportionality factor (which is different for each orifice meter)

Δm is the orifice meter pressure drop, inches of water

T_m is the upstream gas temperature to the meter, °R

P_m is the absolute upstream meter pressure, inches of mercury

M_m is the molecular weight of the gas entering the meter.

For a given orifice meter the K_m must be determined by calibration. Actually K_m is a function of Reynolds number and hence it will not be constant over the total flow rate range of the meter. However, for a small flow rate range, such as is used for most source sampling applications K_m is constant. It is therefore very important that the orifice be calibrated for the range of flow rates it will be measuring during source tests.

Orifice meters which are used in sampling trains are sized so that a pressure drop of up to 10 inches of water occurs for the desired flow rates. The design and construction details of an orifice meter are given in Reference 16, and a calibration procedure is presented in Reference 17. This calibration procedure is outlined in the following section on total sample volume.

TOTAL SAMPLE VOLUME

The total volume, or integrated volume, must be determined in most sampling trains. This provides the sampled volume required for calculation of pollutant concentration. Dry gas meters which are available from 0.2 to 150 cfm capacities are generally used.[5] These are operated internally as shown in Figure 4.15.

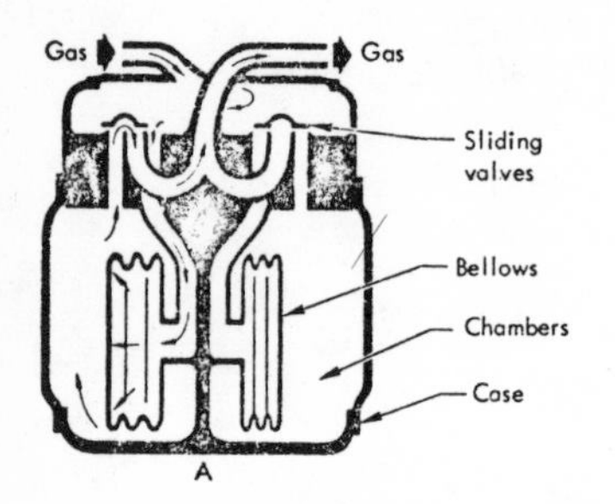

Figure 4.15. Idealized cross section of a dry gas meter. The operation of this instrument is described in the text (from G. Nelson, Controlled Test Atmospheres, Ann Arbor, Michigan: Ann Arbor Science Publishers, Inc., 1971).

> The gas enters the meter and passes the left-hand sliding valve, filling the left-hand bellows and forcing gas out of the chamber on the left. When this bellows is fully extended, the sliding valves shift and direct the incoming gas to the right-hand bellows. This forces the gas from the right-hand chamber as shown in Figure 4.15b. When both bellows are fully extended, the sliding valves again shift and the bellows empty separately as shown in Figures 4.15c and 4.15d. This cycle of alternately filling and emptying the bellows is linked to the volume dials, which register the corresponding volume changes.[5]

For sampling, the smallest dial face divisions should be 0.01 cfm because the meter movement is not smooth over one revolution and if an error is made in reading, the maximum error that can occur is 0.1 cf, one revolution. If one revolution represents 1 cf, then a significant error can be made in reading the measured volume.[5,7]

The dry test meter is calibrated using a wet test meter. As shown in Figure 4.16 the gas displaces water in the wet test meter chamber and causes the rotor to revolve.

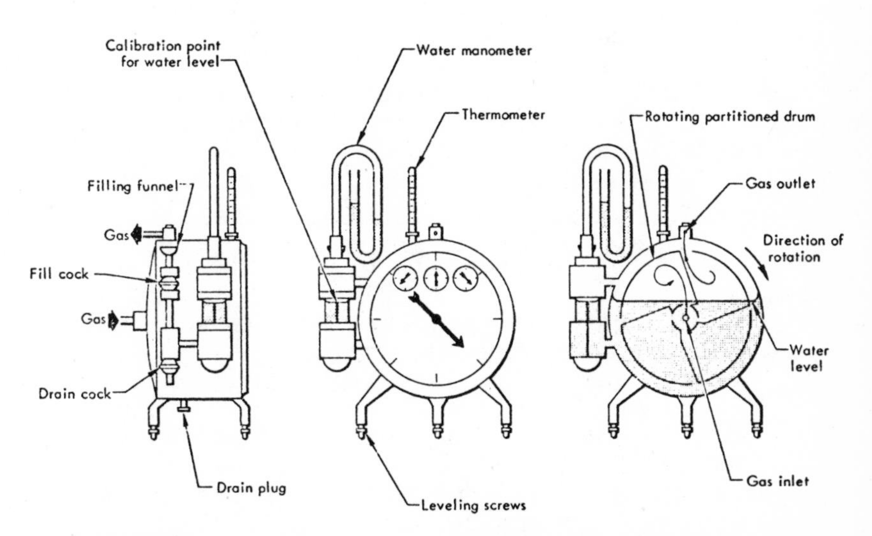

Figure 4.16. Side, front, and cross-sectional views of a wet test meter (from G. Nelson, Controlled Test Atmosphere, Ann Arbor Science Publishers, Inc., Ann Arbor, Michigan, 1971).

> A wet test meter consists of a container which houses a revolving drum that is about two-thirds submerged in water. The drum is divided into four sections, each of which has an inlet and an outlet. When a gas enters the meter, it exerts a buoyant force that turns the drum clockwise. As one quadrant becomes full, a new quadrant rotates into the filling position and the full one begins expelling gas through its outlet. The drum rotates smoothly if the gas is fed to the meter at a steady rate. The total volume for a given time interval is recorded by the decade dials provided. Attached to the meter are a water manometer for measuring the internal and external pressure differential and a thermometer for measuring the temperature of the incoming gas. The filling funnel and the drain and fill cocks are used to keep the water level at the calibration point.[5]

It should be noted that the gas leaving this meter is saturated with water vapor. The set-up for calibrating the dry gas meter and the orifice meter is shown in Figure 4.17. The wet test meter should be one which has a scale of one cubic foot per revolution. The calibration test is performed and the needed information recorded for Figure 4.18. The following equation is used to calculate how well the dry gas meter performs relative to the wet test meter. This should be done for each set of data. If the ratio makes $\gamma < 0.99$ or $\gamma > 1.01$, then the dry gas meter should be adjusted internally as needed and then recalibrated. If wide variation occurs then possibly there are leaks or the dry gas meter needs repair.

$$\gamma = \frac{V_w P_b \, (t_d + 460)}{V_d \, (P_b + \frac{\Delta m}{13.6}) \, (t_w + 460)} \qquad (4\text{-}15)$$

where,

- γ is the ratio of accuracy of the wet test meter to the dry gas meter
- V_w is the gas volume passing through the wet test meter, ft^3
- P_b is the absolute barometric pressure, inches Hg
- t_d is the average temperature of the gas in the dry gas meter, °F
- t_w is the temperature of the gas in the wet test meter, °F
- Δm is the orifice meter pressure drop, inches H_2O

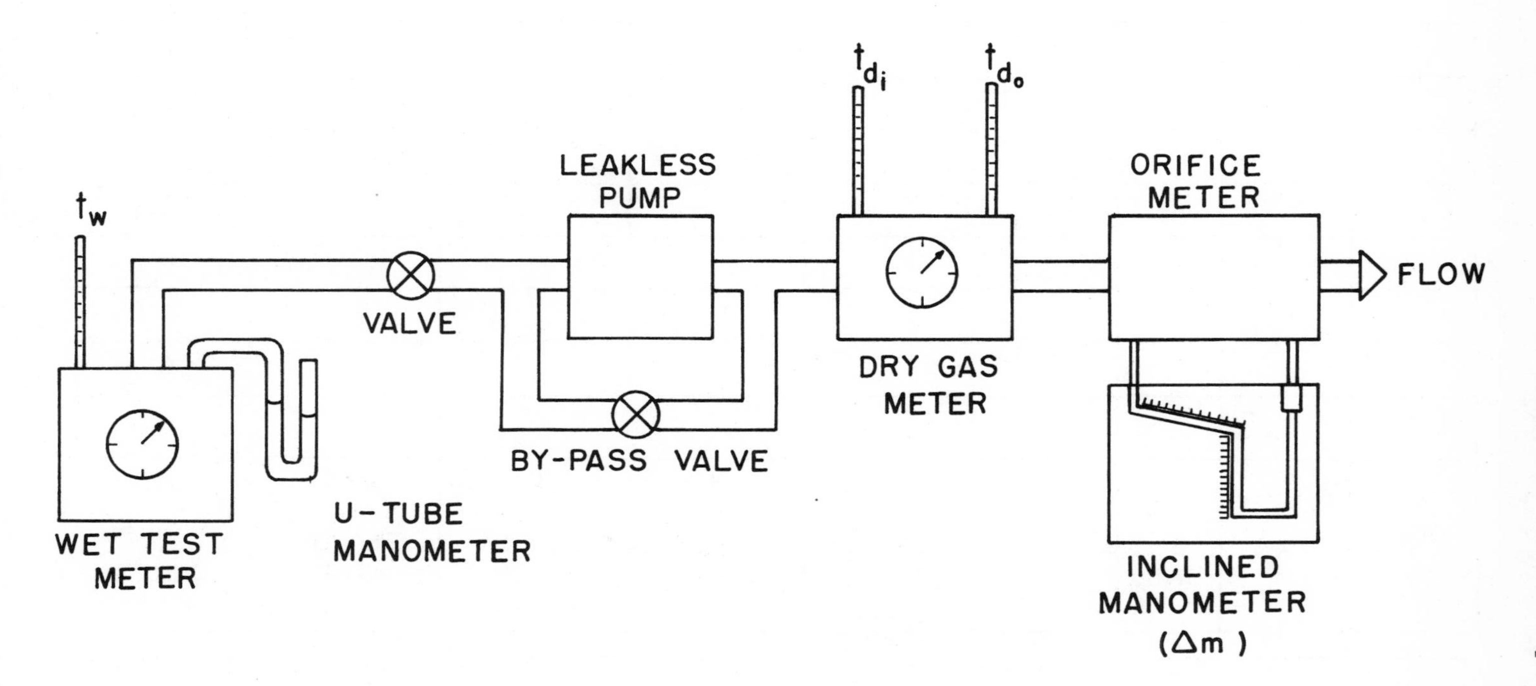

Figure 4.17. *Set-up for calibration of dry gas meter and orifice meter.*

DRY GAS METER NO. ______ ORIFICE METER NO. ______

DATE ______ WET TEST METER NO. ______

BAROMETRIC PRESSURE P_b = ______ IN Hg CALIBRATED BY ______

ORIFICE SETTING Δm IN H_2O	WET TEST METER VOLUME V_w FT^3	DRY TEST METER VOLUME V_d FT^3	TEMPERATURE				TIME Θ MIN.	γ	K_m
			WET TEST	DRY TEST					
			t_w °F	t_{d_i} °F	t_{d_o} °F	t_d °F			
0.5									
1.0									
1.5									
2.0									
2.5									
3.0									
3.5									
4.0									
							AVERAGE		

Figure 4.18. Data sheet for calibrating dry gas meter and orifice meter.

V_d is the volume of gas passing through the dry gas meter, ft^3, and
θ is the time, min.

The data in Figure 4.18 can be used to calibrate the orifice meter. In making these calculations and applying Equation 4.14, the wet test meter data is used for the flow rate. Then

$$Q_m = \frac{V_w}{\theta}$$

Also use

$$T_m = t_{d_o} + 460°$$

$$P_m = P_b + \frac{\Delta m}{13.6}$$

$$M_m = 29 \quad \text{(for air)}$$

GAS CONDITIONING

Often the gas sample must be conditioned or treated before or while it is passed through the sampling train components. This is done either to preserve the sample or to prevent damage to the sampling train. Typical gas conditioning operations include condensing, drying, heating and dilution.

Condensing

Condensers are used to remove the water vapor from the gas sample. They work on the principle that the partial pressure of water vapor decreases with a decrease in sample temperature. For example, as indicated in the steam tables[9] the partial pressure at 32°F is only 0.15% of the partial pressure at 300°F. Thus the use of an ice bath type condenser is an effective way to remove water vapor from a gas sample.

The ice bath condenser is used in source sampling to protect other components from damage and to determine the moisture content of the gas stream. The deposition of water vapor and water soluble constituents in such components as dry gas meter and pump can damage them almost beyond repair. The moisture content must be determined in order to calculate the molecular weight of the stack gases. The details of these procedures are the subjects of Chapters 6 and 7.

The ice bath condenser usually consists of several wet and dry impingers connected in series, but it may be as simply as coiled copper tubing. A measured initial amount of water is put into the impinger-type condenser to assist in the condensation process. The sampling train is operated and a known amount of gas sample is passed through the system. By observing the pressure and temperature operating conditions, the amount of water condensed, and the amount of gas passing through the system, the moisture content of the gas stream can be calculated (see Chapter 7).

Drying

In source sampling trains, gas drying is used to accomplish the same objectives as condensing. The drying operation is achieved using special chemicals which have a great affinity for water vapor. One such chemical is silica gel. The silica gel strongly adsorbs water and hence its change in weight can be used to calculate the moisture content of the gas stream. Commercial silica gel, which is granular and has a bright blue color, can be obtained. As it becomes saturated with water vapor its color changes to a light pink. If this method is to be used to determine the moisture content of a gas sample, care must be taken to insure that all particulate matter is removed first and that there is no other major constituent in the gas stream which is also adsorbed by the silica gel. The silica gel releases (desorbs) the adsorbed water vapor upon heating to 350°F and it can be reused.

Often the condensor and drying tube are used in series to increase collection efficiency. Large mesh silica gel (6-16) is used to prevent the possibility of entrainment of small particles which might damage other components.

Heating

The heating operation is used solely to preserve the gas sample prior to passing it through the pollutant collector. This is an effective method of preventing condensation of water vapor and high molecular weight substances. Therefore it is common practice to heat sampling probes and pollutant collectors such as filters to prevent deposition. If such a condensation process were allowed to occur, it would cause loss of gaseous and particulate constituents from the gas sample. In addition to causing a sampling error these materials could be be deposited in inaccessible areas of the sampling system and lead to later malfunctions.

Ideally it is a good policy to try to maintain stack gas temperatures throughout the sampling train preceding the filter. However, high temperatures favor chemical reactions such as oxidation of hydrocarbons in a gas stream containing appreciable amounts of oxygen. Low temperatures, as mentioned, are conducive to condensation of water vapor and high molecular weight hydrocarbons. Hence a compromise is required and most probes and heated filter boxes operate at about 250°F. Heat sensitive sampling train components will not be affected by this temperature. Water vapor will not condense and some of the safety problems involved with the handling of hot equipment will be alleviated.

Dilution

Addition of a dry gas can be an effective method for preventing condensation. When this gas is added the sample is diluted. Condensation is prevented because the dry gas is capable of supporting a part of the water vapor from the gas sample even though the temperature of the mixture is reduced. The dry gas must be added in such a manner that the original sample constituents are not altered. This could be a problem with respect to particulate matter because the dilution (mixing) process could cause such events as particle agglomeration and deposition.

GAS MOVERS

The purpose of the gas mover is to draw the sampled gas through the sampling train. The detailed requirements of such a component are different depending upon its location in the sampling train, the sampling strategy used, and the pollutant tested for. The types of gas movers most commonly employed are pumps and evacuated containers. Othe rmethods of less interest for moving the gas sample include ejectors and liquid displacement. Extensive coverage of gas movers is presented in Reference 6.

Selection Criteria

A gas mover must (a) provide adequate flow rate, (b) be durable, and (c) be portable.

<u>Flow Rate</u>. The gas mover must be able to overcome the head loss of the other sampling train

components and thereby provide the desired flow rate. It must be able to provide a wide range of flow rates as required by proportional (isokinetic) sampling conditions. Often the head loss across the pollutant collector, such as a filter, increases through the sampling test. This puts an added burden on the gas mover to be able to maintain the constant sampling conditions at the nozzle tip within the stack.

The gas mover must be leakless when it is located ahead of the gas meter in the sampling train. If it isn't, then the metered volume will be greater than the sampled volume and hence the measured pollutant concentration will be less than the true pollutant concentration. The EPA sampling train shown in Figure 2.5 falls in this category. All of the other sampling trains shown in this figure have the gas mover located after the gas meters and therefore no error is involved if a leak exists.

Durability. The gas mover must be durable in that it is exposed to the corrosive environment of the sample gas. During most source tests it is in constant operation and should be a long life component. The design should enable this component to be maintained easily; the key components should be accessible and replaceable with a minimum amount of time.

Portability. The need for portability becomes readily apparent when performing source tests, and consequently the gas mover must be small and lightweight.

Pumps

There are several types of pumps which make suitable gas movers for source sampling trains. All are of the positive displacement type which means that there is a direct linear correspondence between the flow rate and pressure. The pressure represents that amount of suction which must be available to overcome the resistance of the gas to movement through the sampling train. In the commercial source sampling equipment, discussed in Chapter 16, the reciprocating diaphragm and rotary vane pumps are commonly used. Schematics of these pumps are shown in Figure 4.19.

The diaphragm pump operates on the moving diaphragm principle. Gas is drawn into the chamber

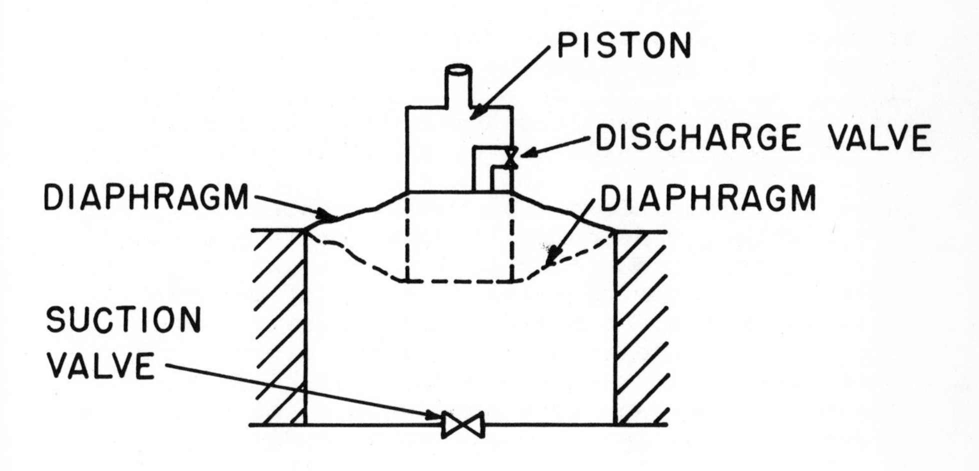

(a) Diaphragm pump

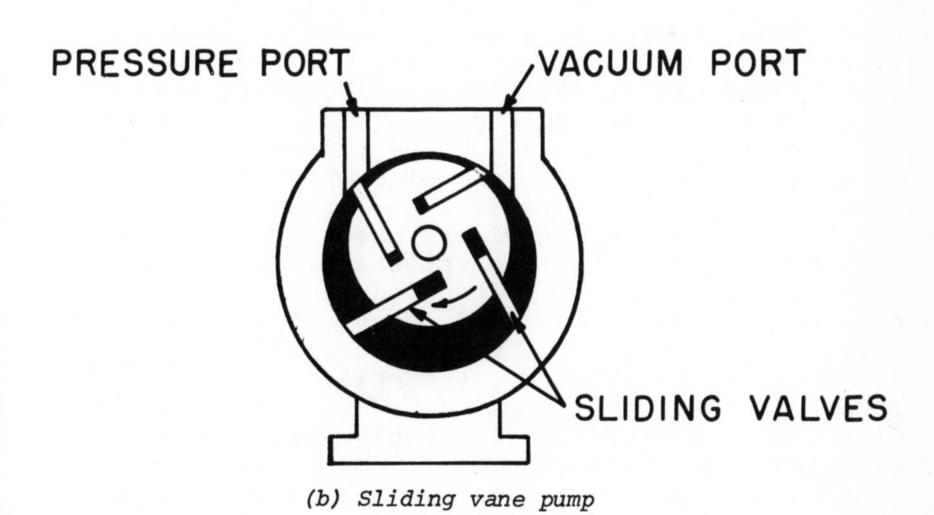

(b) Sliding vane pump

Figure 4.19. *Schematic of pumps used in source sampling.*

on a suction stroke and pushed out on the discharge stroke. On the suction stroke the suction valve is open, allowing gas to flow in. On the discharge stroke the suction valve closes and the discharge valve opens allowing the gas to flow out. This intermittant operation can cause some flow fluctuation (pulsation) in the sampling train. However, this problem can be somewhat reduced by running two such pumps in parallel or by a specifically designed surge chamber in the flow line. The diaphragm in these pumps is made out of metal, rubber or plastic.

The rotary vane pump is one rotor in a casing, which is machined eccentrically in relation to the shaft. The rotor contains a series of movable vanes which seal against the pump casing. The vanes are free to slide in and out of the slots as the rotor turns. If the pump must be leakless, then only the fiber vane type pump with an oiler should be used. The oiler may have to be modified so that no ambient air leaks into the system through the oil bowl. Oilless and carbon vane pumps should be avoided because they have a tendency to leak.

Evacuated Containers

The use of an evacuated container incorporates the gas mover and pollutant collector into one unit. This procedure is used for taking grab samples. The container can be evacuated with a vacuum pump either in the laboratory or the field. The initial conditions, temperature and pressure in the container must be known so that the total volume of sampled gas can be calculated. A good example of this type gas mover for the determination of oxides of nitrogen from combustion operations is described in Chapter 13. A schematic of such a sampling train is also shown in Figure 2.4.

The grab sample containers used are either glass with ground glass fittings or stainless steel. The glass containers must be handled carefully; they are usually encased in foam plastic to protect the flask from breakage and the personnel from hazard.

Long term integrated gas samples are sometimes necessary in source sampling. The accurate determination of the average CO_2 content in incinerator emissions is a prime example. When integrated samples are necessary flexible gas sampling bags or stainless steel gas cylinders are usually employed. The sample containers are evacuated prior to sampling and gradually filled over a long period of time.

Typical volumes and flow rates of these samples are 2-3 cubic feet and 0.025 cfm respectively. Rigid containers under an initial vacuum will fill themselves. Flexible bags must be filled with the aid of a pump. The bags can be pumped full directly or indirectly by drawing a vacuum on an air tight container in which the bag is placed. Figure 4.20 shows a commercially available integrated sampling system.

FLOW CONTROL

Flow regulation for most sampling trains is accomplished by using a throttling valve preceding the gas mover. This valve varies the vacuum the gas mover must work against and thereby changes the flow rate. A more sophisticated arrangement uses two valves. One precedes the pump and provides a coarse control while a second one is installed in a recycle (by-pass) loop to protect the pump and provide a fine control. This latter arrangement is used in the EPA particulate sampling train which is shown in Figure 2.5.

The major requirements of the flow control valve are: (a) it allows sensitive flow rate adjustment to meet proportional sampling (isokinetic) conditions and (b) it does not allow any leakage. Both of these requirements depend upon the valve construction. The leakage problem poses the same potential error as was discussed for the pumps. Good quality needle valves are required in most source sampling applications.

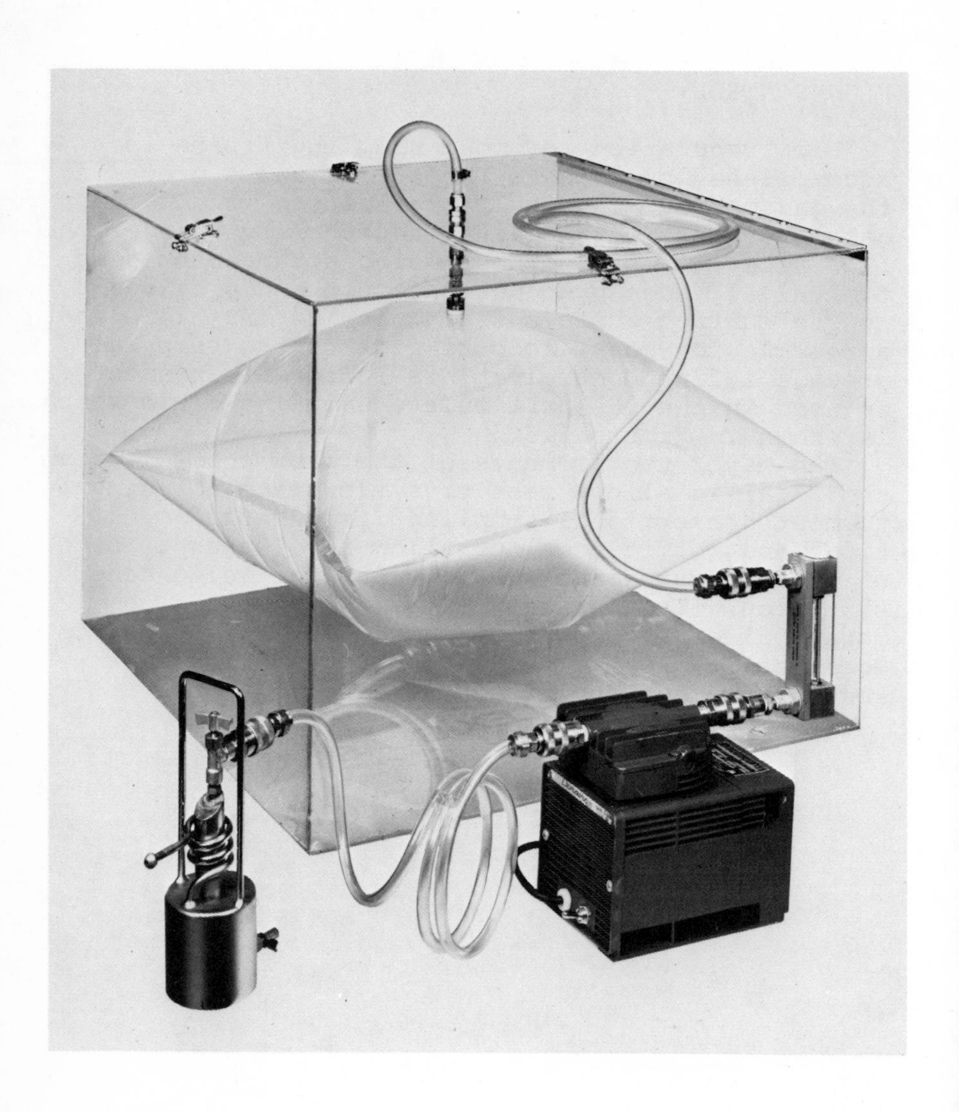

Figure 4.20. Integrated gas sampling system (courtesy of Scientific Glass Inc., Houston, Texas).

REFERENCES

1. Beckwith, T. G., and Buck, N. L. Mechanical Measurements, Addison-Wesley, Reading Mass. 1964.
2. Performance Test Codes, American Society of Mechanical Engineers, New York, New York.
3. Doebelin, E. O. Measurement Systems: Application and Design, McGraw-Hill Book Company, New York, New York. 1966.
4. Streeter, V. L. Fluid Mechanics, McGraw-Hill Book Company, New York, New York. 1966.
5. Nelson, F. O. Controlled Test Atmospheres, Ann Arbor Science Publishers, Ann Arbor, Michigan. 1971.
6. Air Sampling Instruments for Evaluation of Atmospheric Contaminants, American Conference of Governmental Industrial Hygienists, Cincinnati, Ohio. 1972.
7. Holmes, R. G. editor. Air Pollution Source Testing Manual, Los Angeles County Air Pollution Control District, Los Angeles. 1965.
8. Messersmith, C. W., Warner, C. F., Olsen, R. A. Mechanical Engineering Laboratory, Wiley and Sons, New York, New York. 1958.
9. Steam, Its Generation and Use, The Babcock & Wilcox Company, New York, New York. 1963.
10. "Performance Standards for New Stationary Sources," Federal Register, Vol 36, No. 247, December 23, 1971.
11. Morrow, N. L., Brief, R. S., and Bertrand, R. R. "Sampling and Analyzing Air Pollutant Sources," Chemical Engineering, January 24, 1972.
12. Dennis, R., Samples, W. R., Anderson, D. M., and Silverman, L. "Isokinetic Sampling Probes," Industrial Engineering Chemistry, Vol 4-9, No. 2, February 1957, pp. 294-302.
13. Toynbee, P. A., and Parkes, W. J. S. "Isokinetic Sampler for Dust Laden Gases," International Journal of Air and Water Pollution, Vol 6, Pergamon Press, Great Britain. 1962.
14. Cooper, H. B. H., and Rossano, A. T. Source Testing for Air Pollution Control, Environmental Science Services Division, Darian, Connecticut. 1971.
15. Perry, R. H., Chilton, C. H., and Kirkpatrick, S. D. Chemical Engineers' Handbook, McGraw-Hill, New York, New York. 1963.
16. Martin, R. M. Construction Details of Isokinetic Source-Sampling Equipment, Publ. No. APTD-0581. Air Pollution Control Office, EPA, Research Triangle Park, North Carolina. 1971.
17. Rom, J. J. Maintenance, Calibration, and Operation of Isokinetic Source Sampling Equipment, Publ. No. APTD-0576. Office of Air Programs, EPA, Research Triangle Park, North Carolina. 1972.

CHAPTER 5

CONDUCTING A SOURCE TEST

Source testing should be viewed as a scientific experiment requiring an organized and methodical approach. The source test is conducted to provide an answer to a question concerning the emissions from a specific source. If no other means can provide the answer, then a source test, as a last alternative, must be conducted. This chapter presents a systematic approach for organizing and conducting a source test. Figure 5.1 provides a check list for conducting a successful source test. It begins with the identification of the problem and the establishment of relations between the parties involved. Following this, the details of the test must be developed and carried out.

IDENTIFY PROBLEM

Purpose

The end results of any source test are numbers which will quantitatively prove or disprove a hypothesis concerning emissions from a source. For example, it may be hypothesized that a particular source emits significant quantities of several different pollutants. If the test results show this to be false, then an extensive effort was expended in conducting a test which produced unusable results. Source testing is too expensive, time consuming, difficult, and in many cases hazardous to allow this to happen. Therefore proper identification of the purpose is fundamental to a successful and useful source test.

With few exceptions, the parties who require the testing will not be those who conduct the test.

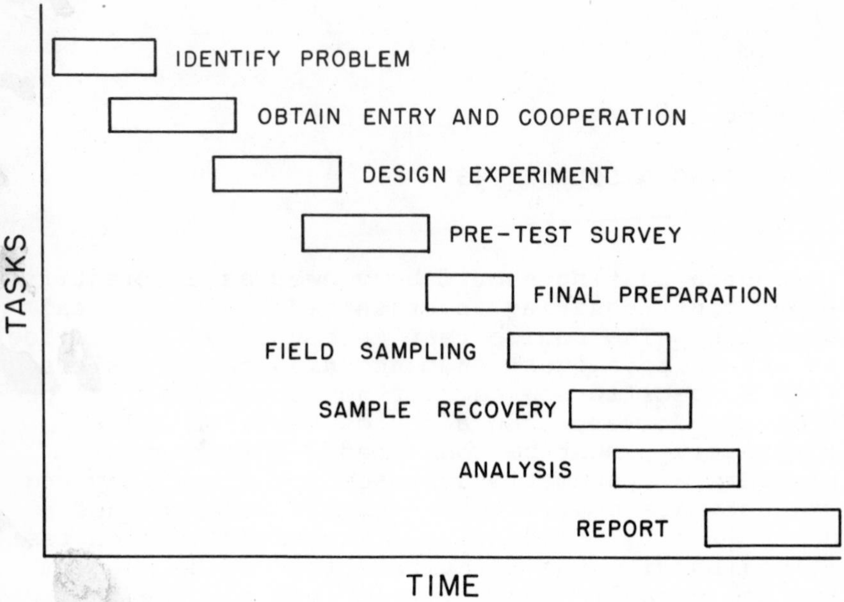

Figure 5.1. Planning a source test.

Thus the first party must adequately describe the needs and problems to the second party, who will then perform the test. Communications thus assume a major role in the successful completion of a source test. For air pollution control agency and company situations this communication will be between administrator, tester, and industrial personnel. It will be between the consultant and industrial personnel in the consulting situation. In most cases the tester or consultant is an engineer. The remainder of this section will be directed to the engineer.

The engineer must have a clearly defined statement of purpose. This begins with the specific question raised. He must know what is required.

If the party requiring the test cannot clearly identify the purpose, then the engineer must work to accomplish this. In certain situations this clarification may be in the form of a contract. Unfortunately the engineer will not always have adequate first hand information as to what is involved with the specific test. This situation is illustrated by a consultant who must prepare a bid for a contract to do sampling but will be able to see the site only after the contract has been awarded. Under these conditions he must rely on past experience and other sources of information to make the decisions and estimates. The ability of the engineer to do this hinges on his knowledge or background. Of course the easiest and most exact way to generate the background information for formulating a concise statement of the problem is to make a site visit.

Process Information

The development of a complete background must include several specific areas. The engineer must learn the process. What are the raw materials used? What are the operations performed to produce the finished product? Is it a continuous or batch process? How much product is produced? What is the pollutant of concern? What is the source strength for this pollutant? The answers to most of these questions often can be obtained by investigating the literature and then making a few calculations. The calculations involve making material balances and/or using pollutant emission factors such as presented in Chapter 3.

In the case where an initial plant visit cannot be used to gather the background information, several sources of information exist as an alternative. First a phone call to the plant manager may still provide important information. Remember, however, many people will choose not to give out such process information over the phone. A second approach would be to conduct a literature survey to determine if pertinent data from other similar opeations had been reported. References 1 through 8 are particularly useful in this respect. These references may provide such important information as process diagrams, descriptions of raw materials, materials handling procedures, emission factors and in some cases even hints on how to make the desired source test measurements.

A rather unique source of literature information is provided by the Environmental Protection Agency.[9] The Air Pollution Technical Information Center (APTIC) has developed a computerized literature search system whereby only a list of key words is needed as input. The computer output literature search is a series of abstracts which may be scanned rather rapidly to determine which are of possible interest. This service saves many hours in the library and at the present time is provided at no charge.

Laws

Before conducting any source test the engineer must determine what laws are involved. These laws may dictate the measurements to be made and how they are to be made. Questions to be considered are: Exactly what laws apply to this source? Are they local, state or federal laws? Who is the enforcement official to receive the filed report? Are there any prescribed test procedures and equipment which must be used? Are there requirements covering what capacity the process must be operated at during testing? How are the results to be reported? What type of process records must be kept during the test? The answers to these questions can be found most easily by contacting the local air pollution control authority. They will have copies of the various regulations and most likely will be able to discuss the pertinent details with you. In most cases it is also advisable to contact the state authority as well; this might be the State Board of Health or the State Environmental Protection Agency.

ENTRY AND COOPERATION

Obtaining permission of entry and the cooperation of plant personnel is of paramount importance. This can best be illustrated by the situation in which a sampling team arrives at the gate and finds the guard has no knowledge of their arrival. After several phone calls, the team is allowed to enter but the three people who could authorize equipment passes are not available. Therefore the team must return and start at least a day late. To prevent this situation, the engineer must have the details of entry worked out before hand. Cooperation begins on a contract in the form of an agreement as to what each party will supply or do. However, for a successful test, the needs of the source test engineer must also be understood by the maintenance and electrical shops. This is discussed in further detail in the section on the Pre-Test Survey. The plant officials will have to plan the process operation for the test if there are such requirements. The equipment operator is also of importance to the test. He may be required, or requested, to maintain certain levels of operation and also keep records of operation parameters.

Much time can be saved in testing if all the individuals involved external to the team are thoroughly familiar with what is going on and what is required of them. An example of failure to do this might be the loss of a particulate sample on a power plant because the boiler tubes were inadvertantly blown during the middle of a test run. A directive through management channels to prevent this would be much more cumbersome than an arrangement worked out directly with the operators.

Be sure to determine available insurance coverage for the test personnel, though you neither want nor expect an accident. Clarification of this point will protect everyone involved.

DESIGN EXPERIMENT

After accomplishing the first two steps, the experiment must be designed to answer a specific question. For example, "Does the newly constructed Dawson Powerplant comply with the federal regulation[10] for emission of particulate matter?" In this case the background information has been b brought together for a clear statement of the question. Similarly for other types of sources different statements of purpose might be formulated regarding control efficiency of specific control devices and product loss from manufacturing operations.

Once it is established that a source test is clearly needed then the test must be designed. The test design must answer What, How, Where, and When. The approach to answering these varies, depending upon the ability to make a site visit to the facility.

What

The engineer must know which pollutant or pollutants must be measured. The expected emission rates or concentrations should be estimated in the background survey. However, are there any other constituents which must be measured? Are any of the constituents expected to be hazardous and thus must be handled with care?

How

The next consideration is how to conduct the test. What type of sampling strategy is needed? What is the length of sampling time based on the categories discussed in Chapter 2? What are the sample size restrictions? What special procedures are required by law? How many runs will be required to give a reliable number? If the specific testing procedure is not specified by law, then an appropriate method will have to be selected. As discussed in Chapter 2 a number of these have been developed. Each procedure selected will have to be evaluated in terms of its ability to allow proper sampling and analysis of the constituents. A chemist should be consulted in this type of selection.

Where

The term location includes knowing both where the plant is located and where within the plant the sampling is to be done. For the first plant visit, directions are of course important. If, in the test design, a plant visit has already been completed then the location of sampling site and the numbers of ports and sampling points can be decided. If no site visit has been made then these considerations can be conducted during the Pre-Test Survey phase.

A method for determining the number of sampling points required to give a representative sample has been prescribed by EPA.[10] The details of this method are presented in Chapter 9. The requirement on upstream and downstream distances from a flow obstruction places a restriction of the location of sampling sites. The best location will therefore be in a long vertical, straight section of duct. This location may not always be the safest location for personnel, however. In the selection of sampling sites, there will be many considerations beside the number of upstream and downstream diameters from obstructions. If sampling ports are to be constructed, then a design must be readied for the plant personnel. In most sampling trains, the important dimension is the diameter of the port. Some trains require a 3-inch pipe coupling which must be welded to the stack. Other trains require supports that may be either welded or bolted to the stack. Possibly the ports can be made by the sampling personnel but whatever the case, these details must be worked out before the sampling date.

When

The decision as to when to take the sample involves a few considerations. If the process operates continuously, then the choice of a sampling date can be a matter of convenience for all concerned. However, if the process operates only 1.5 hours a day, then time also becomes a restraining factor in the experimental design. The sample time for the method selected cannot exceed the operation time of the process. If three runs constitute a test, then most likely three days must be allowed for testing in this latter case. The testing dates must be planned with the plant personnel so as not to conflict with construction and maintenance operations which would produce nonrepresentative test conditions. Furthermore, source tests may also have to be scheduled with respect to the availability of other personnel (*e.g.*, federal or state inspectors) who might want to witness the source test.

PRE-TEST SURVEY

The heart of the Pre-Test Survey activities is a site visit. All contractual agreements must be finalized. At this time those who will be involved directly with the test or the site preparation will have to be informed of what is required and the time of completion or performance. The sampling strategy must be finalized and the sampling sites selected.

The advantage of making a site visit after researching the process is that the engineer is less apt to overlook important factors. He is more cognizant of expected problems and can ask pertinent questions of plant personnel to seek proper solutions.

For the Pre-Test Survey it is a good idea to make up a check list of important information needed to conduct the source test properly. An example of such a check list is given in Chapter 9.

During the site visit it is sometimes a good idea to take a Polaroid camera and battery-powered tape recorder along. This way you can obtain pictures of the sampling sites. At the end of the visit these pictures can be reviewed with plant personnel and you can then show them that you have pictures pertinent only to your work. The tape recorder allows you to record supplemental information not included on the Pre-Test Survey form.

Process Information

During the Pre-Test Survey a complete description of all processes should be obtained. This should take the form of process diagrams, operating conditions and operating procedures. This information is needed to confirm the sampling strategy selected to adequately estimate pollutant mass emission rates.

All air pollution control devices should be noted and also described. If the amount of pollutant collected is known or the efficiency is known, then this information should be recorded. This information along with the estimate of the uncontrolled emissions, such as those reported by Reference 7, can be used to estimate the mass emission rate of the pollutant.

Another point to be investigated is the use of process monitors. What types of monitors are used and where are they located? Have they been calibrated recently? How many are operating and what are the units reported? Will this information be available during the source test? These measurements are important because they provide a record of the process operation during a time when the source test will be performed. Also if a part of a test has to be repeated, it may be necessary to try to reproduce the process conditions and this could be done on the basis of the process monitor information.

Site Selection

During the Pre-Test Survey the sampling site must be readied. The logical place to start is at existing ports if they are present. However, they must meet the requirements set forth by the pertinent regulations. In other words, what and where are the obstructions in terms of upstream and downstream diameters from the ports? What are the stack dimensions and will the present ports allow the stack to be divided into equal areas? What is the expected number of sampling points at these ports? Is there a better place for sampling which would require a fewer number of sampling points? If so, can the ports be installed at this location and will there be adequate room for equipment and personnel to operate safely? Chapter 9 considers these problems in detail.

If the decision is made to install ports for sampling, plant personnel must be notified. If possible this should be done while all parties involved are at the location. Suitable sketches or drawings must be made and given to the person who will do the work. If the decision regarding the installation of ports is not to be made until after the site visit, then preliminary arrangements should still be made with plant personnel.

Logistics

The logistics of the sampling operations in the plant must be worked out during the Pre-Test Survey visit. A final list of necessary equipment is compiled from information gained during the site visit.

The first consideration should be getting equipment and personnel to the sampling site. What type of passes or authorizations are required? Where will equipment be unloaded? Where are the parking facilities? Are elevators available? If not, how many flights of stairs are there and can all of the equipment be moved to the sampling site?

The next consideration is the site itself. Will scaffolding be required? Must ladders be used? All equipment, procedures and working environments should meet OSHA standards.[11] If existing ports are to be used, can the plugs or caps be removed or are they firmly corroded into place? These should be removed during the site visit since this allows inspection of the ports and determination of the approximate wall thickness of the duct. Another consideration is the weather: will the test crew and equipment be exposed to adverse weather?

It should also be determined if the testing personnel will be exposed to any hazardous gases. If this is the case, hooding arrangements must be used or in some cases the testing team may have to wear environmental suits. In any event, these arrangements must be made during the site visit.

Since almost all source sampling equipment requires electricity, the availability of power should be checked for the sampling site. A discussion with the maintenance man will determine if voltage and amperage requirements can be met.

If a laboratory exists in the plant, permission to use it would be a great convenience. The source testing procedures require many items which could be routinely obtained or used from the laboratory. These include ice, distilled water, a pan balance, and such solvents as acetone.

Knowing the location of many other items, like vending machines, cafeteria and restrooms, within the plant is a necessity. In addition there should be a sample recovery area as close to the sampling site as possible. It is unadvisable and almost impossible to properly recover samples on windy, drafty, or dusty scaffolding. If any equipment is to remain at the plant overnight, a lockable storage area is needed. In conjunction with the overnight arrangements, the location and rates of motels or hotels should be investigated. Similarly the location of local restaurants is helpful. If ice is not available within the plant, a local vendor must be found. Local transportation facilities should be investigated if they seem necessary.

Pre-Test Survey Measurements

To assist in the final preparations for the test, several simple measurements may be made to obtain an approximate idea of process conditions and pollutant concentrations. For example, there are length-of-stain detector tubes[12] available commercially for many different gases. These may be useful in identifying the order of magnitude concentration. Also portable combustion test kits are available which use filter spot method to provide some visual indication of process performance.

Several stack gas measurements should be made: pressure, temperature and velocity. A conventional pitot tube can be used to approximate the stack gas velocity. If the velocity is very low or very high, some special procedure may have to be used. Knowledge of the gas temperature is important because high values may necessitate the use of a water-cooled probe. A conventional pitot tube can also be used to measure the approximate static pressure within the stack.

FINAL PREPARATION

Final preparation should be done carefully so that the success of the sample is guaranteed because it is expensive and time-consuming to have to repeat the sample. Each sampling procedure to be used should be reviewed and a list of materials developed. Along with thisan estimate of sampling time requirements should be prepared. These procedures should be reviewed to insure that they will give the desired numbers to test the original hypothesis.

This also includes any supporting tests such as moisture and molecular weight determinations.

The Pre-Test Survey information should be used to indicate the sampling points within the stack cross-section, and these points should be labeled. The desired sampling time at each point should then be chosen. A time sequence should be developed for the sampling itself. This will allow the assignment of specific duties to individuals of the stack team, thereby allowing the sampling to be done in a minimum of time.

Each procedure should be reviewed and a data sheet prepared with all the required parameters listed. For example, it is useless to record gas meter readings if the temperature and pressure of the meter are not also recorded. At this point a process parameter data sheet should also be developed. Specific procedures for sample handling and processing upon sample return to lab should be developed.

The equipment necessary to conduct the source test must be prepared. Is there any special equipment needed? This may require the construction of a special length probe or possibly a water-cooled probe. The ordering of chemicals may be required at this time also. If stack sampling is to be done repeatedly, these chemicals should be stocked. All weighing and labeling of filters must be done at this time. Also chemical reagent containers must be labeled. All equipment should be packaged for transportation to the site. This should include extra or spare equipment which is needed because (a) glassware is breakable, (b) pumps burn out, and (c) dry gas meter diaphragms wear out.

Other functions which must be performed during preparation of equipment are calibrations and operational checks. The gas meters, flow meters, and pitot tubes must be calibrated. This should be done routinely prior to each test. Less routine calibrations include thermometers, probe and area heaters, and vacuum gauges. The sampling train can be checked for leaks at this point. However, this should be done again prior to sampling.

A tool kit with tools adequate for minor repairs on the sampling train should be included with the equipment going to the site. If something goes wrong perhaps it can be fixed in the field and little sampling time will be lost.

All safety equipment must be taken. This includes hard hats, safety glasses, safety belts, ropes, asbestos gloves, overalls, respirators, chains, etc. as the situation demands.

One of the major steps in the final preparation is to reconfirm all arrangements with plant personnel. Are the scaffoldings ready? Have the ports been prepared? Have provisions been made so that the process operation will be at proper conditions? Which gate is to be used for entry? Depending on the situation, it may be a good idea to insure all personnel and equipment for liability or damage. The lodging reservations should be made, along with those for transportation.

SAMPLING AND ANALYSIS

The mechanics of the sampling procedure should be almost routine because of the extensive work done in preparing for it. The first step, after entry to the plant, is to unload all materials in the predetermined areas. Someone must then get the ice. All equipment is unpacked and sampling train supporting devices are installed at the site. The sampling train is then assembled and all performance checks and leak tests are conducted. Any last minute modifications that might be needed are then performed. All preliminary tests are run. The final preparations, the selection of a nozzle and the marking of the probe length, are concluded and the sampling is begun. Pertinent process and test data are recorded on the data sheet at the required intervals.

Sample recovery and handling should receive major attention. After 3 to 6 hours of sampling, the sample recovery should not be done hastily. There is no point in sampling if the collected material is later altered or lost by improper handling or lack of labeling and recording. All collected samples should be quantitatively transferred to sample containers. The collected samples should be recovered in an area protected from wind or dust. Samples should not be put into dirty containers or dropped or mixed. If there is some doubt about a sample, discard it and perform the test again.

Attention should be paid to the chemical stability of the collected sample. Analyses must be completed before any errors are introduced due to deterioration of the sample. Are any of the reagents sensitive to heat or sunlight and therefore require protection? Are the sample containers inert with respect to the collected samples? If there are any questions regarding these points a chemist should be consulted.

Sample recovery forms should be used to insure that the proper data is recorded. Chapters 10 through 13 covering specific test procedures contain examples of such forms.

Analysis of the collected source samples is a laboratory activity. It is done as any routine lab analysis and according to the procedures decided upon in the design of the experiment. Any deviations from these procedures should be recorded and explained. A laboratory data sheet should be used which maintains the labelling system used during the sampling. Any calculations done during the analysis should be included in the record.

THE REPORT

The report presents the test results and recommendations. Future decisions regarding compliance, selection and design of control equipment, etc. will be based upon this report. A number of people who read this report may use it for decision making, yet most of them probably were not involved in any way regarding the test work itself. For this reason the report is the most important phase of source testing. All the work invested to properly define, prepare for, conduct a source test and analyze the resulting data will be wasted if a poor report results.

Thus, the report must provide certain basic information. It must answer why the test was done, who did it, how it was done, where it was done, when it was done, and what results were obtained. It must present all documentation of the results. A suggested report format is included in Table 5.1, and the following subsections discuss the report content.

Introduction

The purposes of the test must be stated. This may be done simply by indicating what hypothesis was tested. This section should also include a brief description of type of process and control methods used, pollutants sanpled, and basic sampling methods used.

Table 5.1

Suggested Report Format

Introduction
1. Statement of purpose for the test
2. Location, dates, personnel
3. Type of process and control methods
4. Pollutants emitted
5. Any other important background information

Summary of Results
1. Emission results
2. Process operation
3. Statement of conclusions and recommendations

Process Information
1. Description of process and control devices
2. Flow diagram
3. All process results including example calculations
4. Presentation of allowable emissions
5. Any required operation demonstrated

Sampling Procedures
1. Sampling port location and dimensioned cross-section
2. Sampling point location and dimensions
3. Symbols and labeling system
4. Sampling train diagram
5. Brief description of sampling procedures; any deviations discussed and justified
6. Analytical procedures briefly discussed; any deviations discussed and justified

Results
1. Complete emission results with example calculations
2. Evaluation of results.
3. Any additional conclusions or recommendations

Appendix
1. Raw field data
2. Laboratory report
3. Test log
4. Raw process operation data
5. Calibration procedures and results
6. Standard procedures
7. Applicable regulation
8. Any related reports
9. Project participants and titles

Summary of Results

A brief discussion of test results including tables is necessary. These tables should include emission results, process operating data, and any auxillary data such as analysis of products and waste from the process. Remember, this may be as far as the plant manager reads! Thus, the summary should also include conclusions concerning the original purpose of the test.

Process Description

A detailed description of the process is essential. All pertinent details regarding specific equipment must be included. A flow diagram must be given showing operating conditions and materials and energy balances.

Location of Sampling Points

A brief description of the location and number of sampling points used in the cross section is needed. The location and points should be identified with the same symbols that are used on data sheets. The reason for selecting these points should be explained.

Process Operation

The process conditions which existed during the sampling must be described. This section should include any other information which is needed to describe the process.

Sampling and Analytical Procedures

A brief description of sampling and analytical procedures used is required. All modifications from the standard procedures and the reasons for them should also be noted. The complete procedures should be included in the appendix of the report.

Calculations and Results

Thissection should show all example calculations and the complete results for each of the tests. This may also involve a statistical evaluation of the results.

Discussion

In this section the validity of the results is discussed. What is the percentage of error? Why did it occur? Here, any errors or questions that could be raised about the report should be discussed by the reporter. It is better that these points be brought up now rather than by a cross examining attorney or a plant manager.

Appendix

The appendix should include all data sheets, the sampling procedures, related reports, and a listing of all participants in the test.

REFERENCES

1. Stern, A. C. editor. Air Pollution, Academic Press, New York. 1968.
2. "Process Flow Sheets," Chemical Engineering, McGraw-Hill, New York.
3. Shreve, N. Chemical Process Industries, McGraw-Hill, New York. 1967.
4. Kirk, and Oppenheimer. The Encyclopedia of Chemical Technology
5. Devorkin, H. Source Testing Manual. Air Pollution Control District, County of Los Angeles, Los Angeles, California. 1965.
6. Journal of the Air Pollution Control Association, Pittsburgh, Pennsylvania.
7. Compilation of Air Pollutant Emission Factors, Publ. No. AP-42. U. S. Environmental Protection Agency, Research Triangle Park, North Carolina. 1972.
8. Cooper, H. B. H., and Rossano, A. T. Source Testing for Air Pollution Control, Environmental Science Services Wilton, Connecticut. 1971.
9. Air Pollution Technical Information Center, U.S. Environmental Protection Agency, Research Triangle Park, North Carolina.
10. "Performance Standards for New Stationary Sources," Federal Register, Vol 36, No. 247, December 23, 1971.
11. "Occupational Safety and Health Act," Federal Register, Vol 36, No. 105, May 29, 1971.
12. Air Sampling Instruments, American Conference of Governmental Industrial Hygienists, Cincinnati, Ohio. 1972.

CHAPTER 6

THE CALCULATION OF MOLECULAR WEIGHT

The molecular weight of the source gas stream must be known in order to properly determine the stack gas velocity and volumetric flow rate during a source test. The average stack gas velocity, $\overline{V}_s$, is determined by applying Equation 4-11 at a number of different traverse points as shown in Figure 2.2.

$$V_s = 85.48\ C_p \left(\frac{T_s \Delta P}{P_s M_s} \right)^{1/2} \tag{4-11}$$

where

M_s is the molecular weight of the stack,
ΔP is the velocity head, in H_2O
T_s is the stack gas temperature, °R
P_s is the absolute stack gas pressure, in Hg
C_p is the pitot tube coefficient.

The average stack gas velocity, $\overline{V}_s$, is computed by taking the average of all measurements. The volumetric stack flow rate, $\overline{Q}_s$, is then calculated using Equation 2-2.

$$\overline{Q}_s = \overline{V}_s A_s \tag{2-2}$$

When proportional sampling (sampling in proportion to stack gas flow rate) is required, the flow rate of the sampled gas has to be measured and controlled accurately. An orifice meter is often used to do this.

Equation 4-14 shows the dependence of the orifice meter flow rate on the molecular weight of the metered gas.

$$Q_m = K_m \left(\frac{T_m \Delta m}{P_m M_m} \right)^{1/2} \tag{4-14}$$

where

M_m is the molecular weight of the gas entering the meter
K_m is a proportionality factor
T_m is the upstream gas temperature to the meter, °R
Δm is the orifice meter pressure drop, in. of H_2O
P_m is the absolute upstream meter pressure, in. of Hg.

This chapter will outline the computation methods for determining the molecular weight of the gas stream. In some instances the source may be well defined and hence nomographs have been developed to aid in estimating the molecular weight.

The calculation of the average molecular weight of a gas stream is performed using the ideal gas laws. For most source sampling situations the conditions are such that significant error is not introduced by using these laws. In general the laws do not hold for conditions of extremely high pressures and temperatures. The two laws used are the Ideal Gas Law and Dalton's Law of Partial Pressure.[1]

IDEAL GAS LAW

The Ideal Gas Law is one which describes the state of a gas as a function of the physical parameters

$$P\ V = \frac{m\ R\ T}{M} \qquad (6-1)$$

where

P is the absolute gas pressure
V is the volume of the gas
m is the mass of the gas
R is the universal gas constant
T is the absolute gas temperature
M is the molecular weight of the gas.

DALTON'S LAW

Dalton's Law of Partial Pressure states that the pressure exerted by a gaseous component of a gas-mixture at a given temperature is the same as it would exert if it filled the whole space alone. This of course assumes that no chemical reactions or other similar processes are occurring. The significance of this law then can be expressed as

$$P = \Sigma P_i \tag{6-2}$$

where

P is the total pressure of the gas mixture
P_i is the partial pressure of the associated with each component.

Partial presure is simply the pressure which one gaseous component exerts; therefore the total pressure is the sum of all the partial pressures.

COMPUTATION METHOD

The Ideal Gas Law and Dalton's Law of Partial Pressure can be used to develop a method of computing the molecular weight of a gaseous mixture. Consider a mixture of gases in a closed container. Applying the Ideal Gas Law for component "i" with partial pressure P_i we get

$$P_i V_{mix} = \frac{m_i R T_{mix}}{M_i} \tag{6-3}$$

where

V_{mix} is the volume of the gas mixture container
T_{mix} is the absolute temperature of the gas mixture.

Similarly for the mixture we get

$$P_{mix} V_{mix} = \frac{m_{mix} R T_{mix}}{M_{mix}} \tag{6-4}$$

Dividing Equation 6-3 by Equation 6-4 and cancelling terms we obtain

$$\frac{P_i}{P_{mix}} = \frac{\frac{m_i}{M_i}}{\frac{m_{mix}}{M_{mix}}} \tag{6-5}$$

If the same mixture is at constant T and P, then we can obtain

$$\frac{m_i}{M_i} = \frac{P_{mix} V_i}{R T_{mix}} \tag{6-6}$$

$$\frac{m_{mix}}{M_{mix}} = \frac{P_{mix} V_{mix}}{R T_{mix}} \tag{6-7}$$

Substituting into Equation 6-5 we obtain

$$\frac{P_i}{P_{mix}} = \frac{V_i}{V_{mix}} \tag{6-8}$$

But, by definition

$$\frac{V_i}{V_{mix}} = B_i$$

where B_i is the mole fraction and thus

$$B_i = \frac{V_i}{V_{mix}} = \frac{P_i}{P_{mix}} \tag{6-9}$$

Rearranging Equation 6-3 we obtain

$$P_i \; V_{mix} \; M_i = m_i \; R \; T_{mix} \tag{6-10}$$

Summing over all components in the gas mixture we obtain

$$V_{mix} \; \Sigma(P_i \; M_i) = R \; T_{mix} \; \Sigma \; m_i \tag{6-11}$$

But since the whole is equal to the sum of its parts

$$\Sigma \; m_i = m_{mix} = \frac{P_{mix} \; V_{mix} \; M_{mix}}{R \; T_{mix}}$$

Substituting this into Equation 6-11 and cancelling terms we obtain

$$M_{mix} = \frac{\Sigma(P_i \; M_i)}{P_{mix}} \tag{6-12}$$

But recalling Equation 6-9, this can be expressed

$$M_{mix} = \Sigma(B_i \; M_i) \tag{6-13}$$

where

M_{mix} is the molecular weight of the ideal gas mixture

B_i is the mole fraction of each component gas $\left(B_i = \frac{V_i}{V_{mix}} = \frac{P_i}{P_{mix}} = \text{\% by volume}\right)$

M_i is the molecular weight of each component ideal gas

Equation 6-13 when applied to a fossil fuel combustion process takes the following form.

$$M_{mix} = \left\| \begin{array}{l} B_{CO_2} \; (44) + B_{O_2} \; (32) + B_{CO} \; (28) \\ + B_{N_2} \; (28) + B_{H_2O} \; (18) \end{array} \right. \tag{6-14}$$

The measurement methods for determining the "B" terms in Equation 6-14 are presented in Chapter 7. The following examples, 6-1 and 6-2, illustrate how Equation 6.13 is used to calculate the molecular weight of the process waste gas from an incinerator and a sulfuric acid plant.

Example 6.1. The calculation of the stack gas molecular weight for an incinerator.

Stack Gas Analysis

Gas	M_i	*% by Volume* (B_i)	$B_i\ M_i$
N_2	28	72%	0.72 x 28 = 20.20
O_2	32	5%	0.05 x 32 = 0.16
CO_2	44	8%	0.08 x 44 = 3.51
CO	28	0%	0.00 x 28 = 0.00
H_2O	18	15%	0.15 x 18 = 2.70
Total		100%	M_{mix} = 26.57

Example 6.2. The calculation of the tail gas molecular weight for H_2SO_4 production.

Tail Gas Analysis

Gas	M_i	*% by Volume* (B_i)	$B_i\ M_i$
N_2	28	96.0%	0.98 x 28 = 27.40
O_2	32	3.5%	0.035 x 32 = 1.12
SO_2	64	0.5%	0.005 x 64 = 0.32
Total		100.0%	M_{mix} = 28.84

NOMOGRAPHS

For some sources the operational parameters are known so that nomographs can be constructed to assist in determining the molecular weight of the gases. Combustion processes are an example of this, with Figure 6.1 showing such a nomograph. While the nomographmay not be as accurate as the previously discussed methods, it does provide a first estimate

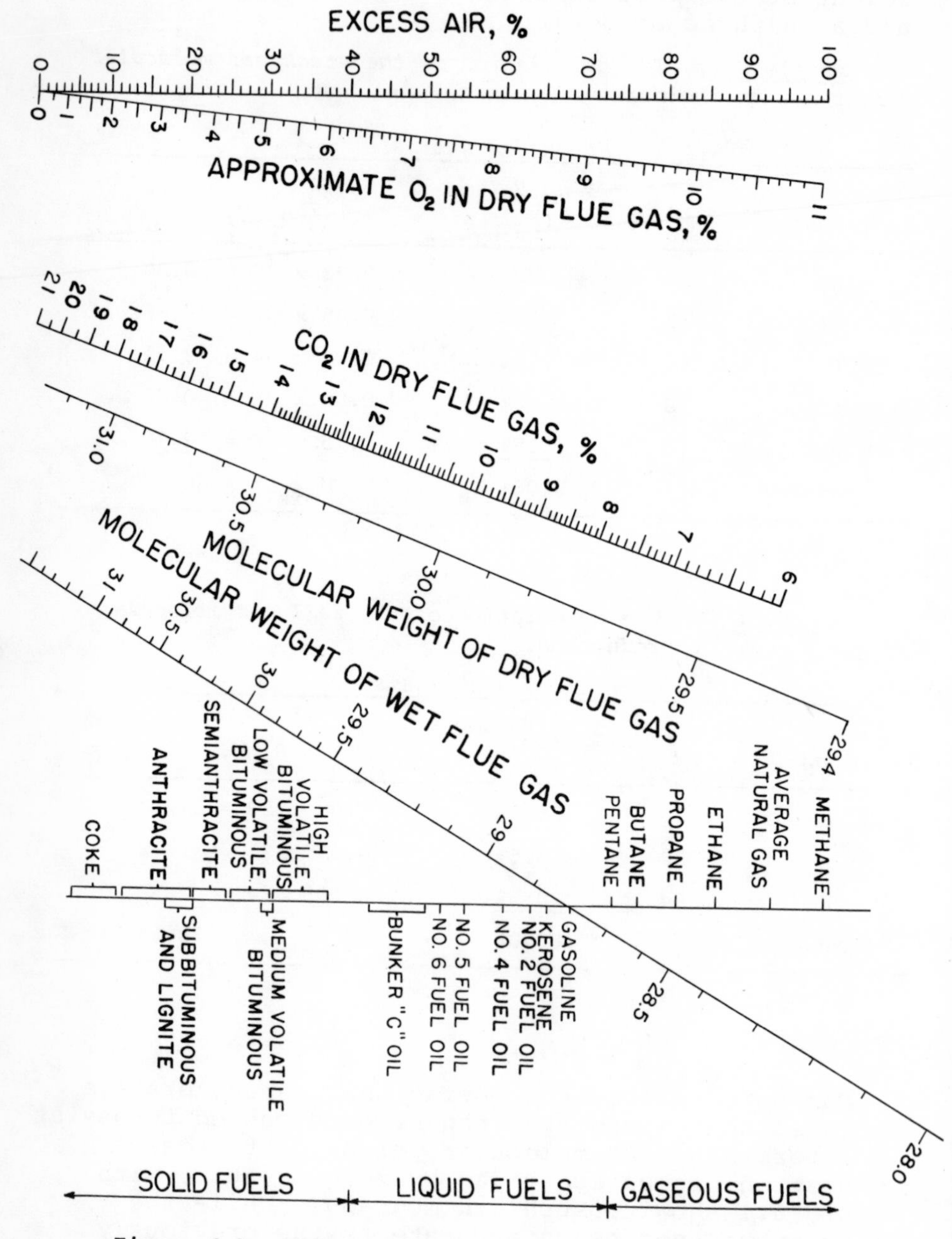

Figure 6.1. Nomograph to estimate the molecular weight of the dry flue gas.[2]

of the molecular weight. This in itself is useful since it must be used in properly setting the proportional sampling conditions for obtaining a representative sample. It is important to notice that for a wide range of fuel types and excess air the molecular weight varies theoretically between 29.4 and 31.0.

REFERENCES

1. Mooney, D. A. Mechanical Engineering Thermodynamics, Prentice-Hall, Inc., Englewood Cliffs, N.J. 1959.
2. Atmospheric Emissions from Coal Combustion--An Inventory Guide, Publ. No. 999-AP-24. U.S. Public Health Service, Division of Air Pollution, Cincinnati, Ohio. 1966.

CHAPTER 7

CARRIER GAS MEASUREMENTS

This chapter will outline the methods used to make molecular weight and velocity measurements of the pollutant carrier gas. Knowledge of the carrier gas characteristics is essential for attaining proper sampling conditions, with information such as gas flow rate and composition needed. As indicated in Chapter 6 some of this basic information must be available in order to calculate the average molecular weight of the waste gas stream. The pollutant mass emission rate, PMR_s, cannot be calculated unless the stack gas flow rate is known.

The constituents often present in substantial amounts and thus having the greatest influence on molecular weight depend on the process being studied. However CO_2, CO, O_2 and N_2 are often major constituents. Other gaseous constituents such as hydrocarbons, oxides of nitrogen and oxides of sulfur are also usually present but their proportion is quite small.

STACK GAS VOLUME FLOW RATE

As discussed in Chapter 4 a pitot tube is used to determine the volume flow rate of the pollutant carrier gas during a source test. However, pitot tubes merely measure instantaneous velocity at a point and cannot be used to directly measure gaseous volume flow rates. The equation expressing instantaneous gas velocity at a point when a pitot tube is used is

$$V_{s_1} = K_p \; C_p \left(\frac{\Delta P_1 \; T_{s_1}}{P_{s_1} \; M_{s_1}} \right)^{1/2} \qquad (7\text{-}1)$$

where

V_{s_1} is the instantaneous gas velocity at some point 1 in a flow stream, fps.

K_{p_1} is 85.48 $\frac{ft}{sec} \left(\frac{lb}{lb\text{-}mole\text{-}°R}\right)^{1/2}$

C_p is the dimensionless pitot tube coefficient ($C_p = f(\Delta P)$)

ΔP_1 is the velocity head measured at point 1, in H_2O

T_{s_1} is the absolute temperature measured at point 1, °R

P_{s_1} is the absolute stack pressure at point 1, in Hg

M_{s_1} is the molecular weight of the sampled gas (ideal) at point 1, $\frac{lb}{lb\text{-}mole}$.

Thus to determine the true average velocity of a gas stream, pitot velocity measurements have to be made simultaneously and continuously at all points within the cross-sectional flow area. Practically speaking this is an impossible task. However, practical methods have been developed to estimate the average velocity through a cross-sectional flow area. The federal performance standards describe such procedures.[1] EPA Method 2 is a procedure for determinine average stack gas velocity and volumetric flow rate for stationary air pollutant sources.

EPA Method 1 provides a procedure for approximating the true gas velocity over the cross-sectional flow area. Providing that the source gas velocity is not a function of time but only of location within the flow area, Method 1 can be applied. Method 1 is designed to help pick a sampling location where the cross-sectional variations of velocity pressure, ΔP, and aerosol mass concentrations are at a minimum. To aid in the extraction of representative samples, the method requires sampling or traverse points as a function of distance from upstream and downstream flow disturbances as shown in Figure 7.1. The details of the method are further discussed in Chapter 9.

An S-type pitot is used for velocity measurements. The tendency of S-type pitot tubes to become plugged by dust-laden air streams is less than that of standard pitot tubes. The method requires that the average stack gas velocity be determined by applying Equation 4-11 as follows

$$V_{s(avg)} = K_p \; C_p \; (\Delta P)^{1/2}_{avg} \left(\frac{(T_s)avg}{P_s \; M_s} \right)^{1/2}$$

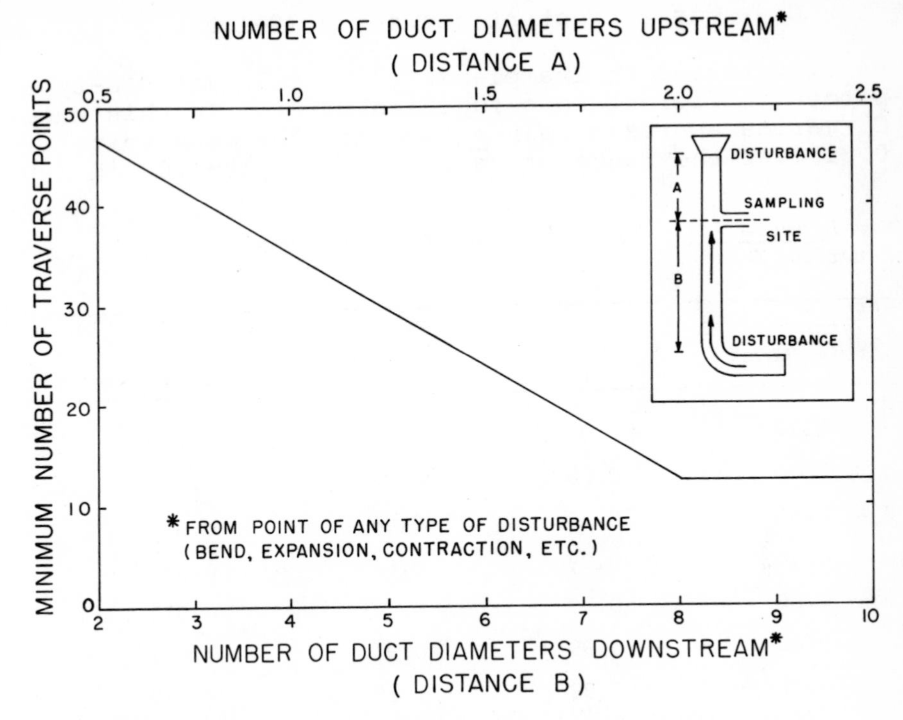

Figure 7.1. Minimum number of traverse points (EPA Method 1).

It assumes that the molecular weight and static pressure of the sampled gas do not vary within the sampling time, or sampling area. Method 2 also assumes that the stack gas temperatures and velocity pressures at the individual points within the stack do not vary with time. This method, however, does consider that the pitot coefficient, C_p, for S-type pitot tubes is a function of ΔP and the physical conditions of the pitot tube. Thus the method requires that all S-type pitot tubes used be calibrated after each field use and that the pitot coefficient, C_p, calculated be within $\pm$ 5% over the working range of velocities expected (usually 0-1 in H_2O). A typical calibration of an S-type pitot tube is shown in Example 7-1. The S-type pitot tube is calibrated against a standard pitot according to the following equation:

$$C_{p_{S\text{-}type}} = C_{p_{std}} \left(\frac{\Delta P_{std}}{\Delta P_{S\text{-}type}} \right)^{1/2} \qquad (7\text{-}2)$$

where

$C_{p_{std}}$ is a constant between 0.98-1.00.

As illustrated in Example 7.1, the calculated S-type pitot coefficient is a function of ΔP. In this case C_p varies approximately $\pm$ 1.2% from the mean value of 0.81 over the range of calibration (0.10-1.00 in H_2O).

Example 7.1. Typical S-type pitot calibration using Equation 7-2.

$\Delta P_{standard}$ in H_2O	$C_{p_{standard}}$	$\Delta P_{S\text{-type}}$	$C_{p_{S\text{-type}}}$
0.10	0.99	0.15	0.818
0.20	0.99	0.31	0.804
0.30	0.99	0.46	0.808
0.40	0.99	0.61	0.810
0.50	0.99	0.76	0.812
0.60	0.99	0.91	0.812
0.70	0.99	1.10	0.798
0.80	0.99	1.20	0.817
0.90	0.99	1.35	0.817
1.00	0.99	1.50	0.818
		ave	0.81
		range	0.80 - 0.82

As calibrated, this pitot meets the requirements of EPA Method 2 over the entire range and the average C_p value can be used without significant error. A more accurate method, however, would be to use the multiple point calibration generated rather than the average C_p value for all velocity calculations. Table 7.1 is an example of a velocity traverse data sheet for field use.

Table 7.1

Velocity Traverse Data Sheet

Source ______________ P_{bar} _____ in Hg P_s _____ in H_2O

Firm ______________ Tester ______________

Location ______________

Test No. ______________

Date ________ Time ________

Upstream Dist. _____ D

Downstream Dist. _____ D

Molecular Weight _____

Stack X-Section

Sample Port	Traverse Point	ΔP in H_2O	C_p	$\sqrt{\Delta P}$	T_s, °R	Remarks
Average						

MOLECULAR WEIGHT DETERMINATION

The molecular weight of the pollutant carrier gas depends on the process being studied. However, most processes using air as raw material usually have considerable amounts of N_2 and O_2 in their gaseous emissions. For instance, combustion process gases usually consist of N_2, O_2, CO_2, H_2O vapor and possibly CO. The tail gases from acid manufacturing plants (HNO_3 and H_2SO_4) are mixtures of N_2 and O_2 with less than 1% SO_x or NO_x. Detector tube methods are available for SO_2 and NO_x determinations that are accurate enough for molecular weight determinations. Orsat analyzers are available for the determination of CO_2, O_2, and CO concentrations. While many other more accurate methods exist, the orsat method still provides adequate combination of accuracy and flexibility needed.

Orsat Analyzer

The Orsat analyzer uses a wet chemical method which allows the volumetric determination of carbon dioxide, oxygen and carbon monoxide under controlled sample conditions. If no other major constituents are present, then the nitrogen content is estimated by a subtraction process. Since the gases are collected over water, the measurement of the concentration of these components allows only for the determination of the dry molecular weight of the sample. The Orsat operates under constant pressure and temperature conditions and the Ideal Gas Laws are used. Thus Equation 6-9 applies and the test measures the most fractions, B_i, of CO_2, O_2, and CO.

$$B_i = \frac{V_i}{V_{mix}} \qquad (6\text{-}9)$$

where

V_i is the each sample component
V_{mix} is the volume of the gas sample mixture.

Apparatus. A typical Orsat apparatus is shown in Figure 7.2. A measured volume of sample is collected over water and hence is always saturated. The sample burette, located on the right side, usually has a volume of 100 ml (50 ml burettes are also available). The percentage by volume of the constituents is determined by successive absorption of carbon dioxide, oxygen, and carbon monoxide from the sample using successive absorbing burettes filled

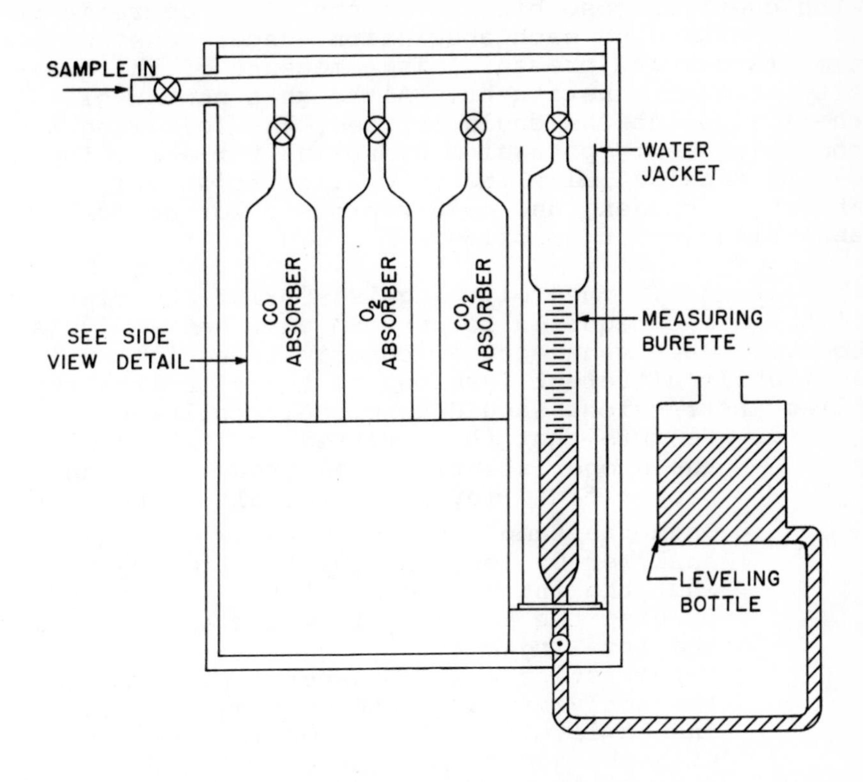

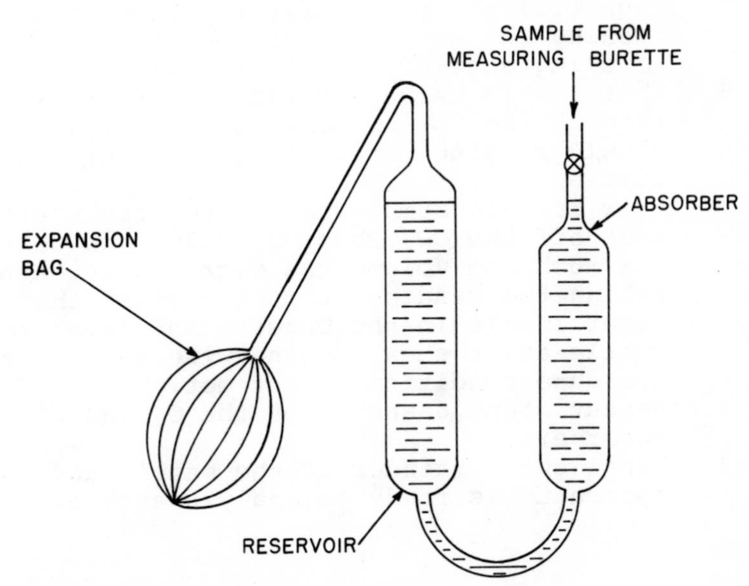

Figure 7.2. *Schematic of Orsat analyzer.*

with chemical absorbing solutions. The decrease in volume caused by each absorption (under constant temperature and pressure) is a measure of the quantity of each constituent. Although a number of chemical solutions could be used, the following are commonly used: potassium hydroxide for absorbing carbon dioxide, alkaline pyrogallic acid for absorbing oxygen, and acid cuprous chloride for absorbing carbon monoxide.

Procedure. The following is a brief description of the steps required to perform an Orsat analysis. However, each manufacturer constructs equipment in a slightly different fashion and therefore the supplied instructions should be closely followed.

1. Be sure that the burettes are filled with the proper solutions and brought to the proper fluid levels. Shut all valves.
2. Open the sample valve and draw gas into the analyzer by lowering the leveling bottle.
3. Shut the intake and expel the sample by opening the atmospheric valve and raising the leveling bottle.
4. Repeat steps 2 and 3 several times so that the analyzer is flushed out and the water in the leveling bottle becomes saturated with the gas to be sampled.
5. Then fill the glass burette with sample by opening the sample valve and lowering the leveling bottle.
6. Shut sample valve and flush gas sample through the CO_2 absorber several times by slowly raising and lowering the leveling bottle.
7. Draw the gas sample back into the burette, shut off the CO_2 absorber valve and observe how much the volume has decreased. When taking the reading it is important that the water levels in the burette and leveling bottle are the same; otherwise the constant pressure condition is not met.
8. Repeat steps 6 and 7 for the O_2 and CO absorbers.
9. Expel the remainder of the sample and the apparatus is ready to analyze another sample.

Operational Problems. A number of operational problems can occur and may cause significant error in the measurement. All fittings and valves must be leak proof so that no ambient air is allowed to enter the analyzer during the test. It is important that the gas sample always be passed in sequence through the CO_2, O_2 and CO absorbing solutions. Otherwise an error will occur. The O_2 absorbing solution also has a great affinity for CO_2 and the CO absorbing solution will also absorb O_2. The water in the burette must be saturated with sample gas or else it will absorb part of the sample prior to the use of the absorbers and thus cause an error. When flushing the sample through the absorbing solution, care must be taken not to force any of the liquid into the capillary column manifold; this will ruin a test and may require disassembly of the Orsat for cleaning. The test should be performed at ambient (70°F) temperatures. Gas samples should therefore be allowed sufficient time to come into temperature equilibrium with the water jacket (which should be at 70°F); this should require only a few minutes at most. The absorbing solutions should be replaced frequently when conducting source tests. If a large number (about 10) of passes are required through one of the absorbers before a stable reading is achieved, this is a sign that the solution should be replaced.

The Orsat analysis data is used to calculate the dry molecular weight of the gas sample. The volume of water vapor in the sampled gas does not enter into the analysis. The sample gas is collected over a water column and therefore is always saturated with H_2O vapor. Thus the H_2O vapor pressure of the sampled gas is adjusted and always is a constant (0.72 in Hg at 70°F and 29.92 in Hg). Applying Equations 6-13 Orsat data:

$$M_{mix} = \Sigma B_i M_i \qquad (6\text{-}13)$$

$$M_{dry\ gas} = B_{CO_2}(44) + B_{O_2}(32) + B_{CO}(28) + B_{N_2}(28)$$

When the mole fraction of H_2O vapor, B_{H_2O}, is known in conjunction with an Orsat analysis, the actual molecular weight of the sampled gas can be determined:

$$M_{mix} = (1-B_{H_2O})\ M_{dry\ gas} + B_{H_2O}\ (18)$$

SAMPLING TRAINS FOR CARRIER GAS CONSTITUENTS

The federal EPA Method 3 describes procedures for taking grab samples and integrated samples.[1] Figures 7.3 and 7.4 show these sampling trains.

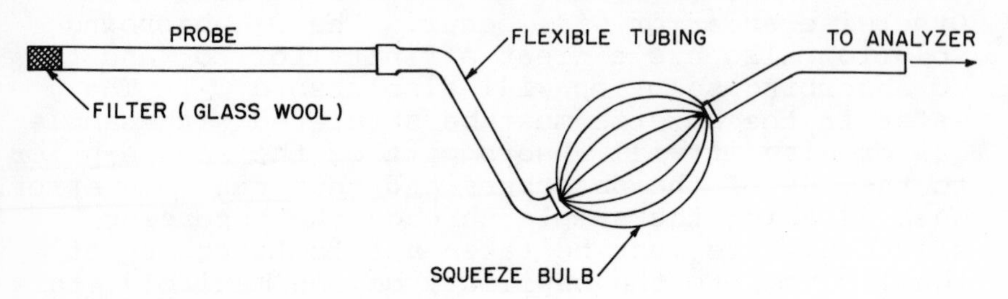

Figure 7.3. Grab sampling train for Orsat analysis.

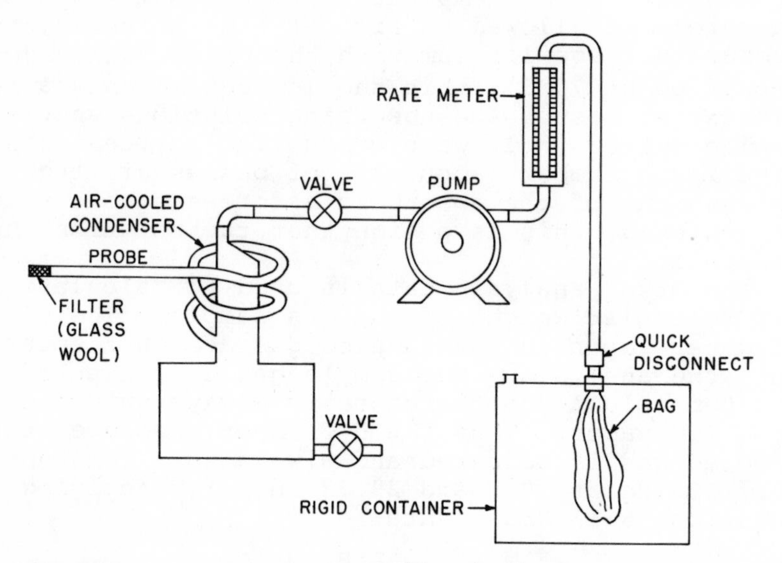

Figure 7.4. Integrated sampling train for Orsat analysis.

These methods are prescribed basically to provide the necessary data for calculating the molecular weight of the gas stream.

Grab Sample Method

This method is used where the source is *uniform* and *steady*. As discussed in Chapter 2, if this method is used for any other conditions, a representative sample is not obtained. Glass wool is used to remove particulate matter. The probe may be either glass or stainless steel, and the squeeze bulb is a leakless one-way type which transports the gas sample to the Orsat analyzer.

Integrated Sample Method

The integrated sample method is most appropriate for sources which are *uniform* but *unsteady*. This method provides the capability of sampling proportionally for periods up to about 150 minutes. A composite sample is analyzed for the various constituents by the Orsat method. The air condenser is used to remove the excess moisture from the gas sample stream. The pump must be leakless so that no ambient air is brought into the system.

Equipment and Reagents. The following items are necessary to conduct the test.

1. stainless steel or Pyrex glass probe
2. glass wool
3. air-cooled condenser or equivalent
4. needle valve
5. leakless diaphragm-type pump
6. rotameter with range of 0-0.035 cfm
7. Tedlar or equivalent plastic bag with volume of 2 or 3 cubic feet
8. pitot tube apparatus
9. Orsat analyzer apparatus
10. rigid container for sample bag
11. necessary tubing and fittings to connect components.

Field Procedures. Set up the sampling train as shown in Figure 7.4. The bag should be checked for leaks. It does not matter if a small portion leaks out, but an error will result if ambient air is allowed to leak into the bag. The sample bag should be evacuated by drawing a vacuum on it; the sampling train pump can be used to do this. Place the probe in the line and then purge the system with sample gas. Use the needle valve to set the flow rate somewhere near midrange on the rotameter. The exact value is not important as it will have to be adjusted periodically to maintain proportional sampling conditions. Connect the sample bag and make sure that all connections are leakproof. Observe the initial stack gas velocity and sample flow rate. Throughout the test adjust the sample flow rate to correspond to the changes in stack gas velocity which is proportional to the $(\Delta P)^{1/2}$ manometer reading.

Calculations

The Orsat data is used in Equations 6-9 and 6-13 to calculate the mole fractions and dry molecular weight respectively. The data can also be used to calculate the amount of excess air supplied to combustion processes. This is done by averaging the test values and substituting into the following equation

$$\% \text{ EA} = \left(\frac{(\% \text{ O}_2) - 0.5\ (\% \text{ CO})}{0.264\ (\% \text{ N}_2) - (\% \text{ O}_2) + 0.5\ (\% \text{ CO})}\right) 100 \tag{7-4}$$

where

% EA is the percentage of excess air

% O_2 is the percentage of oxygen by volume on a dry basis

% N_2 is the percentage of nitrogen by volume on a dry basis

% CO is the percentage of carbon monoxide by volume on a dry basis

0.264 is the volume ratio of oxygen to nitrogen in the air.

Other Gaseous Constituents

When it is obvious that some other gas is a major constituent then it must be included in the analysis. This situation would be governed by the source and should actually be identified during the initial phases of planning for the source test. Special sampling and analysis procedures must be identified. The reader is referred to References 2, 3, and 4 for assistance should this problem arise.

MOISTURE CONTENT DETERMINATION

Four basic methods can be used to measure or estimate the moisture content of a gas stream: condensation, adsorption, saturation pressure, and psychrometry. Of these, condensation and adsorption are used in source sampling. The condensation method employs an ice bath and is quite effective if the moisture content is over 1%. The adsorption method works well for lower moisture contents, but does not work well for temperatures above 300°F.

If the gas stream is saturated with water vapor then the steam tables[5] can be used; then only the temperature of the gas stream need be measured in order to make the moisture content determination. At saturation, the gas is holding the maximum amount of water vapor possible for a given temperature. If the gas stream is not saturated, the psychrometric methods developed by heating and ventilation engineers can be used. Basically the method involves measuring the wet-bulb and dry-bulb temperatures of the gas stream and then using psychometric charts.[5]

Condensation

The condensation method of determining moisture content uses an ice bath arrangement. The gas sample is drawn from the stack at a known flow rate and passed through the condenser. The moisture content is determined by observing the amount of condensate collected for a known volume of gas sample. In making this type of measurement it is very important that none of the moisture condenses prior to reaching the bath. Of course some moisture leaves the condenser but this quantity can be calculated because the gas becomes saturated at the ice bath temperature. Thus the total volume of moisture in the sample is

$$V_{wo} = V_{wc} + V_{wm} \tag{7-5}$$

where

- V_{wo} is the volume of extracted water from the source at gas meter conditions
- V_{wc} is the volume of water that condensed at the condenser referred to gas meter conditions
- V_{wm} is the volume of water passed through the gas meter at meter conditions.

V_{wm} can be calculated as follows:

$$V_{wm} = \frac{e_m}{P_m} V_m \tag{7-6}$$

where

- e_m is the water vapor pressure for saturated conditions and meter temperature
- P_m is the absolute pressure at the meter
- V_m is the total volume of gas passed through the meter at the meter conditions.

The proportion of the moisture (by volume) existing in the duct at the sampling point is expressed

$$B_{wo} = \frac{V_{wo}}{V_{wc} + V_m} \qquad (7\text{-}7)$$

substituting Equation 7-5 into Equation 7-7 we obtain

$$B_{wo} = \frac{V_{wc} + V_{wm}}{V_{wc} + V_m} \qquad (7\text{-}8)$$

Solving Equation 7-5 for V_{wm} and substituting into Equation 7-8, we obtain

$$B_{wo} = \frac{V_{wc} + \left(\frac{e_m}{P_m}\right) V_m}{V_{wc} + V_m} \qquad (7\text{-}9)$$

Multiplying the numerator and denominator by P_m

$$B_{wo} = \frac{P_m V_{wc} + e_m V_m}{P_m [V_{wc} + V_m]} \qquad (7\text{-}10)$$

Applying the Ideal Gas Law, Equation 7-5, and solving for V_{wc} and substituting into Equation 7-10, we obtain

$$B_{wo} = \frac{m_{wc} R T_m + e_m V_m M_w}{m_{wc} R T_m + P_m V_m M_w} \qquad (7\text{-}11)$$

where

m_{wc} is the mass of water collected at the condenser
T_m is the absolute gas temperature at the meter
M_w is the molecular weight of water
R is the universal gas constant.

The value determined by Equation 7-11 is used directly to determine the molecular weight of the stack gas.

Adsorption

The moisture content of a gas stream may be determined by adsorbing the water vapor onto a solid phase. Solids such as silica gel and Drierite R have a high affinity for water vapor and may be used for this purpose. These adsorbents indicate the presence of water by changing from blue to pink color. A measured weight of dry silica gel is placed in a drying tube and the sample gas is passed through it. The water content is measured by the weight change. As long as the silica gel does not become completely saturated, the partial pressure of water vapor leaving the drying tube is negligible. This is true as long as the gas temperature remains low; the adsorption process is less effective at higher temperatures. In fact, adsorbents are usually regenerated by heating. With these considerations, then, it is important that the gas stream temperature contacting the adsorbent be below 300°F. This method works best for low temperature and highly saturated gas streams. The adsorption method cannot be used for all types of waste gas streams. Other constituents may be also adsorbed by the adsorbent and cause a weight increase; this would cause the calculated moisture content to be too high.

SAMPLING TRAIN FOR MOISTURE

Federal EPA Method 4 is a condensation-adsorption technique to determine the moisture content of a gas stream.[1] The method is intended for use in those situations where the gas stream does not contain water droplets. If this latter situation does occur, then the gas stream can be assumed to be saturated and the psychrometric method applied. It is also possible to determine the moisture content by condensation at the same time in which the particulate sampling test is performed (EPA Method 5 described in Chapter 10).

Figure 7.5 shows the federal EPA moisture train. A heated probe is used to prevent condensation prior to the condensor. Some glass wool is put into the

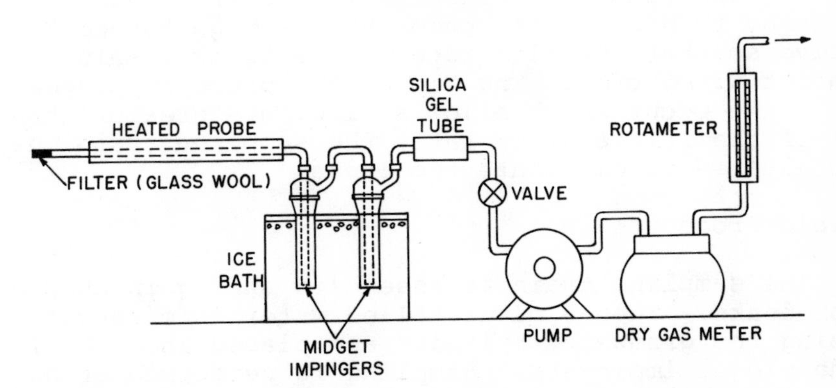

Figure 7.5. Federal EPA moisture train (Method 4).

end of the probe to act as a particulate matter filter. Two midget impingers, which act as condensers, are connected in series within an ice bath. The impingers are usually followed by a silica gel drying tube to collect the moisture carry-over due to physical entrainment and the moisture content at saturated conditions. For example if the gas conditions are 45°F and 29.92 in. Hg, about 1% water vapor exists.

Equipment and Reagents

The following items are needed for this method:

1. heated stainless steel or Pyrex glass probe
2. two midget impingers each with 30-ml capacity
3. ice bath
4. silica gel (optional; used to protect pump)
5. needle valve
6. diaphragm type pump (leakless)
7. rotameter with flow rate range of 0 to 0.1 cfm
8. 25-ml graduated cylinder
9. barometer which can be read to 0.1 in. Hg
10. pitot tube apparatus
11. distilled water
12. dry gas meter which can be read to 0.01 ft^3
13. tubing and fittings to connect components
14. thermometer for gas meter.

Laboratory Calibration

The dry gas meter and other components should be checked and calibrated as per methods presented in Chapter 4. The sampling train should be assembled in the laboratory and checked for leaks. This is done by turning on the pump and setting the needle valve so that the flow rate is the desired value (about 0.075 cfm). The tip of the probe is plugged; if leaks occur which allow a flow rate greater than 1% of the initial flow rate, the sampling train must be altered to eliminate the leaks.

Field Procedure

The sampling train is assembled and again checked for leaks. Then 5 ml distilled water is measured using the graduated cylinder and placed in each of the midget impingers. Sampling is performed at one point within the stack and the sampling rate is held

Table 7.2

Field Moisture Determination Data Sheet

Firm ____________________ Source Description

Location ____________________

Test No. ____________________

Date ____________ Time ____________

Tester ____________________

Barometric Pressure ____________

Clock Time	Gas Volume Through Meter, (V_m), ft^3	Rotameter Setting ft^3/min	Meter Temperature °F	Pitot Data		Remarks
				ΔP	$(\Delta P)^{1/2}$	

constant at about 0.075 cfm. This method assumes that *uniform* and *steady* conditions exist as discussed in Chapter 2. However, if wide variation in flow rate occurs, it is suggested that proportional sampling be used which implies *uniform* but *unsteady* conditions. The proportional sampling is accomplished by using the pitot tube apparatus. Using the rotameter, a sampling rate is selected arbitrarily for a given pitot tube manometer reading. During the test the needle valve is used to change t the sampling flow rate. Relative changes are made according to $(\Delta P)^{1/2}$; this is done because the pitot tube responds to stack gas velocity changed according to Equation 7-1. Sampling is continued until droplets begin to carry over from the first to the second impinger or until the dry gas meter indicates that 1 cubic foot of gas has been sampled. After completing the test the amount of condensate collected should be measured to the nearest 0.5 ml. All test data should be recorded on a form similar to the one shown in Table 7.2.

Calculations

The moisture content of the gas stream is calculated using Equation 7-11. The actual measurement obtained from the condenser is m_{wc}. This amount of condensate is measured to the nearest 0.5 ml, and m_{wc} is known to the nearest 0.5 gm. The value of e_m is determined by using steam tables.

REFERENCES

1. "Performance Standards for New Sources," Federal Register, Vol 36, No. 247, December 23, 1971.
2. Air Sampling Instruments, American Conference of Governmental Industrial Hygienists, Cincinnati, Ohio, 1972.
3. Methods of Air Sampling and Analysis, American Public Health Association, Washington, D.C., 1972.
4. Ruch, W. E. Chemical Detection of Gaseous Pollutants, Ann Arbor Science Publishers, Ann Arbor, Michigan, 1966.
5. Mooney, D. A. Mechanical Engineering Thermodynamics, Prentice-Hall, Inc., Englewood Cliffs, N.J. 1959

CHAPTER 8

ERRORS IN SOURCE TESTING

The previous chapters have indicated that a source test contains many different steps. The sampling train itself is comprised of a multitude of components. Each component has a specific purpose, *i.e.*, measurement of or control of pressure, temperature or flow rate. Poor accuracy and/or precision of these components and their subsequent use compounds the overall error of the estimate of pollutant concentration, $\overline{C}_s$, and pollutant mass emission rate, $\overline{PMR}_s$.

The final use of the source test data often is to determine emission compliance. As mentioned in Chapter 3 the regulations are given on the basis of pounds per million BTU's for combustion sources and pounds per ton of product for such processes as acid plants. Thus for fuel-fired boilers the fuel consumption rate and heating value must also be determined before compliance can be ascertained. These determinations then are also sources of error. However, in this chapter we consider only those errors associated with the gas sampling procedures. Only these latter actions are under the direct control of the source test team. In the gas sampling portion of source testing, it is important to identify which step(s) has the greatest influence on the overall error. A test team must strive to improve its precision and accuracy; source testing is a costly operation and it is not feasiable to perform many tests on a single source. Often very important and costly decisions must be made on the results of a single source test.

This chapter will discuss types of errors and then identify the probable errors associated with proportional and grab sampling procedures. Before doing this, however, we must first review some basic statistical concepts. The works of Li[1] and Schenck[2] provide a greater detailed coverage of the statistical concepts used in this chapter.

TYPES OF ERRORS

There are two broad classes of errors which must be recognized: (1) determinate or systematic errors and (2) indeterminate or random errors. Systematic errors are due generally to poor procedures but the magnitude of the error is often constant for each case or condition. Also it is of a unidirectional nature, which means it is always either positive or negative. Thus if the magnitude and direction of the error are known, their effect can be taken into account. Random errors are due to the effects of uncontrolled variables. They are as likely to be positive as negative and their specific magnitude and direction cannot be predicted for each measurement. They occur because in each test there are a large number of experimental variables (each of which causes a small error) which are left uncontrolled.

Systematic errors may be caused by equipment and personnel. The systematic equipment errors may be taken into account by proper calibration techniques as discussed in Chapter 4. Systematic human errors are reduced by using standard procedures. The attempt must be made to evaluate or eliminate the bias by such methods. Otherwise the alternatives are (1) to forget about the bias completely and accept the answer as being accurate or (2) to guess at the magnitude of the bias for each individual or piece of equipment. Usually neither of these latter alternatives is considered desirable. It is important to note that a mistake and an error are not the same. A mistake is simply a human blunder and no amount of statistical consideration can take it into account.

Random errors may be attributed to uncontrolled environmental conditions and human errors of observation. Usually it is difficult to differentiate between systematic and unavoidable human errors. However, their effect is quite different and thus if at all possible they should be separated. The variation due to unavoidable human errors can be treated by statistical methods.

When dealing with errors, three terms are often used: precision, accuracy, and bias. *Precision* is the tendency of measurements to be reproducible from test to test. *Accuracy* is the nearness of a measurement to the true value. *Bias* is a measure of the magnitude of the deviation of the measured (predicted) value from the true value. Example 8.1 illustrates the meaning of these terms.

Example 8.1. Four different sampling teams were sent out to determine the mass emission rate of a number of different air pollution sources. In each case the true emission rates were known beforehand but were not revealed to the teams. The results of the tests are shown in Figure 8.1. Team A demonstrated good precision and accuracy. They were accurate because the predicted values were all close to the true values and they were precise because all values for a given condition were close together. Team B was able to achieve good precision but their accuracy was poor. Essentially, for a given condition, the predicted values were close together but there existed a bias. Team C achieved poor precision and poor accuracy. Their predicted values were neither consistently together nor close to the true values. Team D definitely had poor precision and accuracy was also doubtful. In this latter case it should be noted that the average or mean of all of D's observations is actually the same as the true value for each case. Thus it appears that while team D may be able to estimate the true pollutant emission rate it would have to conduct many tests because its precision is poor.

All of the teams except A would have a difficult time making a living. Team B definitely has a systematic error. This is shown as a bias on Figure 8.1b. Such a bias could be caused by faulty or noncalibrated equipment.

Teams C and D have problems with precision and accuracy. Further tests would have to be made to determine whether their problems were due to systematic errors, random errors or both.

Systematic Errors

Systematic errors are determined by using standard calibration methods. The calibration methods for various source sampling train components are presented in Chapter 4. Figure 8.2 shows several calibration curves. Assuming that the random error is insignificant when compared to the systematic error, the measured values can be converted to the true values. The following expression applies

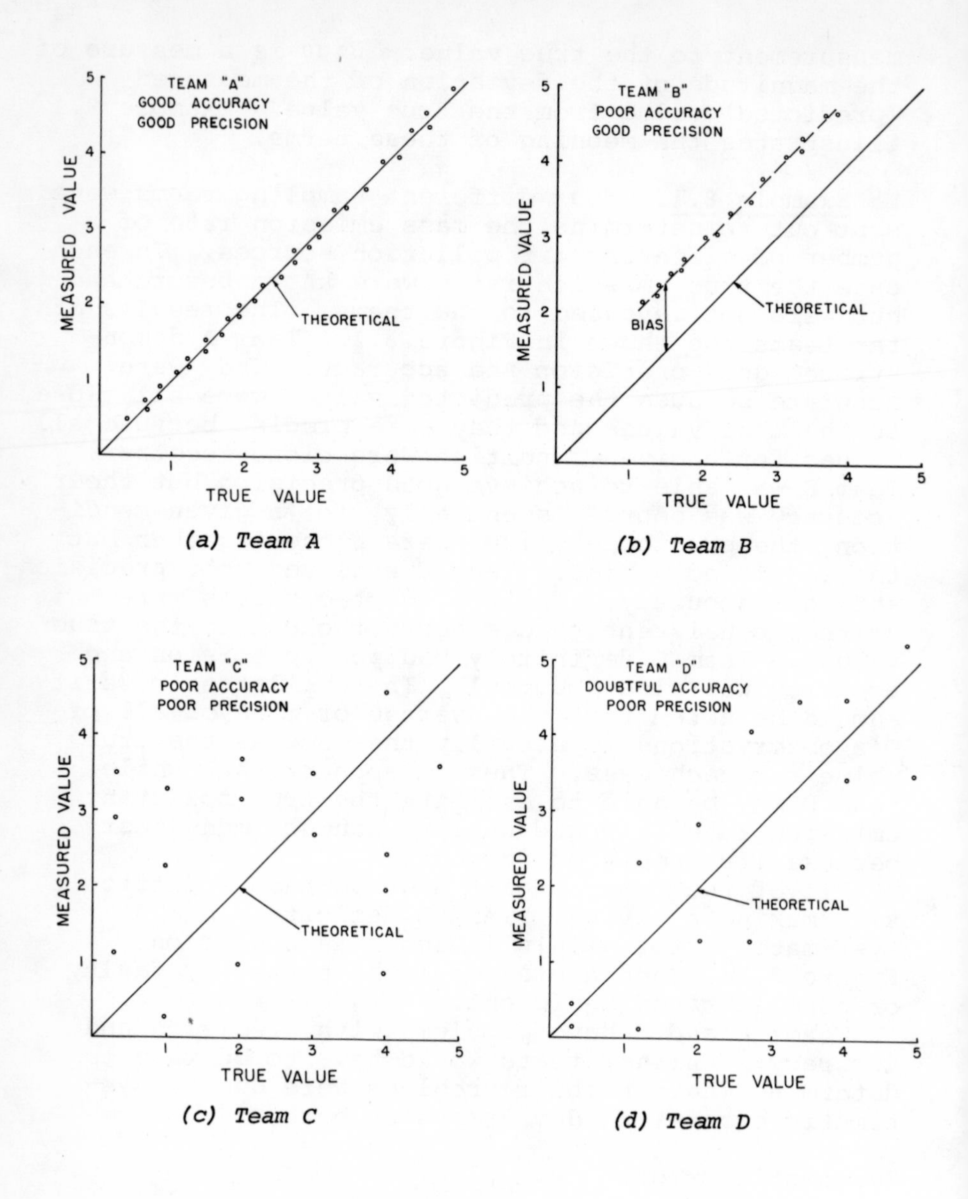

Figure 8.1. *Results of source sampling tests (Example 8.1).*

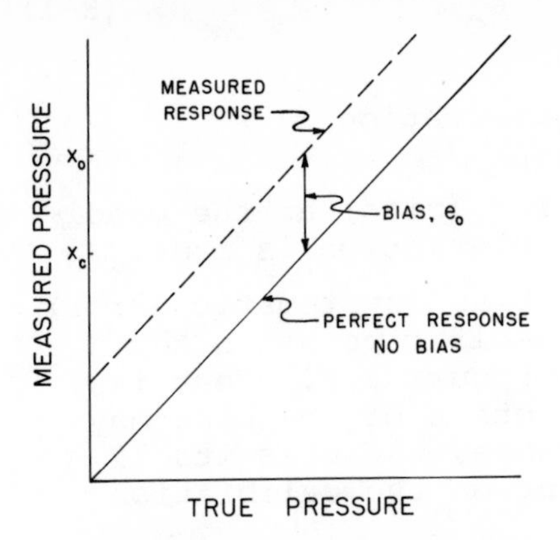

(a) Bias is constant and unidirectional over the whole test range.

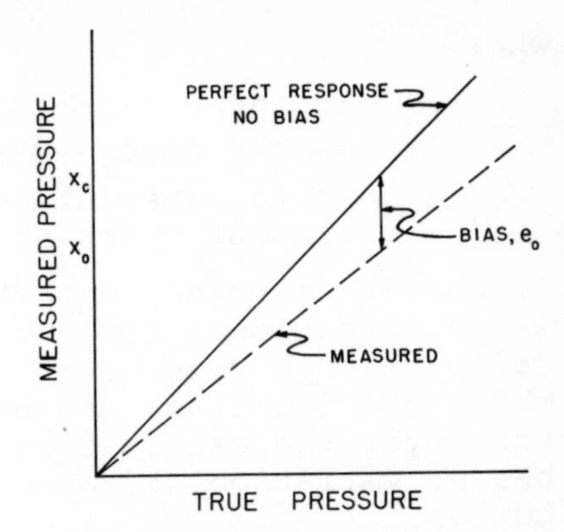

(b) Bias is unidirectional but not constant over the whole test range.

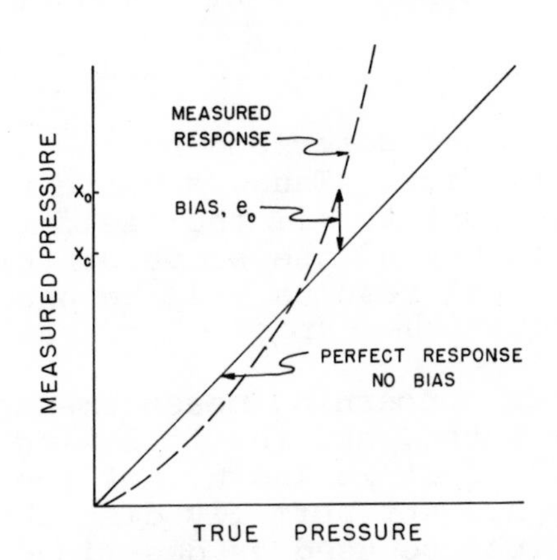

(c) Bias is neither constant nor unidirectional over the whole test range.

Figure 8.2. Evaluating systematic errors by calibration of pressure measuring devices.

$$X_c = X_o \pm e_o \tag{8-1}$$

where

X_c is the corrected observation
X_o is the observed value
e_o is the absolute error (bias) at the condition at which the observation is made.

It should be noted that the bias (systematic error) is unidirectional for any specific point on the calibration curves shown in Figure 8.2. That is, while the magnitude and direction of the bias may vary over the calibration range, the bias itself has no variation at any point on the calibration curves.

Instrumentation bias is not necessarily bad. If the bias can be evaluated by proper calibration techniques, there is no reason why the instrument cannot be used. Thus, when accounting for systematic errors by calibration the instrument should be calibrated under the conditions at which it will be used.

Random Errors

Random errors are described by the laws of probability and statistics. Thus, a basic understanding of both is important to all the field measurements. However, this is beyond the scope of this text and hence the principal results will be used mainly as a basis to discuss the errors of source sampling procedures.

Random errors concerning measurements are usually normally distributed, and the following statistical methods apply. If it is known that the observations from a process are not normally distributed, then other methods must be used as described in Reference 1. Sometimes it is possible to perform a transformation on the data and the resultant is then normally distributed. Examples of such transformations would be square root and logarithmic transformations. The term normal identifies how the readings or measurements group around the mean. Normally distributed data is described by its population mean and standard deviation.

The equation describing a normal distribution is given

$$y = \frac{1}{\sigma\sqrt{2\pi}} \exp\left(-\frac{(x-\mu)^2}{2\sigma^2}\right) \tag{8-2}$$

where

- y is the relative frequency of occurrence
- x is the independent variable
- σ is the standard deviation of the population
- μ is the population mean.

The probability that an observation lies between x_1 and x_2 is determined by integrating Equation 8-2 as follows:

$$P(x_1 \leq x \leq x_2) = \frac{1}{\sigma\sqrt{2\pi}} \int_{x_1}^{x_2} \exp\left(-\frac{(x-\mu)^2}{2\sigma^2}\right) dx \quad (8\text{-}3)$$

The integration process simply calculates the area under the function described in Equation 8-2. Most statistics and engineering handbooks present tables for these calculations. As shown in Figure 8.3, 68.3% of the population lies with 1 σ of the mean, 95.5% within 2 σ and 99.7% within 3 σ.

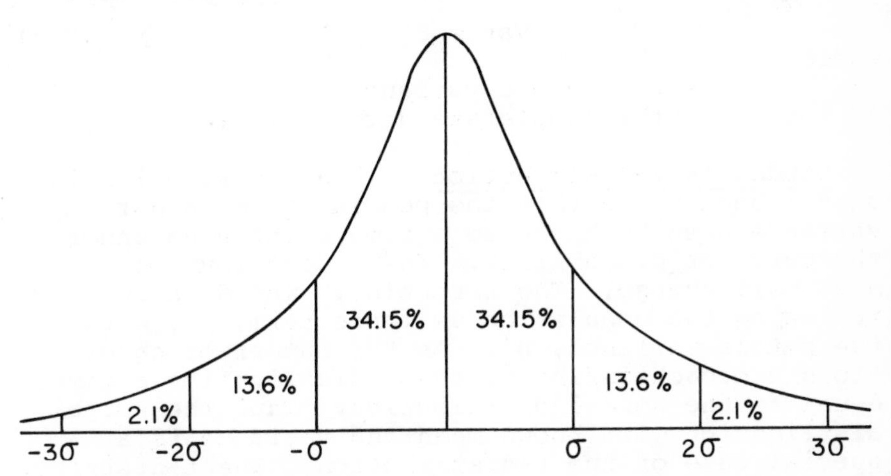

Figure 8.3. Relative areas under a normal distribution.

The standard deviation determines the spread of the population. In source sampling neither the mean nor standard deviation are actually known. In fact, it is the whole purpose of source testing to accurately estimate the mean mass emission rates for various pollutants. Estimates of these parameters can be made from test data. Basically, it is desirable to have as small a standard deviation as possible because this reduces the number of samples required to obtain a reliable estimate of the mean.

Mean. The mean of a population can be estimated from the sample observations using the following expression:

$$\bar{x} = \frac{\Sigma x_i}{N} \tag{8-4}$$

where
- $\bar{x}$ is the mean
- x_i is the individual observation
- N is the total number of observations.

Standard Deviation. The standard deviation of a population can be estimated from the sample observations using the following expression:

$$S = \left(\frac{\Sigma (x_i - x)^2}{N-1}\right)^{1/2} \tag{8-5}$$

Variance. The term variance is used sometimes to indicate the spread of data. It is equal to the square of the standard deviation. Thus for samples:

$$\text{Var} = S^2 \tag{8-6}$$

where
- Var is the sample variance
- S is the sample standard deviation.

Student's t-Distribution. This distribution is useful because neither the population mean nor variance need be known to estimate the area under the curve or probability. The t-distribution is also bell-shaped. The t statistic was developed by replacing the population variance term, σ^2, with the sample variance, S^2. As the number of observations approach infinity, the t-distribution becomes equal to the normal distribution. Thus the normal distribution with known mean and variance is a special case of the t-distribution. The t-distribution is useful because it allows making confidence interval estimates of the mean without knowing the population variance, σ^2. The t statistic is given

$$t = \frac{\bar{x} - \mu}{\left(\frac{S^2}{N}\right)^{1/2}} \tag{8-7}$$

The t-distribution tables for this statistic are given by Li.[1] They are also listed in many other texts on mathematical statistics. The usefulness of the t-distribution for source sampling is that it provides a statistical basis for testing hypotheses regarding source emissions. For example, it may be hypothesized that the population mean of the mass emission rate from a source is equal to the compliance standard (see Chapter 5); the alternative hypotheses are that it is less than or greater than the standard. These hypotheses can be written

$$H_o: \overline{PMR_s} = \mu_o$$
$$H_1: \overline{PMR_s} > \mu_o$$
$$H_2: \overline{PMR_s} < \mu_o$$

where μ_o is the emission standard.

In testing a hypothesis, two kinds of errors, Type I and Type II, can be made. A Type I error is the rejection of a hypothesis which is actually true. A Type II error is the acceptance of a hypothesis which is actually false. The probability of committing a Type I error is called the level of significance. Figure 8.4 shows the distribution of t if the hypothesis is true. If a 5% level of significance is used, 2.5% of the area is in each of the tails shown. We reject the hypothesis, $\overline{PMR_s} = \mu_o$, if the observed t falls in the right tail and then accept the hypothesis that $\overline{PMR_s} > \mu$. If the calculated value falls in the left tail, we reject the original hypothesis and accept the hypothesis that $\overline{PMR_s} < \mu$.

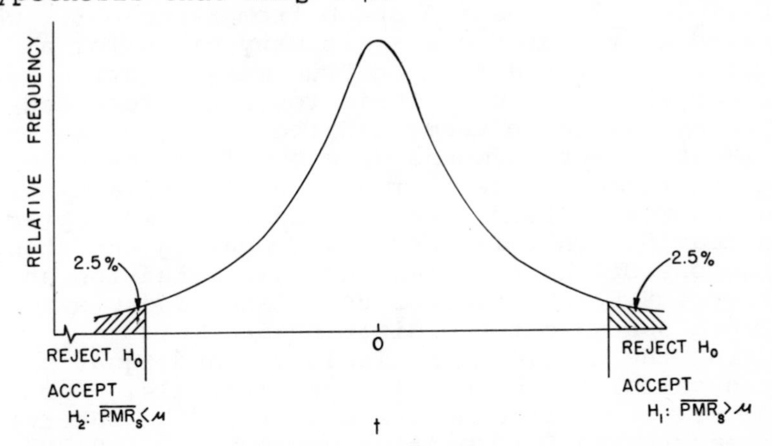

Figure 8.4. t-Distribution for 5% level of significance.

Confidence Intervals. In source testing it is often desirable to provide a confidence interval for the predicted pollutant mass emission rate. This is done by rearranging Equation 8-7, giving the following inequality:

$$\bar{x} - t_\alpha \left(\frac{S^2}{N}\right)^{1/2} < \mu < \bar{x} + t_\alpha \left(\frac{S^2}{N}\right)^{1/2} \quad (8\text{-}8)$$

where t_α is the value of t associated with the desired confidence coefficient and known degrees of freedom, α. If the level of confidence used is 5%, the confidence coefficient is 95%. Using such a value gives a 95% level of confidence regarding the estimate of the interval. For the t-distribution the number of degrees of freedom is (N-1). It is evident from Equation 8-8 that small values of the sample variance, S^2, and large values of N both contribute to decreasing the interval. The value of t_α also decreases as N increases. It must be realized that the confidence interval given by Equation 8-8 is not a single interval but rather a collection of intervals. For each sample of N observations, $\bar{x}$ and S^2 can be calculated, and using a t-table a confidence interval of $\overline{PMR_S}$ can be calculated.[1] The practical aspect of all this is that if a sampling team has an inherently high variance, it must increase its number of tests to achieve the desired confidence interval; this latter action while statistically commendable simply costs too much money. Example 8.2 illustrates how the confidence interval of the mean can be calculated.

Example 8.2. Teams A and D from Example 8.1 were required to repeat their tests many times for a condition which had a known true mass emission rate of 2.0. The results of their tests are recorded in Table 8.1. Both teams had the same average or mean. The observed means were equal to the true mean and therefore neither team exhibited a testing bias. However, team A was considerably more precise than team D. This is shown by comparing the standard deviations of their test results, a situation shown in Figure 8.5. The larger the standard deviation, the more spread out the distribution.

Using the test data in Table 8.1 and Equation 8-8, we can calculate the confidence intervals. For a 95% level of confidence (t_α = 2.262), the intervals for teams A and D respectively are $\mu = 2 \pm 0.085$ and $\mu = 2 \pm 0.344$. When performing source tests the true mean emission rate is not known. In fact that is the whole purpose for performing the test. Thus when performing the test it is important to have as small a standard deviation as possible. In many cases only three repetitions of the test are performed. It is a time-consuming and costly operation which

Table 8.1

Source Sampling Testing Results for Teams A and D (Example 8.2)

Test Number	*Team A Observed*	*Team D Observed*
1	2.0	2.5
2	1.8	1.9
3	2.1	1.5
4	2.0	1.8
5	2.2	2.1
6	1.9	2.3
7	2.1	2.8
8	2.0	2.2
9	1.9	1.7
10	2.0	1.2
$\overline{X}$	2.0	2.0
S	0.120	0.480
S^2	0.014	0.230

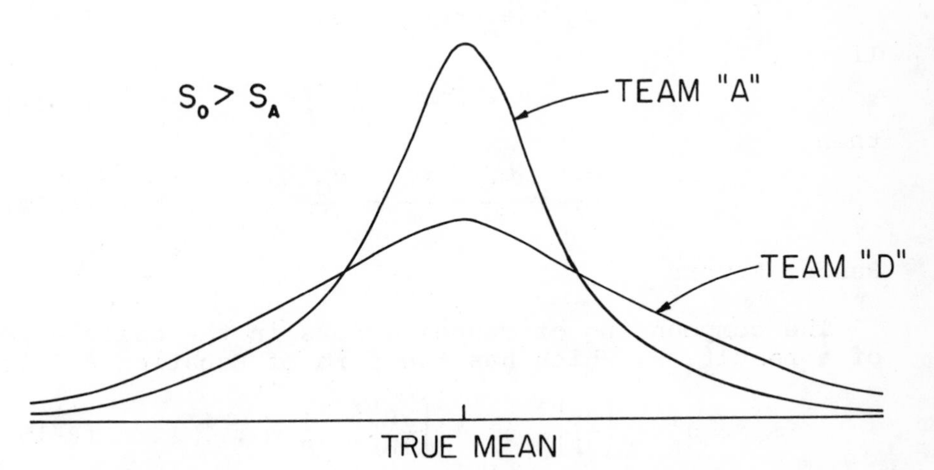

Figure 8.5. Distribution of test results for teams A and D (Example 8.2).

naturally prohibits the use of a large number of observations. In the case of N = 3, the 95% confidence level has a $t_2 = 4.303$ for Equation 8-8.

ERROR PROPAGATION

When using a series of measurements to calculate a result, errors are compounded. The manner in which systematic and random errors affect a result are quite different.[2]

Systematic Errors

The compounding of systematic errors in the calculation of a result, R, follows a general form. If the dependent variable, R, has the form

$$R = f(A,B,C....) \tag{8-9}$$

then

$$e_R = \frac{\partial R}{\partial A} e_A + \frac{\partial R}{\partial B} e_B + ... \tag{8-10}$$

where

e_R is the absolute error of the calculated result
A,B,C etc. are measurements
e_A, e_B, and e_C are absolute errors associated with the measurements.

The following special cases occur for subtraction-addition and multiplication-division. If

$$R = A + B - C \tag{8-11}$$

then

$$e_R = e_A + e_B - e_C \tag{8-12}$$

If

$$R = \frac{AB}{C} \tag{8-13}$$

then

$$\frac{e_R}{R} = \frac{e_A}{A} + \frac{e_B}{B} - \frac{e_C}{C} \tag{8-14}$$

Random Errors

The compounding of random errors in the calculation of a result, R, which has the form of Equation 8-9 is

$$\sigma_R^2 = \left(\frac{\partial R}{\partial A}\right)^2 \sigma_A^2 + \left(\frac{\partial R}{\partial B}\right)^2 \sigma_B^2 + ... \tag{8-15}$$

If the dependent variable has the form of Equation 8-11 then

$$\sigma_R^2 = \sigma_A^2 + \sigma_B^2 + \sigma_C^2 \tag{8-16}$$

If the dependent variable has the form of Equation 8-13 then

$$\left(\frac{\sigma R}{R}\right)^2 = \left(\frac{\sigma A}{A}\right)^2 + \left(\frac{\sigma B}{B}\right)^2 + \left(\frac{\sigma C}{C}\right)^2 \tag{8-17}$$

Maximum Error

Sometimes the term maximum error is used when referring to measurements. Actually it is incongruous with the probability concept and is used mostly by people not familiar with standard deviation. The maximum error is estimated by determining the error of each step and then assuming that they combine in such a way as to provide a maximum error of the result. This approach actually produces estimates of the range. The range is that interval within which all observations fall. The problem with this approach is that it becomes a very unlikely event if very many steps are involved. When this approach is used, Equation 8-10 is applied.

RANDOM ERRORS IN SOURCE SAMPLING

In source sampling we wish to determine the average pollutant mass emission rate from the stack. Recalling Equation 2-1 we have the following relationship

$$\overline{PMR}_s = \overline{C}_s \; \overline{Q}_s \tag{2-1}$$

In order to evaluate the total error associated with this estimate Equation 8-17 is applied, obtaining

$$\left(\frac{\sigma_{\overline{PMR}_s}}{\overline{PMR}_s}\right)^2 = \left(\frac{\sigma_{\overline{C}_s}}{\overline{C}_s}\right)^2 + \left(\frac{\sigma_{\overline{Q}_s}}{\overline{Q}_s}\right)^2 \tag{8-18}$$

Since from Chapter 2

$$\overline{C}_s = \frac{m}{V}$$

and

$$\overline{Q}_s = \overline{V}_s A_s$$

we can also write Equation 8-18 as

$$\left(\frac{\sigma_{\overline{PMR}_s}}{\overline{PMR}_s}\right)^2 \quad \left(\frac{\sigma_m}{m}\right)^2 + \left(\frac{\sigma_V}{V}\right)^2 + \left(\frac{\sigma_{\overline{V}_s}}{\overline{V}_s}\right)^2 + \left(\frac{\sigma_{A_s}}{A_s}\right)^2 \tag{8-19}$$

where

- m is the mass of pollutant collected
- V is the sample volume
- $\overline{V}_s$ is the average stack velocity
- A_s is the cross-sectional area of the stack.

It must be realized that the actual variance estimate calculated from multiple tests consists of three major components: (a) source condition variation, (b) equipment and personnel variation in field procedures, and (c) equipment and personnel variation in the laboratory. The variation due to source conditions is always a significant factor, and is due to both time and spatial variation. Gradients do exist and process fluctuations definitely contribute to the variance. The selection of the proper sampling strategy (as discussed in Chapter 2) will minimize this variation in the $\overline{PMR}_s$. But remember, using the wrong sampling strategy is a mistake, rather than an error. It provides a biased estimate of $\overline{PMR}_s$ and may cause an increase in the variance. Even if the process is a well-controlled one there still will be gradients and local time varying fluctuations due to turbulent conditions.

The total variance, $\sigma^2_{\overline{PMR}_s}$, could be estimated by performing many tests on the source and then applying Equations 8-5 and 8-6. However, this would be a very costly and time-consuming procedure. Also it would have to be done for each source. An alternative method is to try an evaluation of all the terms and variances which go into measuring $\overline{C}_s$ and $\overline{Q}_s$. However, it should be noted that such a variance estimate will always be somewhat less than the variance estimated from field data. The following sections will estimate $\sigma^2_{\overline{PMR}_s}$ for proportional and grab sampling procedures. Example problems will be used to illustrate the utility of the expressions developed.

Proportional Sampling

Proportional sampling is used when there exist time varying fluctuations of the source. This procedure must be used to provide a representative sample and hence an unbiased estimate of the pollutant mass emission rate.

Sample volume. When performing a source test according to proportional sampling procedures the nozzle velocity, V_n, should be some constant proportion of the stack velocity, V_s, at each sampling point. The ability to do this depends upon taking a number of process measurements and making the proper adjustment of the equipment. Throughout the test this procedure must be maintained if the correct sample volume, V, is to be obtained.

Therefore the stack velocity, V_s, must be measured and the nozzle velocity, V_n, adjusted appropriately. The variance associated with the sample volume, σ^2_V and that associated with maintaining proportional conditions, $\sigma^2_{(V_n:V_s)}$ are the same. Our ability to achieve and maintain proportional sampling conditions reflects directly in σ^2_V. We will consider here a special case of proportional sampling which is called isokinetic sampling. For the special case of isokinetic sampling the stack and nozzle velocities must be equal at all times, giving $V_n:V_s = 1$. We shall now develop an expression for $V_n:V_s$. Following Equation 4-6 we have

$$V_n = \frac{Q_n}{A_n} = \frac{4\ Q_n}{\pi\ D_n^2}$$

From Equation 4-11 for a pitot tube we have

$$V_s = K_p C_p \left(\frac{T_s\ \Delta P}{P_s\ M_s}\right)^{1/2} \qquad (4\text{-}11)$$

But Q_n is determined by using an orifice meter and taking into account the different gas conditions. Thus

$$Q_n = \left(\frac{1-B_{wm}}{1-B_{ws}}\right) Q_m \qquad (8\text{-}20)$$

which accounts for difference in moisture contents of the two gas conditions.

Also from Equation 4-14 we obtain

$$Q_m = K_m \left(\frac{T_m\ \Delta m}{P_m\ M_m}\right)^{1/2} \qquad (8\text{-}21)$$

Thus combining Equations 8-20 and 8-21 we get

$$V_n = \frac{4\ K_m}{\pi\ D_n^2} \left(\frac{1-B_{wm}}{1-B_{ws}}\right) \left(\frac{T_m\ \Delta m}{P_m\ M_m}\right)^{1/2} \qquad (8\text{-}22)$$

Combining Equation 4-11 and 8-22 we obtain

$$\frac{V_n}{V_s} = \frac{4\,K_m}{K_p C_p D_n^2 \pi} \left(\frac{1-B_{wm}}{1-B_{ws}}\right) \left(\frac{T_m}{T_s} \; \frac{\Delta m}{\Delta P} \; \frac{P_s}{P_m} \; \frac{M_s}{M_m}\right)^{1/2} \qquad (8\text{-}23)$$

where

- V_n is the gas velocity at the tip of the nozzle, fps
- V_s is the gas velocity measured by the pitot tube, fps
- D_n is the diameter of the nozzle, ft
- B_{wm} is the mole fraction of moisture in the gas at meter conditions
- B_{ws} is the mole fraction of moisture in the gas at stack conditions
- T_m is the absolute gas temperature at meter, °R
- T_s is the absolute gas temperature in the stack, °R
- P_s is the absolute stack gas pressure, in Hg
- P_m is the absolute gas pressure at the orifice meter, in Hg
- Δm is the manometer deflection of orifice meter device, in H_2O
- ΔP is the manometer deflection of pitot tube device, in H_2O
- M_s is the molecular weight of the stack gas, lb_m/lb_m-mole
- M_m is the molecular weight of the gas at the orifice meter, lb_m/lb_m-mole
- K_m is the orifice meter coefficient including appropriate constants to make Equation 4-12 dimensionally consistent
- K_p is the pitot tube equation constant with approximate units to make Equation 4-11 dimensionally consistent and
- C_p is the pitot tube coefficient which is dimensionless.

Thus Equation 8-23 describes the result required for the field phase of obtaining a representative sample and properly estimating the true mass emission rate. By applying Equation 8-15 to Equation 8-23 the desired error expression is obtained:

$$\left[\frac{\sigma(V_n:V_s)}{V_n/V_s}\right]^2 = K_m^{-2}\ \sigma_{k_m}^2 + K_p^{-2}\ \sigma_{k_p}^2 + C_p^{-2}\ \sigma_{c_p}^2 + D_n^{-4}\ \sigma_{D_n^2}^2 +$$

$$(1-B_{wm})^{-2}\ \sigma_{1-B_{wm}}^2 + (1-B_{ws})^{-2}\ \sigma_{1-B_{ws}}^2 +$$

$$1/4\ \Big(T_m^{-2}\ \sigma_{T_m}^2 + T_s^{-4}\ \sigma_{T_s}^2 + \Delta m^{-2}\ \sigma_{\Delta m}^2 +$$

$$\Delta P^{-4}\ \sigma_{\Delta P}^2 + P_s^{-2}\ \sigma_{P_s}^2 + P_m^{-4}\ \sigma_{P_m}^2 + M_s^{-2}\ \sigma_{M_s}^2 +$$

$$M_m^{-4}\ \sigma_{M_m}^2\Big) \qquad (8\text{-}24)$$

Average Stack Gas Velocity. The average stack gas velocity is determined by taking measurements at multiple points within the duct as discussed in Chapter 2. Equation 4-11 is applied and the average values for ΔP and T_s are used. Thus

$$\overline{V}_s = K_p C_p \left(\Delta P_{avg}\right)^{1/2} \left(\frac{(T_s)_{avg}}{P_s M_s}\right)^{1/2} \qquad (8\text{-}25)$$

and applying Equation 8-17 we obtain

$$\sigma_{\overline{V}_s}^2 = C_p^2\ \Delta P_{avg}\ T_{s_{avg}}\ P_s^{-1}\ M_s^{-1}\ \sigma_{k_p}^2 +$$

$$K_p^2\ \Delta P_{avg}\ T_{s_{avg}}\ P_s^{-1}\ M_s^{-1}\ \sigma_{c_p}^2 +$$

$$1/4\ \left(C_p^2\ K_p^2\ \Delta P_{avg}^{-1}\ T_{s_{avg}}\ P_s^{-1}\ M_s^{-1}\ \sigma_{\Delta p_{avg}}^2\right) +$$

$$1/4\ \left(C_p^2\ K_p^2\ \Delta P_{avg}\ T_{s_{avg}}^{-1}\ P_s^{-1}\ M_s^{-1}\ \sigma_{T_{s_{avg}}}^2\right) +$$

$$1/4\ \left(C_p^2\ K_p^2\ \Delta P_{avg}\ T_{s_{avg}}\ P_s^{-3}\ M_s^{-1}\ \sigma_{p_s}^2\right) +$$

$$1/4\ \left(C_p^2\ K_p^2\ \Delta P_{avg}\ T_{s_{avg}}\ P_s^{-1}\ M_s^{-3}\ \sigma_{m_s}^2\right) \qquad (8\text{-}26)$$

Pollutant Mass Measurement. The laboratory methods used generally are small sources of error when compared to the field operations. Of course the actual magnitude is dependent upon the procedure itself. As an example consider the particulate sampling outlined in further detail in Chapter 10. The mass measurement of the particulate matter requires four basic measurements. The result follows the form of Equation 8-11

$$m = (W_2 - W_1) + (W_4 - W_3) \tag{8-27}$$

where

- m is the collected weight of particulate matter
- W_1 is the tare weight of the filter
- W_2 is the gross weight of the filter after sampling
- W_3 is the tare weight of a beaker
- W_4 is the gross weight of the beaker and the particulate material collected in the probe.

Thus Equation 8-16 applies and

$$\sigma_m^2 = \sigma_{W_1}^2 + \sigma_{W_2}^2 + \sigma_{W_3}^2 + \sigma_{W_4}^2 \tag{8-28}$$

If an analytical laboratory balance is used, each term in Equation 8-28 has a value of about 0.01 mg^2. Thus the variance of the result is 0.04 mg^2 and the standard deviation is 0.2 mg. If the resultant weight of material collected is greater than 100 mg, the laboratory error will be less than 0.2%.

Stack Area. The cross section area of a stack can be calculated by determining the internal dimensions. These dimensions are measured through the sampling ports or possibly determined from available construction drawings. A reasonable variance for the linear dimension is about 0.01 feet. Equation 8-17 applies for variance estimates of the area determination. The actual value of $\sigma_{A_s}^2$ depends upon the dimensions of the stack.

Probable Error. The probable error associated with proportional sampling is determined by evaluating the terms in Equations 8-19, 8-24, 8-26, and 8-28. Table 8.2 gives a compilation of values for variances. *These variance estimates are based upon the experience and judgment of the authors.* At present this must be done since adequate information is lacking from component manufacturers and the general literature. These values are applied in Example 8.3.

Table 8.2
Estimated Error Terms Associated with Proportional Sampling Procedures

Term	*Use and Units*	*Variance,* σ^2
K_m	orifice meter coefficient	0.0004
K_p	$\frac{ft}{sec}\left[\frac{lb}{lb\ mol°R}\right]^{1/2}$	0 (constant)
C_p	pitot tube coefficient, dimensionless	0.0001
D_n^2	nozzle diameter, in^2	0.000001
$1-B_{wm}$	dry gas at meter, mole fraction	0.0001
$1-B_{ws}$	dry gas at stack, mole fraction	0.0001
T_m	gas temperature at meter, °R	4
T_s	gas temperature at stack, °R	4
Δm	orifice meter pressure drop, in. H_2O	0.0004
ΔP	pitot tube velocity pressure, in. H_2O	0.0004
P_s	stack gas pressure, in. Hg	0.01
P_m	gas pressure at meter, in. Hg	0.01
M_s	molecular weight of stack gas	0.05
M_m	molecular weight of gas at the meter	0.05
A_s	area of stack, ft^2	0.02
ΔP_{avg}	average pitot tube velocity pressure, in. H_2O	$\Sigma\ \sigma^2_{\Delta P}$
$T_{s_{avg}}$	average gas temperature at the stack, °R	$\Sigma\ \sigma^2_{T_s}$
W_1	tare weight of filter, mg	0.04
W_2	gross weight of filter and sample, mg	0.04
W_3	tare weight of beaker, mg	0.04
W_4	gross weight beaker and sample, mg	0.04

Example 8.3. A source test for particulate matter was conducted on a coal-fired boiler. Table 8.3 lists the operational data. Samples were taken at 12 points within the stack according to federal EPA procedures; this is discussed in Chapter 9. To estimate the variance of the pollutant mass emission rate, $\overline{PMR_s}$, see Equation 8-19. Since the particulate matter test is conducted under isokinetic conditions, Equations 8-24, 8-26 and 8-28 should be used. Values for the individual variance terms such as $\sigma^2_{T_s}$, $\sigma^2_{\Delta P}$ and $\sigma^2_{P_m}$ are obtained from Table 8.2.

Applying Equation 8-24 we obtain

$$\frac{\sigma^2_{\frac{V_n}{V_s}}}{\left(\frac{V_n}{V_s}\right)^2} = 2.7 \times 10^{-3}$$

Thus

$$\frac{\sigma_{\frac{V_n}{V_s}}}{\left(\frac{V_n}{V_s}\right)} = 5.2 \times 10^{-2} = 5.2\%$$

For isokinetic conditions $V_n:V_v = 1$ and recalling the argument that $\sigma_{\frac{V_n}{V_s}} = \sigma_V$ we have

$$\sigma^2_{V_i} = 2.7 \times 10^{-3}$$

But, the total sample volume is obtained by taking smaller sample volumes at 12 points within the stack. Applying Equation 8-16 we obtain

$$\sigma^2_V = (2.7 \times 10^{-3})\ 12 = 3.2 \times 10^{-2}$$

We obtain

$$\left(\frac{\sigma_V}{V}\right)^2 = \frac{3.2 \times 10^{-2}}{100^2} = 3.2 \times 10^{-6}$$

Applying Equation 8-26 we obtain

$$\sigma^2_{\overline{V}s} = 12.74$$

Table 8.3

Operational Values for Source Test for Particulate Matter (Example 8.3)

Term	*Value*
K_m	1.8
K_p	85.48
C_p	0.81
D_n^2	0.060
$1-B_{wm}$	0.975
$1-B_{ws}$	0.940
T_m	560
T_s	800
Δm	6.00
ΔP	0.50
P_s	29.63
P_m	30.10
M_s	28.8
M_m	30.0
A_s	30.0 (5 x 6)
ΔP_{avg}	0.54
$T_{s_{avg}}$	800
$\overline{V}_s$	48.7
V	107
W_1	202
W_2	380
W_3	159403.3
W_4	160273.3

and therefore

$$\left(\frac{\sigma_{V_s}}{\overline{V}_s}\right)^2 = \frac{12.74}{48.7^2} = 4.38 \times 10^{-3}$$

From Equation 8-28 we obtain

$$\sigma_m^2 = 0.16$$

and thus

$$\left(\frac{\sigma_m}{m}\right)^2 = 1.05 \times 10^{-7}$$

Finally, considering the stack area and applying Equation 8-17,

$$\left(\frac{\sigma_{A_s}}{A_s}\right)^2 = \frac{0.01}{5^2} + \frac{0.01}{6^2} = 7.2 \times 10^{-4}$$

Now applying these calculated values to Equation 8-19

$$\left(\frac{\sigma_{\overline{PMR}_s}}{\overline{PMR}_s}\right)^2 = 61.03 \times 10^{-4}$$

and therefore

$$\frac{\sigma_{\overline{PMR}_s}}{\overline{PMR}_s} = 7.8 \times 10^{-2} = 7.8\%$$

Thus with reference to Figure 8.3 this means that 68.3% of the measured $\overline{PMR}_s$ values can be expected to fall within 7.8% of the true $\overline{PMR}_s$. Also we see that 95.5% of the observations should fall within 2σ or in this case 15.6%. These of course are based upon the assumptions that the sampling strategy and equipment used will give an unbiased estimate of the true $\overline{PMR}_s$. Note that the largest source of error in this test is the pitot tube method for measuring stack gas flow rate. As expected the laboratory procedures are only a small source of random variation (error).

Maximum Error. Shigehara *et al.*[3] have applied the maximum error approach to the isokinetic sampling problem. They actually consider only the isokinetic velocity ratio (Equation 8-23) and eventually argue that the maximum error is 15.4%. However, this is not really as meaningful as it might first seem because the probability of the event is only 1 in 10^{22}. Table 8.4 shows the maximum relative errors associated with each measurement. Note that the ΔP, Δm and C_p terms were significant contributors to the total error.

Table 8.4

Maximum Relative Errors for Isokinetic Source Sampling Operations[3]

Measurement	*Maximum Error, %*
Stack temperature, T_s	1.4
Meter temperature, T_m	1.0
Stack gauge pressure, P_{gs}	0.42
Meter gauge pressure, P_{gm}	0.42
Atmospheric pressure, P_{atm}	0.21
Dry molecular weight, M_d	0.42
Moisture content, B_{ws}	1.1
Pitot tube pressure head, ΔP	10.0
Orifice pressure differential, Δm	5.0
Pitot tube coefficient, C_p	2.4
Orifice meter coefficient, K_m	1.5
Diameter of probe nozzle, D_n	0.8

Grab Sampling

The grab sampling procedure has fewer steps and hence fewer sources of error. Of course, the major error which can occur is using this method for source conditions which actually require proportional sampling techniques. In the true sense of definitions, this would be a mistake and not an error. In any event it will not be included here. Refer to Chapter 2 to find those source conditions in which a grab sample will provide a representative sample.

Grab sampling procedures, as outlined in Chapter 2, involve obtaining a known volume of sample over a very short period of time. The gaseous constituent is then absorbed into sample liquid reagent to allow its determination. This latter quantitative measurement might be a volumetric reduction as with the Orsat analysis (Chapter 7) or it might involve some colorimetric analysis, such as determination of oxides of nitrogen (Chapter 13).

Once again Equation 8-19 applies. The terms for average stack gas velocity and the area of the stack are the same as for proportional sampling. Only the terms for sample volume and the mass of the pollutant need be evaluated.

Sample Volume. The purpose of the grab sampling process is to obtain a known quantity of gas sample. For example, in the oxides of nitrogen procedure presented in Chapter 13 the following equation occurs:

$$V_{sc} = \frac{T_{std}(V_f - V_a)}{P_{std}} \left(\frac{P_f}{T_f} - \frac{P_i}{T_i}\right) \qquad (13\text{-}1)$$

where

- V_{sc} is the sample volume at standard conditions (dry basis), ml
- T_{std} is the absolute standard temperature, 530°R
- P_{std} is the absolute standard pressure, 29.92 in. Hg
- V_f is the volume of the flask, ml
- V_a is the volume of the absorbing solution, ml
- P_f is the final absolute pressure of the flask, in. Hg
- P_i is the initial absolute pressure of the flask, in. Hg
- T_f is the final absolute temperature of the flask
- T_i is the initial absolute temperature of the flask.

Applying Equation 8-15 to Equation 13-1 we obtain

$$\sigma^2_{V_{sc}} = \left(\frac{T_{std}}{P_{std}}\right)^2 \left(\frac{P_f}{T_f} - \frac{P_i}{T_i}\right)^2 \sigma^2_{V_f} + \left(\frac{T_{std}}{P_{std}}\right)^2 \left(\frac{P_i}{T_i} - \frac{P_f}{T_f}\right)^2 \sigma^2_{V_a} +$$
$$\left(\frac{T_{std}}{P_{std}}\right)^2 \left(\frac{V_f}{T_f} - \frac{V_a}{T_f}\right) \sigma^2_{P_f} + \left(\frac{T_{std}}{P_{std}}\right)^2 \left(\frac{V_a}{T_i} - \frac{V_f}{T_i}\right) \sigma^2_{P_i} +$$
$$\left(\frac{T_{std}}{P_{std}}\right)^2 \left(\frac{P_f V_a}{T_f^2} - \frac{V_f P_f}{T_f^2}\right)^2 \sigma^2_{T_f} +$$
$$\left(\frac{T_{std}}{P_{std}}\right)^2 \left(\frac{P_i V_f}{T_i^2} + \frac{P_i V_a}{T_i^2}\right)^2 \sigma^2_{T_i} \qquad (8\text{-}29)$$

Pollutant Mass Measurement. The laboratory activities for NO_x sampling involve the preparation of the sample and the determination of the absorbance at 420 nm using a spectrophotometer. A blank is also used so that any bias due to systematic error in the preparation of reagents or sample handling may be removed. The mass of nitrogen oxides collected is determined by the following expression which follows the form of Equation 8-11:

$$m_{NO_x} = m_S - m_B \qquad (8\text{-}30)$$

where

- m_{NO_x} is the total mass of NO_x collect
- m_S is the mass of NO_x in the sample as measured by absorbance
- m_B is the mass of NO_x in the blank as measured by absorbance

The resulting error equation is the same as Equation 8-15:

$$\sigma^2_{m_{NO_x}} = \sigma^2_{m_S} + \sigma^2_{m_B} \qquad (8\text{-}31)$$

The magnitude of the error associated with this laboratory measurement depends upon the quality of the spectrophotometer used. However, a realistic variance value for each of the absorbance measurement terms is 10 μg^2, and the standard deviation of the result is about 5 μg. Since the resultant mass has a value of several hundred (> 300) micrograms, the error is on the order of 1.5%.

Probable Error. The probable error associated with grab sampling is determined by evaluating the terms in Equation 8-19. Table 8.5 is a list of the *author's estimates* of these values. The following example applies these values to a grab sampling procedure.

Table 8.5

Estimated Error Terms Associated with Grab Sampling Procedure

Term	*Use and Units*	*Variance,* σ^2
V_f	sampling flask volume, ml	4
V_a	volume of absorbing solution, ml	0.25
P_f	final pressure of flask, in. Hg	0.01
P_i	initial pressure of flask, in. Hg	0.01
T_f	final temperature of flask, °R	4
T_i	initial temperature of flask, °R	4
T_{std}	absolute standard temperature, °R	0 (constant)
P_{std}	absolute standard pressure, in. Hg	0 (constant)

Example 8.4. A source test for oxides of nitrogen was conducted on a coal-fired boiler. This test was performed at the same time as the particulate matter test in Example 8.3. Table 8.6 lists the operational data. Equation 8-19 is used to estimate the variance of the pollutant mass emission rate $\overline{PMR}_s$. Since the oxides of nitrogen test is a grab sample method (see Chapter 13) Equations 8-26, 8-29 and 8-31 should be used. Values for the individual variance terms such as $\sigma^2_{T_f}$ and $\sigma^2_{P_i}$ are obtained from Table 8.5.

Applying Equation 8-29 we obtain

$$\sigma^2_{V_{sc}} = 136$$

and thus

$$\left(\frac{\sigma_{V_{sc}}}{V_{sc}}\right)^2 = \frac{136}{1364^2} = 7.35 \times 10^{-5}$$

Table 8.6

Operational Values for Source Test Oxides of Nitrogen (Example 8.4)

Term	*Value*
V_f	2000
V_a	25
P_f	22.8
P_i	1.9
T_f	535
T_i	535
T_{std}	530
P_{std}	29.92
V_{sc}	1364
m_{NO_x}	400

From Example 8-3 we obtain

$$\left(\frac{\sigma_{\overline{V}_s}}{\overline{V}_s}\right)^2 = 5.38 \times 10^{-3}$$

Applying Equation 8-31 we obtain

$$\sigma^2_{m_{NO_2}} = 10 + 10 = 20$$

Thus

$$\left(\frac{\sigma_{m_{NO_2}}}{m_{NO_2}}\right)^2 = \frac{20}{400^2} = 1.25 \times 10^{-4}$$

From Example 8-3

$$\left(\frac{\sigma_{A_s}}{A_s}\right)^2 = 7.2 \times 10^{-4}$$

Applying these calculated values to Equation 8-19

$$\left(\frac{\sigma_{\overline{PMR}_s}}{\overline{PMR}_s}\right)^2 = 63 \times 10^{-4}$$

and therefore

$$\frac{\sigma_{\overline{PMR}_s}}{\overline{PMR}_s} = 8.0 \times 10^{-2} = 8.0\%$$

Thus, referring to Figure 8.3, we can expect 68.3% of the measured $\overline{PMR}_s$ values to fall within 8% of the true $\overline{PMR}_s$; also from Figure 8.3 we see that 95.5% of the observations should fall with 2σ or in this case 16%. These statements of course are based upon the assumption that the sampling strategy and equipment used will give an unbiased estimate of the true $\overline{PMR}_s$. Note that the largest source of error in this test is the pitot tube method for measuring stack gas flow rate. As expected, the laboratory procedures are only a small source of random variation (error).

SYSTEMATIC ERRORS IN SOURCE SAMPLING

Systematic errors do occur in source sampling procedures. Some can be evaluated and thus the bias can be eliminated from the result. An example of this situation would be calibration of a pitot tube to determine its proper coefficient. Other systematic errors, however, cause biases which are impossible to evaluate. For example, if non-isokinetic conditions are used in sampling for particulate matter, a biased sample is obtained; it is not possible to make a correction after the test to obtain a true estimate of the particulate mass. The magnitude of the error also depends upon the particle size and density. But this usually is not known. The only alternative in this situation is to strive for excellence in maintaining isokinetic sampling conditions.

Equipment and Procedure Biases

There are many cases where equipment causes a sample bias. This definitely occurs if such items as pitot tubes, orifice meters, gas meters, sample flasks, etc. are not calibrated. Leaks in the sampling system will cause a bias which is impossible to account for; the only real alternative is to eliminate them. If the efficiency of the pollutant collector is not known, a bias exists. In the case of reuseable ceramic filters the bias may actually change from test to test. This would occur if the efficiency of collection is not constant.

Test personnel can also cause biased samples. This may occur due to improper procedures for reading manometers, rotameters and other instruments. This sort of bias is minimized by training personnel.

Biased Sampling Conditions

In Chapter 2 a number of the different source categories required that a proportional sampling method be used. If the actual sampling conditions deviate from these requirements, biased samples result. This problem has been considered by Hemeon and Haines,[4] Badzioch[5] and more recently by Smith *et al.*[6] The following discussion closely follows the work of Smith.

There are two basic methods of calculating the pollutant mass emission rate from a stack. The first is known as the *sample concentration method* and is given previously as Equation 2-1.

$$\overline{PMR}_s = \overline{C}_s \overline{Q}_s \qquad (2\text{-}1)$$

by expanding terms we have

$$\overline{PMR}_{s_c} = \frac{m}{V} A_s \overline{V}_s \qquad (8\text{-}32)$$

where

- m is the total mass of the pollutant collected during the sampling period
- V is the actual volume of sample taken at stack conditions
- A_s is the area of the stack
- $\overline{V}_s$ is the average stack gas velocity.

A second method known as the *ratio of area method* can be used to calculate $\overline{PMR}_s$ independently of the total measured sample volume. The method is given as follows:

$$\overline{PMR}_{s_a} = \frac{m\ A_s}{\Theta\ A_n} \qquad (8\text{-}33)$$

where

Θ is the total time of sampling
A_n is the internal cross-sectional area of the nozzle.

If we divide Equation 8-33 by Equation 8-32, noting that

$$\overline{V}_n = \frac{V}{A_n \Theta}$$

we obtain the ratio of proportional (isokinetic) conditions on the average

$$\frac{\frac{m A_s}{\Theta A_n}}{\frac{m}{\overline{V}} A_s \overline{V}_s} = \frac{\overline{V}_n}{\overline{V}_s} \tag{8-34}$$

The significance of this ratio must be understood. Basically if the calculated ratio, V_n:V_s, is appreciably greater than or less than 1.00, the results should definitely be questioned. But, since this is an average, the fact that it equals 1.00 does not insure that the results necessarily are valid. The variance of the estimated mean might be appreciable. It is possible that the average ratio be 1.00 while at all of the traverse (sampling) points the sampling conditions were actually above or below isokinetic. An example of use of this ratio is presented in Chapter 10 for evaluating particulate emission test results. It is called an *isokinetic ratio*, I, and the tests should be rejected and repeated if $I_{avg} > 110\%$ or $I_{avg} < 90\%$. A more statistically desirable approach here would be to use the standard deviation also to measure the deviation from isokinetic conditions. The isokinetic ratio, V_n/V_s, should be calculated at each point and Equations 8-4 and 8-5 used to calculate the mean and standard deviation, respectively. The decision to accept or reject a set of test results would be based upon some allowable deviation of the mean from 1.0 and the standard deviation of the individual isokinetic ratios.

The actual magnitude of error caused by non-isokinetic conditions depends upon the method of calculating $\overline{PMR}_s$ and the particle size distribution (see Figure 2.3). Figure 8.6 shows the error which will occur if the source concentration, C_s, is steady and only gases and small particles are present. In this situation the sample-concentration method provides an accurate estimate of the true

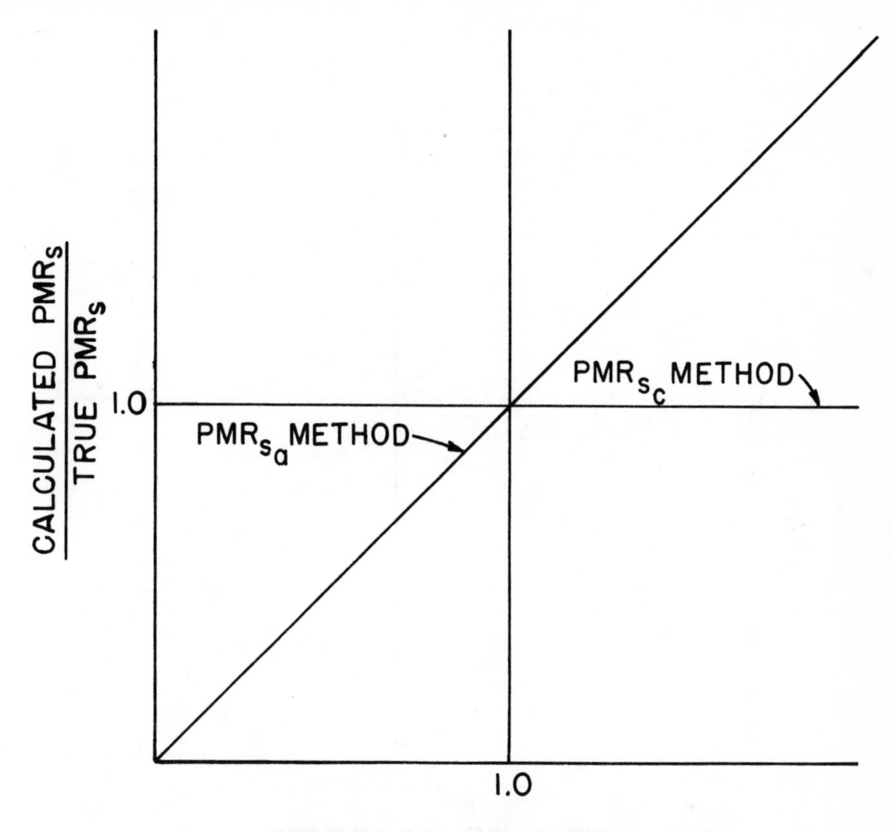

Figure 8.6. Errors due to anisokinetic conditions (gases and small particles).

$\overline{PMR}_s$. But, the ratio of area method is biased; it estimates too low for sub-isokinetic conditions and too high for over-isokinetic conditions. The magnitude of this error is in direct proportion to the deviation from isokinetic. Figure 8.7 shows the error which will occur if only large particles are present. In this situation the ratio of area method is unbiased. The sample concentration method overestimates $\overline{PMR}$ under sub-isokinetic conditions and underestimates $\overline{PMR}$ if over-isokinetic conditions occur.

For large particles the ratio of area method is unbiased. When $V_n:V_s < 1$ a smaller total gas volume

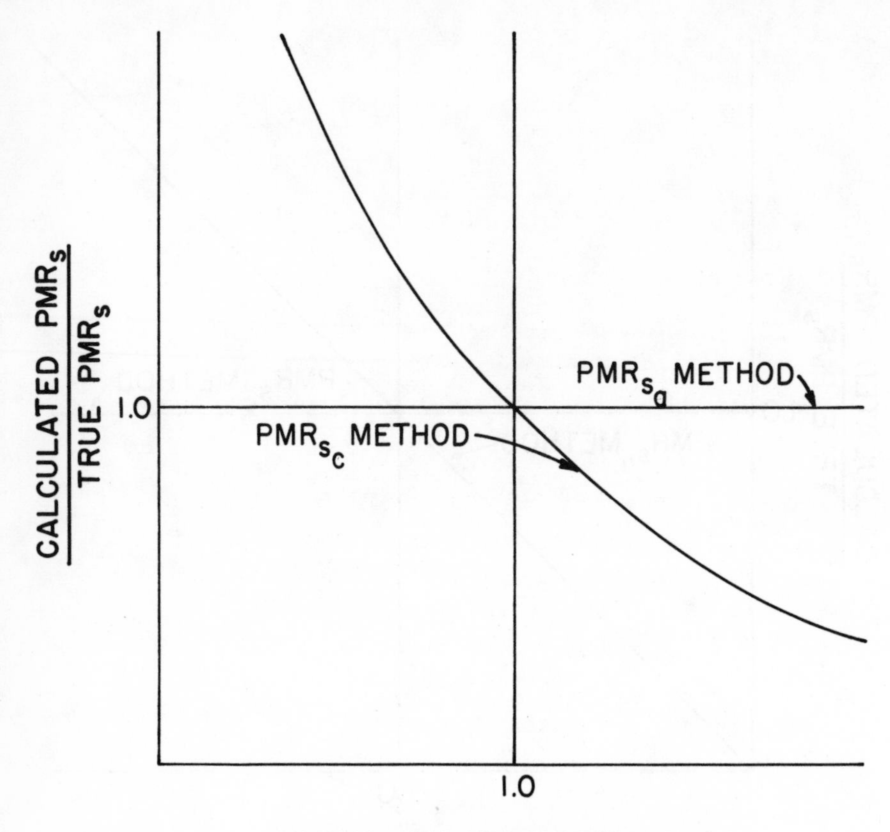

Figure 8.7. Errors due to anisokinetic conditions (large particles).

is taken into the nozzle; the larger particles are unable to follow the streamline and enter the nozzle anyway. Hence the correct mass is collected and $\overline{PMR}_{s_a}$ is unbiased. When $V_n:V_s > 1$ then too much gas is drawn in but the large particles will not be drawn in with it. Hence the correct amount of particulate mass is collected and once again $\overline{PMR}_{s_a}$ is unbiased.

For the situation where only large particles exist, the sample concentration method is biased for any condition other than isokinetic. If $V_n:V_s < 1$, the inertia of the particles causes them to enter the nozzle instead of following the streamlines. Hence the calculated $\overline{PMR}_{s_c}$ is greater than the true value. When $V_n:V_s > 1$ some of the particles go past the nozzle but the gas around them enters the nozzle. Thus not enough particulate matter is collected and the $\overline{PMR}_{s_c}$ is less than the true $\overline{PMR}_s$.

Usually a gas stream does not contain only large or small particles; it has some size distribution which is unknown. Figure 8.8 is derived from Figures 8.6 and 8.7. Basically it shows what errors can be expected. You cannot quantitatively determine the amount of bias due to non-isokinetic conditions. This figure shows qualitative trends only. Nelson[7] and Vitols[8,9] have tried to theoretically quantify this bias. However this can be done only for very simple cases and usually you do not have the information required to make such calculations.

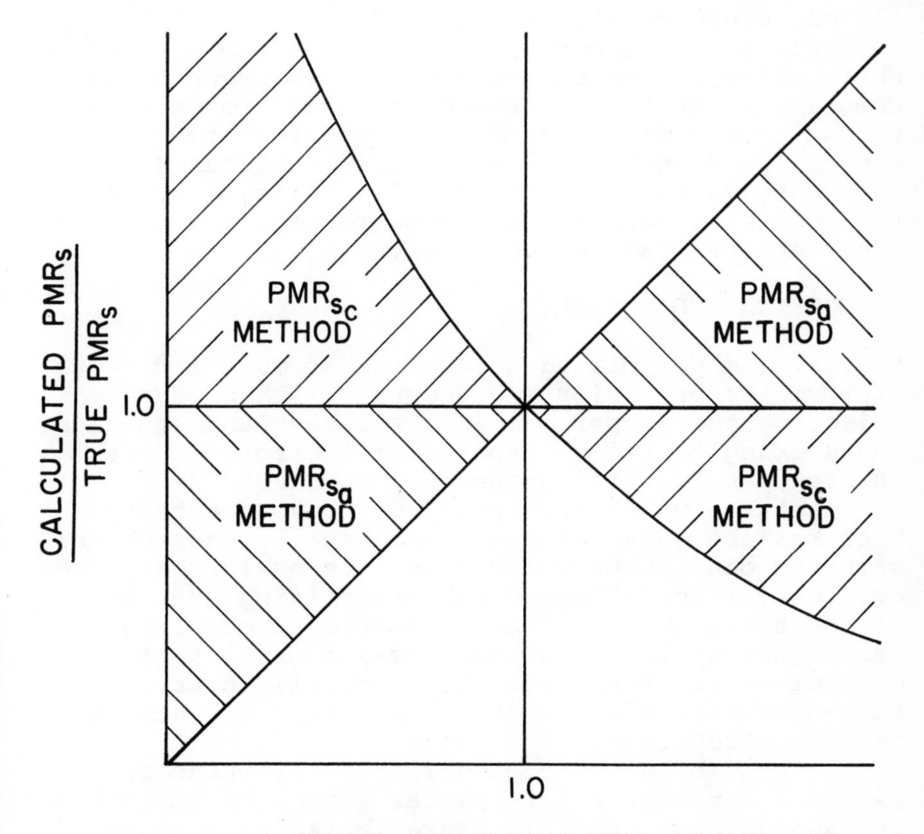

Figure 8.8. *Errors due to anisokinetic conditions.*

MISTAKES IN SOURCE SAMPLING

A mistake is a human blunder. Mistakes occur frequently in source sampling but they cannot be removed or predicted by any statistical manipulation. Some mistakes are easily recognized and can be immediately corrected without any adverse effects on the tests. However, others occur and often are never evident to the test team. In this latter case the only clue may be the wide variation of results from test to test; even this situation is not conclusive since the variation possibly may be explained by source variation.

A serious mistake in source sampling is to have the pitot tube and/or the nozzle improperly aligned with the flow in the duct. This causes biased pitot readings and flow characteristics into the nozzle. Under field conditions it is impossible to determine the magnitude of the bias caused.

Table 8.7 is a compilation of some common mistakes which occur in source testing. It is by no means complete since there are numerous ways to perform tests incorrectly. Seemingly, each test team manages to discover new ways to make mistakes. Fortunately, however, after a few tests, many of the operations and steps become routine and the frequency of mistakes declines.

DISCUSSION AND SUMMARY

The team performing a source test must strive to improve its precision and accuracy. This means selecting the proper sampling strategy and equipment and seeking to reduce experimental errors. This chapter has identified the many sources of errors and mistakes which may occur during source tests. Mistakes are human blunders and usually cause biased sampling conditions which cannot be evaluated. The greatest source of error for determining $\overline{PMR}_s$ by either the grab sample or proportional sampling techniques is the pitot tube measurement of stack gas velocity. Even greater error will occur in this measurement if it is not properly aligned in the gas stream.

The true bias associated with non-isokinetic sampling for particulate matter cannot be evaluated. It depends upon particle characteristics such as size distribution at each traverse point and this is not known. In fact the whole purpose of the test may be to collect particulate matter and then

Table 8.7

Common Mistakes in Source Sampling

Field Work

Mistakes in observing and/or recording of process and test data.
Nonalignment of pitot tube and/or sampling nozzle.
Improper marking of probe and thus sampling at the wrong points.
Breakage of glassware and loss of sample.
Contamination of reagents.
Failure to operate train at correct conditions--sample flow rate, probe and sample box temperatures.
Use of careless sample recovery techniques.
Damaging the filter prior to conducting the test.
Failure to leak test sampling train.
Conducting test while source is not operating at the prescribed conditions.
Using wrong amount of reagents in sampling train.
Overloading electrical circuits and having equipment stop in the middle of a test.
Using sampling probe which has a broken glass liner.
Failure to measure pertinent process parameters.
Leaving important equipment and supplies at the laboratory.
Particulate matter collected inside of buttonhook nozzle and not added to sample catch.

Laboratory Work

Use of noncalibrated or malfunctioning equipment.
Use of impure reagents
Improper preparation and storing of reagents.
Losing samples due to carelessness.
Mistakes in analysis and reduction of analytical data.

Planning, Computations and Reporting

Failure to notify control agency about test.
Selection of wrong sampling strategy.
Selection of wrong analytical method.
Failure to make proper arrangements with plant personnel.

determine the size distribution. Hence quantitatively there is no way that a correction factor can be used to determine the true $\overline{PMR_s}$ once a biased sample has been taken.

Another source of error includes the computational methods used in the field. Chapter 15 discusses computational methods; the use of nomographs in the field is convenient but undoubtedly errors occur in using them. In essence some accuracy is sacrificed for convenience.

Errors occur when determining process information such as full use rate and the fuel heating value. This information is very important when compliance regulations are given on a normalized basis. For example if the standard for a steam generator is 0.1 pounds of particulate per million BTU's input, the measured data is used as follows

$$E = \frac{m A_s \overline{V}_s}{m_c \Delta h V} \qquad (8-35)$$

where

- E is the normalized emission, pounds per 10^6 BTU's
- m is the mass of pollutant collected
- V is the total volume sample
- A_s is the area of stack, ft^2
- $\overline{V}_s$ is the average stack gas velocity, fps
- m_c is the coal use rate, lbs/sec
- Δh is the heating value of the coal, 10^6 BTU's per pound.

Actually Equation 8-17 should be applied to 8-35, thus obtaining

$$\left(\frac{\sigma_E}{E}\right)^2 = \left(\frac{\sigma_m}{m}\right)^2 + \left(\frac{\sigma_V}{V}\right)^2 + \left(\frac{\sigma_{\overline{V}_s}}{\overline{V}_s}\right)^2 + \left(\frac{\sigma_{A_s}}{A_s}\right)^2 + \left(\frac{\sigma_{m_c}}{m_c}\right)^2 + \left(\frac{\sigma_{\Delta h}}{\Delta h}\right)^2 \qquad (8-36)$$

The variance terms for fuel use rate and heating value cannot be estimated unless specific procedures are known. However, the authors feel that both of these measurements when evaluated will be found to be significant contributions to the overall error. In fact for compliance tests a member of the test team should be assigned the duties of monitoring these measurements. It does not seem logical to spend thousands of dollars measuring $\overline{PMR_s}$ if the other terms, m_c and Δh, for example, are determined in a haphazard manner. Thus, if the industrial process instruments must be used for any part of the calculations, they must be calibrated before the test.

REFERENCES

1. Li, J. R. Statistical Inference, Vol I & II, Edwards Brothers, Inc., Ann Arbor, Michigan. 1964.
2. Schenck, H. Theories of Engineering Experimentation, McGraw-Hill, New York, New York. 1961.
3. Shigehara, R. T., Todd, W. F. and Smith, W. S. "Significance of Errors in Stack Sampling Measurements," Paper 70-35, 63rd Annual Meeting of the Air Pollution Control Association, St. Louis, Missouri. 1970.
4. Hemeon, W. C. L., and Haines, G. F. Jr. "The Magnitude of Errors in Stack Dust Sampling," Air Repair 4, 159-164, November, 1954.
5. Badzioch, S. "Correction for Anisokinetic Sampling of Gas-Borne Dust Particles," J. Inst. of Fuel, 106-110, March, 1960.
6. Smith, W. S., Shigehara, R. T., and Todd, W. F. "A Method of Interpreting Stack Sampling Data," Paper 70-34, 63rd Annual Meeting of the Air Pollution Control Association, St. Louis, Missouri. 1970.
7. Nelson, W. G. "Correction of Anisokinetic Sampling Errors," 12th Conference on Methods in Air Pollution and Industrial Hygiene Studies, University of Southern California, Los Angeles, California, April 6-8, 1971.
8. Vitols, V. "Determination of Theoretical Collection Efficiencies of Aspirated Particle Matter Sampling Probes Under Anisokinetic Flow," Ph.D. Dissertation, University of Michigan, 1964.
9. Vitols, V. "Theoretical Limits of Errors Due to Anisokinetic Sampling of Particulate Matter," J. Air Pollution Control Assoc. 16, 2, 1966.

CHAPTER 9

PREPARATION FOR SOURCE SAMPLING

Chapter 5 discusses the conducting of a source test. To some extent, preparation for source testing is also discussed. However, the authors feel that the proper preparation for source sampling is critical and therefore warrants further detailed discussion.

Preparing for a source emission test whether for particulate, SO_2, NO_x or H_2SO_4 most generally requires the following:

1. a pre-test plant survey
2. preparation of the test plan
3. equipment preparation.

Failure to prepare adequately in any of these areas can result in safety hazards, loss of time, undue expense, and inaccurate test results. Thus, the purpose of this chapter is to familiarize the reader with the problems of source sampling preparation.

PRE-TEST PLANT SURVEY

The purpose of a pre-test plant survey is to provide information for the efficient design of the source test. A checklist is most helpful when conducting this survey. An example of one is shown in Table 9.1. The actual source test is a well-defined scientific experiment. In practice the basic methodology of operating the source test equipment becomes almost mechanical. The real problems associated with source sampling are usually involved with the accessibility of the sampling location, the conditions in the working environment, and the variability of the process tested. If conducted properly, a pre-test plant

Table 9.1

Pre-Test Survey Field Report

Firm ______________________ Date of Report ______________

Location ___________________ Date of Test _________________

Process ____________________ Report by ___________________

Process

Describe using box model below.

Approximate Input Rates ______________________________

Approximate Usable Output Rates _______________________

Approximate Waste Output Rates _______________________

Control Equipment Used _______________________________

Operating Conditions for Source Test ___________________

Plant Contacts

	Name	Position
1.	________________	________________
2.	________________	________________
3.	________________	________________
4.	________________	________________
5.	________________	________________

Sampling Location

Show dimensioned sketch

Downstream diameters to nearest flow disturbance _____

Upstream diameters to nearest flow disturbance _____

Rectangular cross section:
Equivalent diameter

$$D = 2\left(\frac{(\text{length})(\text{width})}{\text{length} + \text{width}}\right)$$

Traverse points/port__
Ports_____ Diam _____

Is the working environment in this area extreme in any way?

Explain ____

Table 9.1, continued

Stack Gas Conditions

T_S AVE ______		% H_2O AVE ______	
P_S AVE ______		M_S AVE ______	
ΔP AVE ______	MAX ______	MIN ______	

Storage and Lab Space

Available ______ Where ______

Miscellaneous Equipment Needed

rope	safety belts	respirators
pulleys	safety lines	CO indicators
lift buckets	raingear	ear plugs
hard hats	gas masks	water jugs
safety shoes	(for what gases)	salt tablets

survey will yield many insights that will allow most major problems to be anticipated and dealt with.

Ideally a member of the source testing team should conduct the pre-test survey. The following is an example of the procedure to follow when conducting the survey.

Process

Before traveling to the plant for the pre-test survey, the process of interest should be studied thoroughly. At the time of the survey it is invaluable to know the operating principles of the process and the associated air pollution problems. Familiarity with typical emission rates, control techniques, local emission regulations and sources testing standards is a must. Chapter 3 deals with the general task of gathering process information.

Contacts

Before the plant survey it is advisable to insist that a plant engineer be assigned to oversee the source testing project. Having someone from the plant working directly with the sampling team and the problems encountered will simplify matters considerably. After arriving at the plant for the pre-test survey, spend some time meeting as many people connected with the process as possible. Be sure to meet the plant manager or chief engineer. Knowing people directly responsible for the operation of the process will be of great help, as will meeting a plant chemist and the foremen from the electrical and maintenance shops. It should be obvious that these people can handle routine problems like a blown fuse or the need for some acetone or a 4' pipe wrench much more effectively than any plant engineer.

Sampling Location

The optimum sampling location for the source test is determined with the help of process operating personnel and a plant engineer. This should be done with respect to safety, convenience, and the local required sampling method criteria applicable. Once the location is agreed upon, the following procedure should be followed.

Ports. Measure the diameter and area of the duct at the sampling location, and the distance from the sampling location to the nearest upstream and downstream flow disturbance, such as bends or openings. Using the applicable procedure from the local sampling standards, determine the number of traverse points necessary and design the sampling ports. If the test is to meet federal EPA standards[1] then Figures 9.1 and 9.2 along with Table 9.2 should be used. To use Figure 9.1, measure the distance from the chosen sampling location to the nearest upstream and downstream disturbances. Then use the figure to determine the corresponding number of traverse points for upstream and downstream conditions. Select the higher of the two numbers. For circular stacks the number of traverse points should be increased so that it is a multiple of 4. As shown in Figure 9.2 the sampling points should be located on at least two diameters; when a large number of traverse points are required they can be located on 4 diameters. In every case the traverse axis should divide the stack cross section into equal parts. When a rectangular stack is encountered, the cross section should be divided into equal areas as shown in Figure 9.2. The areas should be of such dimensions that the ratio of the length to width is between one and two. The traverse points should be located at the centroid

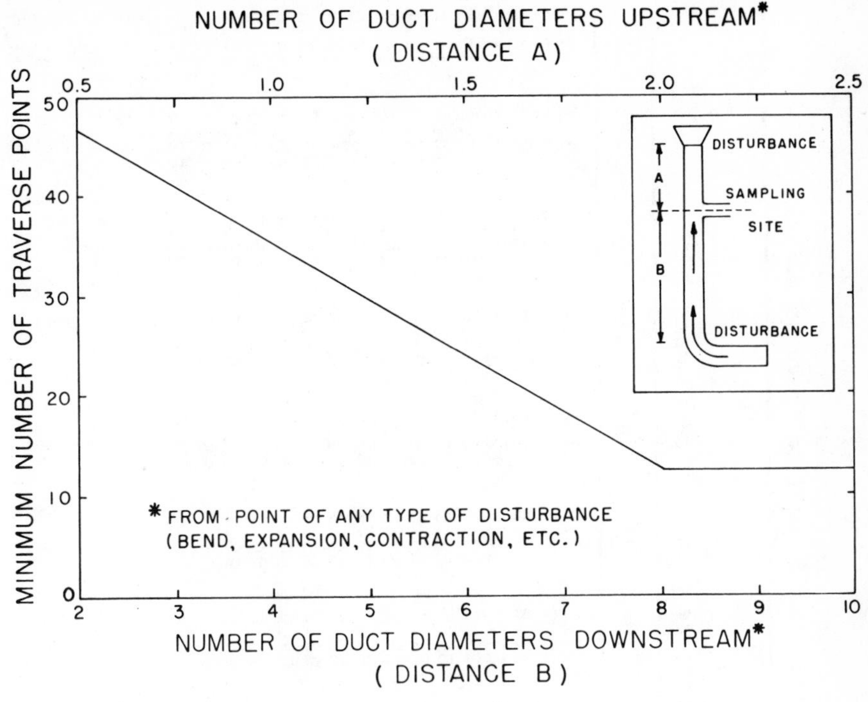

Figure 9.1. Minimum number of traverse points.

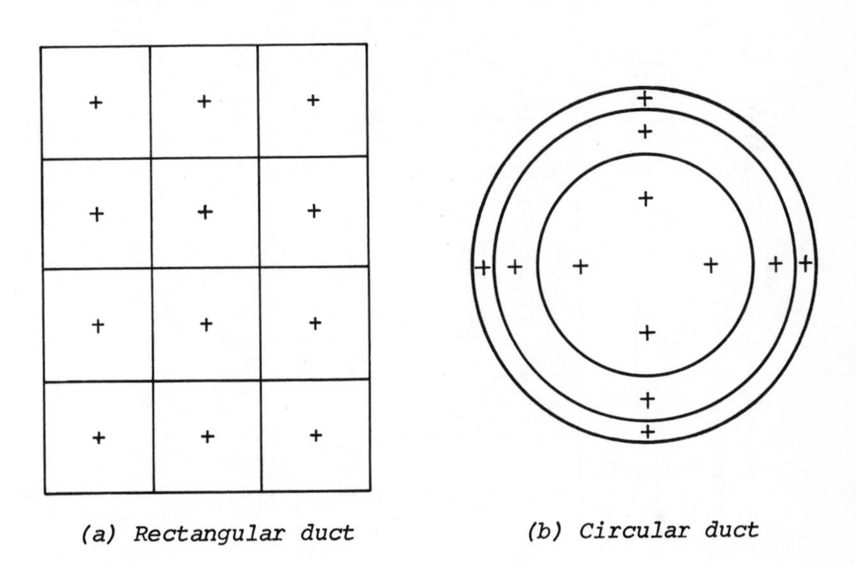

(a) Rectangular duct *(b) Circular duct*

Figure 9.2. Cross sections of circular and rectangular ducts.

Table 9.2

Location of Traverse Points in Circular Stacks
(Per cent of stack diameter from inside wall to traverse point)

Traverse point number on a diameter	*Number of traverse points on a diameter*											
	2	*4*	*6*	*8*	*10*	*12*	*14*	*16*	*18*	*20*	*22*	*24*
1	14.6	6.7	4.4	3.3	2.5	2.1	1.8	1.6	1.4	1.3	1.1	1.1
2	85.4	25.0	14.7	10.5	8.2	6.7	5.7	4.9	4.4	3.9	3.5	3.2
3		75.0	29.5	19.4	14.6	11.8	9.9	8.5	7.5	6.7	6.0	5.5
4		93.3	70.5	32.3	22.6	17.7	14.6	12.5	10.9	9.7	8.7	7.9
5			85.3	67.7	34.2	25.0	20.1	16.9	14.6	12.9	11.6	10.5
6			95.6	80.6	65.8	35.5	26.9	22.0	18.8	16.5	14.6	13.2
7				89.5	77.4	64.5	36.6	28.3	23.6	20.4	18.0	16.1
8				96.7	85.4	75.0	63.4	37.5	29.6	25.0	21.8	19.4
9					91.8	82.3	73.1	62.5	38.2	30.6	26.1	23.0
10					97.5	88.2	79.9	71.7	61.8	38.8	31.5	27.2
11						93.3	85.4	78.0	70.4	61.2	39.3	32.3
12						97.9	90.1	83.1	76.4	69.4	60.7	39.8
13							94.3	87.5	81.2	75.0	68.5	60.2
14							98.2	91.5	85.4	79.6	73.9	67.7
15								95.1	89.1	83.5	78.2	72.8
16								98.4	92.5	87.1	82.0	77.0
17									95.6	90.3	85.4	80.6
18									98.6	93.3	88.4	83.9
19										96.1	91.3	86.8
20										98.7	94.0	89.5
21											96.5	92.1
22											98.9	94.5
23												96.8
24												98.9

of each area. For a rectangular duct, the following equation should be used in relation to Figures 9.1 and 9.2.

$$\text{equivalent diameter} = 2\left(\frac{(\text{length})(\text{width})}{\text{length} + \text{width}}\right) \qquad (9\text{-}1)$$

Usually no less than 12 total traverse points should be used. However, if the stack or duct is less than 2 feet in diameter, the numbers obtained from Figure 9.1 can be multiplied by a factor of 0.67. In no case should a sample be taken at a distance closer than 1 inch from the interior stack wall.

Experience has shown that 4" ports fabricated from standard pipe couplings are the easiest to work with. Generally, the port size is sufficient and pipe plugs are usually easier to remove than pipe caps or flange plates. If the ports to be used for the source test are installed more than two weeks prior to the test time, it is advisable to notify the maintenance department to loosen the pipe plugs before test time.

Sampling Platform. The pre-test plant survey is the time and place to insist on safe scaffolding and ladders. All platforms, scaffolding, ladders, and working environments should meet OSHA standards.[2] Measurements should be taken to insure that the scaffolding or platform can be designed or modified to be compatible with the source testing equipment used. The electrical shop should be notified of the power requirements for the source test. Generally two separate electrical services (at least 115 volt; 20 amp) at each sampling location will be sufficient.

Duct gas conditions. During the pre-test survey the approximate duct gas conditions, temperature, velocity pressure, static pressure, moisture content, and possibly the approximate gaseous constituent concentrations should be estimated. A short 30" standard pitot tube, Magnehelic pressure gauge and dial thermometer can be carried easily in most suitcases. The moisture content and gaseous constituent concentrations of the duct gas usually can be estimated from the process variables. Simplified gas sampling apparatus, allowing approximate concentrations to be determined, is available.

Miscellaneous equipment. Specific equipment necessary for the source test varies because every sampling location is different and requires specialized equipment. Things like ropes, pulleys, winches and safety equipment are examples. If the sampling location is not easily accessible (*e.g.*, no elevator in building), provisions should be made to enlist plant help to simplify the transport of the sampling equipment (sometimes as much as 200-300 lbs) from ground level to the sampling location. This usually results in significant savings in time and money. A stack sampling team costs $20-$25 per man-hour while plant labor costs on the order of $3-$5 per man hour. In addition, a sampling team without broken backs usually does better work, smiles more, and is easier to get along with.

Storage and lab space. During the pre-test survey a request for equipment storage and laboratory space in the proximity of the sampling location should be made. If possible the storage area should be clean and capable of being locked. Laboratory space with a sink for cleaning equipment and a refrigerator for storing unstable reagents such as H_2O_2 is a necessity.

Review

After touring the process and selecting the sampling location, it is a good practice to discuss the details of the upcoming test once again with the plant engineer assigned to the project. At this time it is wise to retire to a quiet office. The process operating conditions for the source emission test should be agreed upon (*e.g.*, for steam boilers--which steam load; maximum attainable or normal operating?). The final contract agreement should be arranged at this time.

PREPARATION OF TEST PLAN

In most states, when source emission tests are performed for the purpose of complying with air pollution regulations or proving compliance with specific regulations, a test plan must be sent to the proper agency. The test plan notifies the agency of the source test. The state and federal environmental protection agencies have the right to observe all source tests for compliance. An outline of a typical source plan is shown in Table 9.3. This general outline indicates what kind of information is usually requested. Some states have chosen to be more specific with their source test guidelines

Table 9.3

Test Plan Outline

I. Source identification
 A. Applicable emission regulations
 B. Applicable test methods and sampling equipment
  1. particulate
  2. gaseous pollutants (*e.g.*, SO_x, NO_x, H_2SO_4)

II. Information to be obtained
 A. Variables to be measured and expected range
  1. process operating data
  2. sampling train data
  3. source emissions
 B. Control equipment operation
  1. design specifications
  2. operating data

III. Analyses required
 A. Particulate
 B. Gaseous pollutants (*e.g.*, SO_x, NO_x, H_2SO_4)
 C. Other (*e.g.*, particle sizing)

IV. Calculations required
 A. Particulate
 B. Gaseous pollutants
 C. Other

V. Preparation needed
 A. By plant personnel
  1. sampling location
  2. process operation
 B. By source test team
  1. calibration
  2. packing equipment

VI. Test schedule

VII. Assignment of responsibilities to individuals

and procedures. A copy of the state of Connecticut's Source Test procedures is given in Appendix B, as an example. The Connecticut Department of Environmental Protection, DEP, requires that an Intent to Test form be completed by the tester, when source tests are being conducted for compliance. This Intent to Test form, shown in Figure 9.3, must be received by the DEP 30 days prior to the anticipated test date. During this time the test plan is reviewed with the intent of protecting the industry involved and the

Date ________
Number
Assigned ________

INTENT TO TEST

Source Test Form #1

I. Source Information

Name ________ Person or Persons Responsible for tests
Address ________ Name ________
________ Telephone ________

III. Premis # ________ (To be assigned by
Registration # ________ D.E.P.)

II. Testing Firm Information

Name ________ Person or Persons Responsible for tests
Address ________ Name ________
________ Telephone ________

Signature ________
Person responsible for tests

IV. Gas Stream Sampling Information. Identify all gas stream components to be sampled.

Component	Sampling Duration minutes per point	total test time	Number of tests (Minimum of 3)	Expected Concentrations	Method Employed to Determine Concentration i.e., material balance, emission factor, reference (specify) etc.
1.					
2.					
3.					
4.					
5.					

V. Stack Information
1. Approx. gas temp. ________ °F
2. Approx. gas flow ________ A.C.F.M.
3. Approx. percent moisture ________ % (percent by volume)
4. Approx. gas density ________ lb/ft (at stack conditions)

VI. Purpose of Test

VII. Identification of Documents Requested on Attached Sheets

VIII ________
IX ________
X ________
XI ________
XII ________
XIII ________
XIV ________

Please attach this form to the information required on Page 2.

Figure 9.3a. Intent to Test form for state of Connecticut, page 1.

Please submit the following information:

VIII. Sampling Train Information

A description of each sampling train to be used. The description should include the following items:

1. A schematic diagram of each sampling train. The name, model number and date of purchase of commercially manufactured trains should be included with the diagram.
2. The type or types of media to be used to determine each gas stream component.
3. Sample tube type, *i.e.*, glass, teflon, stainless steel, etc.
4. Probe cleaning method and solvent to be used (if test procedure requires probe cleaning).

IX. Data Sheets

A sample of all field data sheets to be used in the test or tests.

X. Description of Operations

A written description of any operation, process or activity that could vent exhaust gases to the test stack. This description should include:

1. The composition of all materials capable of producing pollutant emissions used in each separate operation.
2. The feed rate and any operating parameter used to control or influence the operations.
3. The rated capacity of equipment with manufacturer's specification.

The process registration numbers, if applicable, should also be included in the description.

XI. Sampling Area Description

A cross-sectional sketch showing dimensions of the following items:

1. Stack configuration
2. Sampling port locations
3. Sampling point positions for each port.

XII. Stack and Vent Descriptions

A sketch or sketches showing the plan and elevation view of the entire ducting and stack arrangement. The sketch should include the relative position of all processes or operations venting to the stack or vent to be tested. It should also include the position of the sampling ports relative to the nearest upstream and downstream gas flow directional or duct dimensional change. The sketches should include the relative position, type, and manufacturer's claimed efficiency of all gas cleaning equipment.

XIII. Operational Parameters

A statement indicating the types of processes, operations, and operating conditions anticipated during the test or tests, on all equipment venting to the stack.

XIV. Registration

If the equipment or stack to be sampled has not been registered with the Department of Environmental Protection, the applicable registration form or forms must be completed and submitted with the Intent to Test forms.

Figure 9.3b. Intent to Test form for state of Connecticut, page 2.

public health and welfare. The information from the form along with test results and other source information is collected and computer filed.

Once computerized, much additional information is available. Emission factors can be updated and created for similar sources. Control equipment degeneration can also be detected for those sources tested on a routine basis. The progress of all required source tests within the state is also monitored, in this case on a weekly basis. It is also interesting to note that the Connecticut DEP reserves the right to supply the pollutant collection media for the proposed tests. At the conclusion of the field tests the DEP representative may, at his discretion, collect all test data sheets and sample collecting media! Connecticut's approach is a progressive one and may be indicative of future trends for many air pollution control agencies.

EQUIPMENT PREPARATION

To successfully prepare the equipment needed and to insure that the equipment arrives safely at the sample site three things are necessary.

1. an equipment checklist
2. calibration of the equipment
3. proper packing of the equipment.

To insure that all equipment necessary for the source test is prepared for use, checklists should be developed and filled out immediately after returning from the pre-test plant survey. Examples of equipment checklists are given in Chapters 10, 11, 12, and 13. An example of a tool check list is given in Table 9.4. A week or two prior to test time all equipment should be tested for proper operation. Continuity of probe and filter box heaters and the vacuum capacity of the pump are examples. Pitot tubes, dry test meters, and thermocouple-potentiometer systems should be recalibrated before each use. The rough field use incurred during stack sampling necessitates frequent recalibration of these instruments.

After all equipment has been checked, repaired, and calibrated it should be packed immediately for transportation to the sampling site. Care should be taken to protect all fragile glassware and delicate equipment. Polyurethane molded foam packing cases, as shown in Figure 9.4, can be bought for such equipment. When packing for a

Table 9.4

Tool and Supplies Checklist

Tools	*Supplies*
Tool box (*large*)	Hose clamps (1/4-1")
Screw drivers 1 large flat blade 1 medium flat blade 1 very small flat blade 1 large Philips 1 small Philips	Tape 2 rolls electrical 2 rolls fiberglass (hi-temp) 2 rolls filament packing
C-clamps 1 6" clamp 1 3" clamp	Silicone sealer Asbestos cloth Asbestos gloves
Wrenches 2 12" adjustables 1 6" adjustable 1 12" pipe 1 chain wrench (large enough for 4" pipe)	Work gloves Silicone vacuum grease Pump oil Manometer fluid Safety glasses
Pliers 1 pair pump pliers 1 pair electrician's pliers 1 pair needlenose pliers	Anti-seize compound Teflon pipe dope 2-3 Wire elec. adapters 3 Wire elec. triple taps
Tweezers (for handling filters)	
Jack knives	*Spare Equipment*
Tape measure (12')	T-C Extension wire
Tubing cutter	T-C plugs
Hacksaw	Fuses
Flashlight	Electrical connections
Claw hammer	Pipe fittings
Files	Dry cell batteries
Ammeter-voltmeter-ohmeter	Electrical wire
	Valves
Micrometer } for nozzle Inside calipers } diam.	Vacuum gauge Dial thermometers
Slide rule	Manometer parts
Probe holder	SS tubing (1/4, 3/8, 1/2) (small pieces)

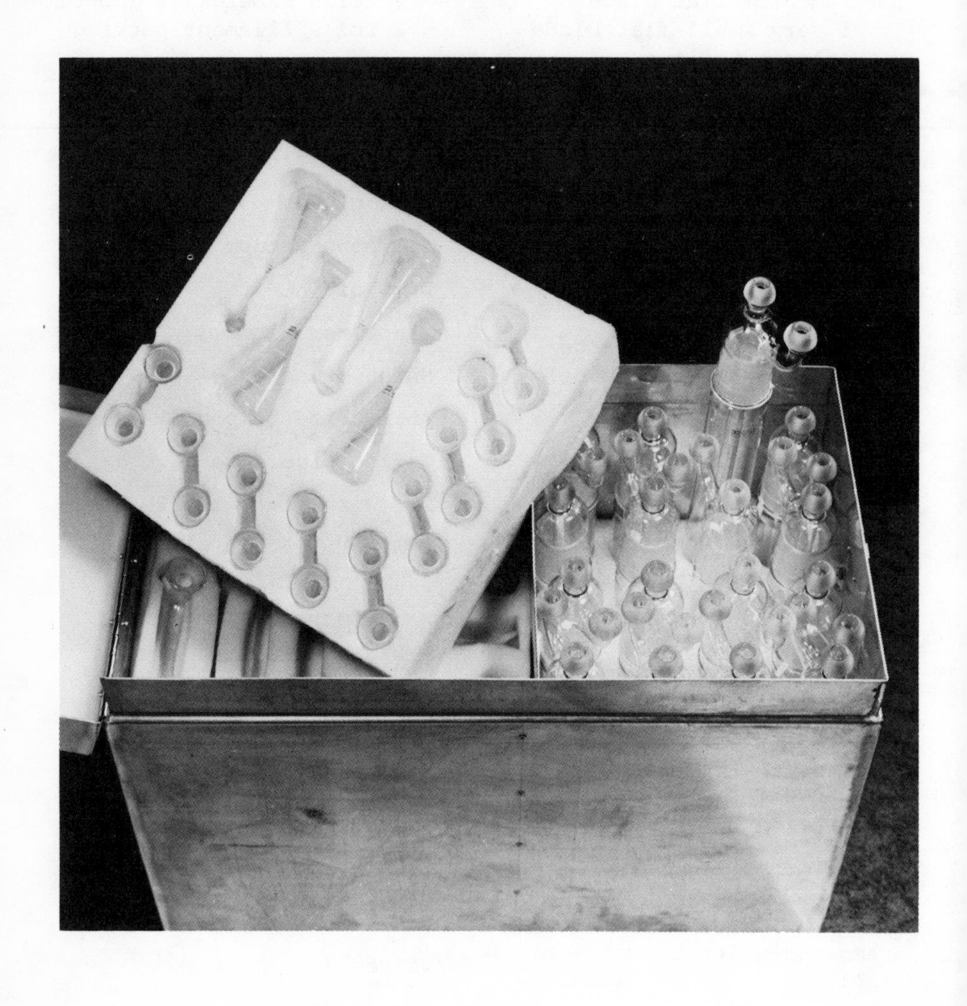

Figure 9.4. Packing case for glassware (courtesy of Glass Inovations Inc., Addison, New York).

source sampling project which requires public air transportation, the airline should be contacted concerning the transport of flammable solvents such as acetone. Long sampling probes (greater than 7') can also cause some problems. Aircraft smaller than DC-9's usually have baggage compartment arrangements that cannot handle these longer probes. Air freight services can usually ship acetone, long probes, and heavy pieces of equipment. However, when using air freight services allow a few extra days for delivery.

REFERENCES

1. "Standards of Performance for New Sources," Federal Register, Vol 36, No. 247, December 23, 1971.
2. "Occupational Safety and Health Standards: National Concensus Standards and Established Federal Standards," Federal Register, Vol 36, No. 105, May 29, 1971.

CHAPTER 10

DETERMINATION OF PARTICULATE MATTER EMISSIONS

This chapter presents EPA Method 5 procedures for particulate matter.[1] This method applies to new steam boilers, incinerators, and portland cement plants. In the future the method is expected to extend to secondary lead smelters, sludge incinerators, brass and bronze foundries, basic oxygen furnace steel mills, and asphalt batch plants. The federal government has also issued source testing guidelines for the testing of incinerators at federal facilities.[2]

Presently the source sampling methods specified by the different states vary considerably. Some states require the inclusion of the condensible particulate fraction, as outlined in the August 14, 1971, *Federal Register*,[3] in the determination of total mass emission rates. The current Method 5 does not include this fraction. Other states have modified the sampling and analytical techniques where they deemed necessary. Although many of the states specify methods that differ from the federal procedures, the general procedure enjoys widespread acceptance. Therefore this chapter presents EPA Method 5.[1]

Presently there is some controversy about EPA Method 5, with some users claiming that the condensation of sulfur trioxide with water into a filterable aerosol adds to the particulate weight. However, the study reporting this relied on an alundum R thimble filter method,[4] and the authors do not share the enthusiasm for the method of verification used. References 5 and 6 also compare EPA Method 5 to the alundum filter method. References 7 and 8 discuss source sampling equipment and data interpretation.

This controversy leads to a discussion of the inherent problem of the definition of exactly what is a particulate. Reference 9 considers this problem in great detail. There are several possibilities:

1. dry particulate at stack temperature
2. dry particulate at 250°F collected on specified filter media
3. dry and condensible aerosols
4. particulates which are in the respirable range
5. particulates and aerosols existing in the ambient atmosphere at the perimeter of the industry.

The first definition is that obtained by the ASME-Thimble method.[10] The second describes the particulate collected by the EPA Method 5. The third is the particulate collected by the LA County method.[11] What is the true definition of a particulate? The one finding usable credance is that which is implied by the existence of the EPA Method 5.

BASIC METHOD

EPA Method 5 relies on the removal of a dust-laden gaseous sample from the duct followed by the subsequent removal of the particulate on a filter media with concurrent measurement of the sample volume to determine particulate concentration. The sample is removed from duct or stack by using a prescribed traversing procedure (see Chapter 9), which in effect is an approximate integration of collected mass and sample volume over the cross-sectional area of the stack or duct. The velocity distribution is also determined during the sample traverse. This information provides the stack gas flow rate, which is used with the particulate concentration to determine mass emission rates. As discussed in Chapter 2, EPA Method 5 requires that isokinetic sampling conditions be maintained. Thus at each traverse point the sample velocity at the nozzle is adjusted to equal the duct velocity. The sampling train required is discussed in detail in the following section.

SAMPLING TRAIN

The sampling train discussed here is based on EPA Construction Details,[12] although this should not necessarily be interpreted as an endorsement of the construction of the equipment. The construction details are used as a basis of discussion because of the differences between available commercial units with different component configurations. The sampling train shown in Figure 10.1 is that specified in EPA Method 5.

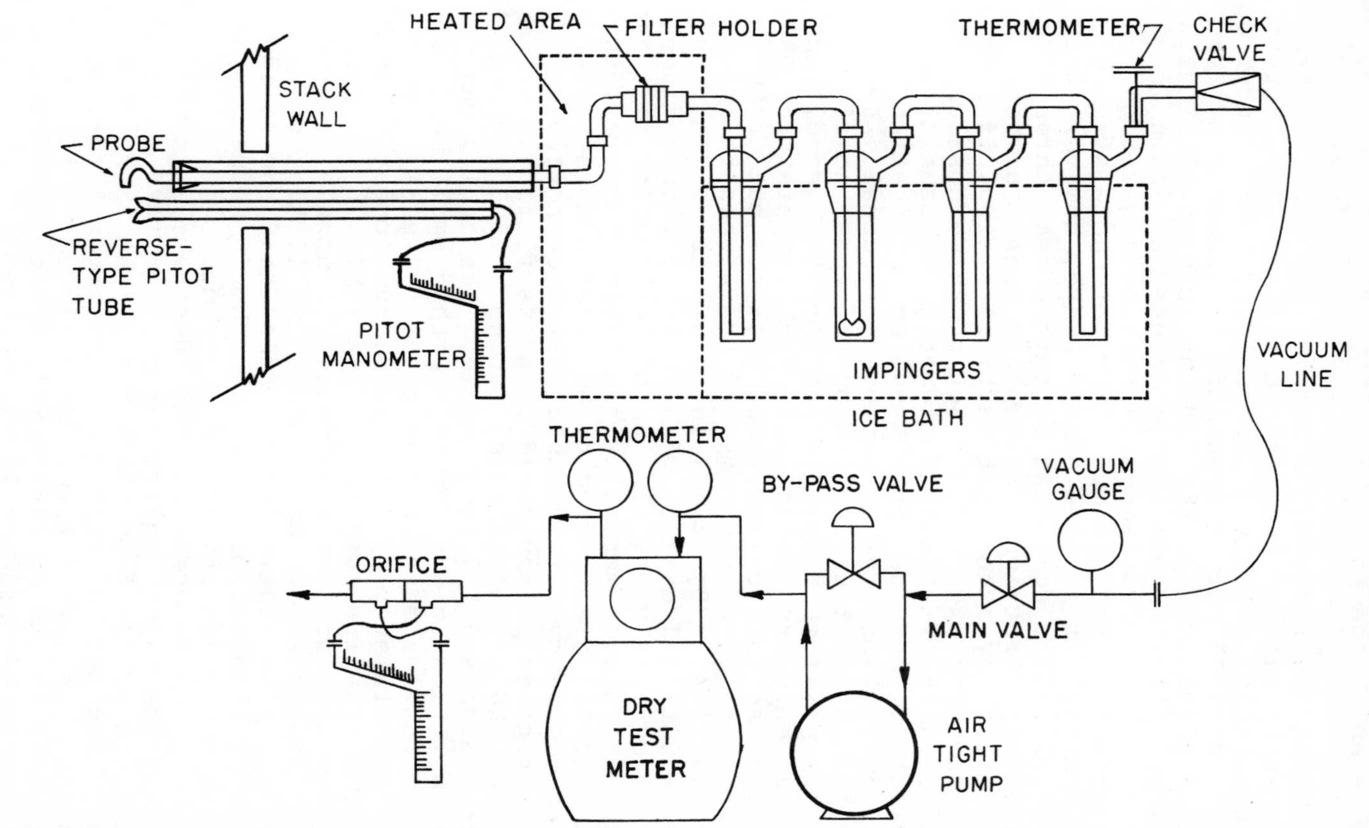

Figure 10.1. Sampling train for particulate matter (EPA Method 5).

Nozzle and Probe

The nozzle removes the sample from the gas stream. It must be aligned parallel to the gas flow, but this alignment is strictly a personal judgment. Incorrect alignment will not be demonstrated directly by the S-type pitot used or the isokinetic variation calculation. A buttonhook nozzle is used as shown in Figure 10.1. Advantageously this configuration requires that the minimum entry port hole size be 2.5 inches. It has the disadvantage, however, that it requires the collected stream to make three 90° changes in direction. This would negate almost completely any attempt at particle sizing, if larger particles are expected. A set of three nozzles, 1/4, 3/8, and 1/2 inch in diameter, is usually sufficient for routine stack sampling applications. Most duct velocities encountered can usually be sampled isokinetically with one of these nozzles. However, extreme velocities and/or stack conditions, such as moisture content and stack temperature, may require isokinetic flow rates that are out of the reliable range of the sampling train when the normal nozzles (1/4, 3/8, or 1/2 inch) are used. In this case additional nozzles of 3/16 and 5/8 inch diameter can be extremely useful. As of yet nozzles of these sizes are not commercially available. However, they can be easily fabricated by any good machine shop. In an emergency a servicable 3/16 inch nozzle can be fabricated in the field by press fitting a piece of 1/4 inch stainless steel tubing (0.035 inch wall) into a standard 1/4 inch nozzle. The tubing can then be filed to a sharp edge with a small hand file. Nozzles made in this fashion usually have a sampling diameter of 0.180 to 0.190 inches. When no measuring instruments are available in the field (micrometer and inside calipers), a nozzle diameter of 0.185 inches can be assumed with little error. However, after returning from the field the inside sampling diameter of the nozzle should be determined before any calculations are performed.

The probe removes the sampled stream from the stack. The major requirement is that it does not alter the sample from stack conditions, a requirement which can be translated into several parameters. The first is temperature. Several compromises have been made in the operation relating to temperature. The sample temperature is allowed to fall below the duct temperature and is maintained at 250°F based on a calibration of the heating element within the probe. If the temperature falls much below this, the water vapor and aerosols in the sample stream begin to condense, which quickly leads to clogging of the filter media. A sample collected in this manner does not conform to the definition of particulate by EPA Method 5. The probe is exposed to both high and low temperatures simultaneously. This places considerable thermal stress on the materials of construction. Method 5 stipulates the use of heat resistant glass sampling liners for probes less than 8 feet in length because of the ease and completeness with which glass can be cleaned following the sample. Stainless steel tubing cannot be completely cleaned and thus some sample will always be lost. However, glass-lined probes longer than 6 or 7 feet are very susceptible to breakage. The mere static deflection of some longer probes can be sufficient to break most glass liners.

Pitot Tube

The pitot tube is a Stausscheibe type, and is also called the S-type or reversed pitot. The use of this pitot is directly specified in Method 5. It is used instead of a standard type because of the clogging tendency of the standard pitot in a dust-laden stream. The S-type tube is also advantageous because it is easy to construct and shows a larger difference in pressure than a standard pitot for the same velocity pressure. An S-type pitot should be constructed out of 3/8-inch thick walled tubing because this design is more applicable to rugged field use and requires less recalibration.

Filter

The fiberglass media used is required to be at least 99.7% efficient for 0.3μ dioctyl phthalate particles. It also must be inert to chemical reaction. This efficiency is calculated on the basis of spherical dioctyl phthalate particles of uniform density and cannot be translated directly to stack particulate. The filters are not reused. Although the exact efficiency for the filters as they are used is not precisely known, the fiberglass filters are considerably more reliable than alundum thimble filters which are reused and therefore of doubtful efficiency.

The filter support media should be a glass frit. Metal-based supports may last only until condensation occurs. Some of the holders have solid hard plastic or teflon shoulders to seal against the filter while other seals are of softer silicone rubbers. There are a variety of filter holder clamps, some having as many as four bolts while others have one huge compression nut. Ease of assembly, durability, filter seal, and handling properties should be considered prior to purchase of filter holders. The filter holder either will be unnoticed or a constant problem. Since the life of the filter media is a function of its surface area, larger filters (4-5-inch diameter) are preferred in stack sampling.

Cyclone

The cyclone, available with many commercial trains, is not required by Method 5. If it is used, the collected material is added to the probe washings. The major reason for using the cyclone is to remove larger particles, thereby increasing filter life. There is no analytical significance to the use of the cyclone other than the uncertainty of efficiently cleaning another piece of glassware.

Impingers

Four impingers are shown in the Method 5 sampling train. These remove the water and allow the moisture content of the gas stream to be determined. At the point when the gas stream enters the impingers, most of the particulate matter has been removed by the filter. All the impingers have 500 ml capacities and are of the Greenburg-Smith design. The first, third, and fourth have been modified by replacing the tip with a 1/2-inch ID glass tube extending to 1/2 inch above the flask bottom (see Figure 4.12). The second impinger is a Greenburg-Smith type with a standard tip. The primary function of the first impinger is to cool the hot gas stream from the filter. Some gas absorption (condensation) will take place in this impinger. The second impinger is for final particulate removal and final gas absorption. The third impinger is a dry trap to catch any carry over from the previous impinger.

The last impinger contains silica gel for the final removal of all water vapor. The silica gel is usually a preweighed amount used in the determination of the vapor content of the sampled volume. It also serves, by the removal of water vapor, as a protection for the working parts of the vacuum pump and dry gas meter.

There are two other items in the impinger train which protect the metering system and sample. The first is a thermometer to enable the exit gas from the bubbler to be maintained at a low enough temperature to protect the metering system. The second item is a check valve which protects the sample. At the higher vacuum encountered as a filter becomes loaded, flow tends to be backward through the bubbler train when the pump is turned off to remove the filter. If the check valve is not used (or sticks), the water in the first bubbler can be drawn back up into the filter holder.

EPA Method 5 does not consider the condensible particulate fraction and hence the impingers are not specifically required. The fragile glass impinger train may be replaced by a suitable condenser, the practicality of this being obvious. It has been found that an ice-packed copper or stainless steel coil (greater than 3/8" ID) followed by a suitable sump (500 ml capacity) can be made workable. The condenser, however, should be followed by a silica gel drying tube to collect the remaining moisture of saturation and protect the vacuum pump and dry gas meter.

Sampling Box

The sampling box holds the probe, the filter holder and the impinger train, and also must supply power to the probe to heat it. It contains a heated area in which the filter holder is located, and the box also contains the ice bath for the impinger train. These areas must be water tight and well insulated. The sampling box must also be light and easy to handle and support since it will be moved with the probe to the required sampling points. There are many choices of sample box supports available with commercial sampling trains. Several use an overhead rail support while others use supports on which the box slides. No one type of support lends itself to all applications. The simpler the support, the easier to fabricate and alter it as the situation requires. In states not requiring that the condensible particulate fraction be determined and considered part of the particulate emissions, the sampling box can be greatly simplified. Where condensibles are of no interest the chore of stack sampling can be made easier by removing the impingers from the sample box and locating them remotely in the same area as the meter box. Method 5 only requires the use of an impinger train to condense and determine moisture. Therefore there is no apparent reason why the filter assembly cannot be connected to the impinger train by a flexible umbilical cord. It is the authors' experience that if care is taken in removing the condensed water vapor from the umbilical cord, the moisture determination will not be significantly affected. Experience has further determined that the filter assembly can be contained in a small light weight aluminum box approximately 10 inches on a side. The use of this smaller and lighter sample box precludes the necessity of sampling support systems such as monorails, etc. A source test conducted in this manner is usually faster, cheaper, and easier than a test conducted with the impingers in a 50 pound sample box swinging precariously at the end of the probe.

Umbilical Cord

The purpose of the umbilical cord is to allow the placement of the heaviest system component, the meter box, at a more convenient location for testing. To do this the cord must carry the sample stream to the meter box, supply power to the probe and filter heater, and transfer the velocity head measurement to the meter box. In some cases, it can provide a communication line or thermocouple extension wire. The maximum length suggested by Method 5 is 100 feet, but umbilical cords have been used up to 150 feet in length.[5] Experience has shown that a 25 foot umbilical cord is sufficient for most routine source tests.

Meter Box

The purpose of the meter box is to enclose and protect the various pieces of the gas flow train which facilitate the taking of isokinetic source samples. The box usually contains:

1. A leakless vacuum pump with suitable control valves to draw the gas sample through the train at the isokinetic rates needed. This pump is quite heavy and causes the meter box to weigh between 50 and 70 pounds. Often it is desirable to have this component separate in order to facilitate handling.
2. A vacuum gauge for measuring the sampling stream pressure. With practice, the vacuum measured during sampling gives some indication of the filter media's life.

3. A dry gas meter equipped with temperature-measuring devices at inlet and outlet for determining the integrated sample volume.
4. A calibrated orifice meter, which is used to monitor the sample flow rate.
5. An inclined dual manometer to measure the velocity head in the source duct and the pressure drop across the calibrated orifice meter.
6. Variable voltage power supplies to maintain the probe and filter box at 250°F by means of their individual heaters.
7. A pyrometer or potentiometer suitable calibrated for thermocouple temperature measurements.

External Equipment

External equipment consists of those items needed to conduct a source test which are not integral parts of the gas flow train itself. A representative measurement of the stack gas temperature must accompany any source test. This measurement is usually taken at regular intervals at some point in the stack. Based on the assumption of a uniform temperature profile and small relative errors encountered with temperature measurement (see Chapter 8), usually it is not imperative to take the stack temperature at each traverse point. It should be noted, however, that while testing uninsulated exterior metal ducts during the winter months, significant errors may be encountered. On occasion under winter conditions the authors have found temperature differences of up to approximately 200°F when the stack temperature was taken at every traverse point.

The local barometric pressure must be measured using an uncorrected barometer. Local pressure is important if the measurement is obtained from the weather service. The final item required is the nomograph, which provides a simple mechanical calculation of what the orifice meter pressure drop should be for a measured velocity head to maintain isokinetic conditions. The use of this nomograph is explained in detail in Chapter 14.

MAINTENANCE AND CALIBRATION

Maintenance and calibration refers only to the work which must be done in the laboratory. The performance of the orifice meter is of primary importance. This is critical to the maintenance of isokinetic conditions, and its calibration should not be ignored. Other critical pieces of equipment are the pitot tube and dry gas meter. Calibration requirements are discussed in the EPA maintenance procedures.[13] Routine maintenance and calibration procedures are outlined in Table 10.1. With several commercial trains, minor additions may be required according to the manufacturer's instructions.

One point should be made about the calibration of the dry gas meter and the orifice. The procedure calls for the use of a wet test meter, which must have the capacity to handle accurately the flow rate required for the calibration. The minimum and maximum flow capacity is usually supplied with each wet test meter.

TEST PROCEDURES

Equipment Set-Up

Once the equipment has been transported to the plant gate the actual field test procedure begins. Generally speaking, the hardest, that is the most physically demanding part of a particulate source test, is getting the equipment from the plant gate to the sampling location. Transporting 200-300 pounds of sampling gear in awkward sizes and shapes from ground level to half way up a stack on top of a seven story boiler house (possibly with no elevator) is no easy task. Since a field sampling team for conducting particulate source tests is usually comprised of only two men, an engineer and a technician, it is advisable to get help at the plant to transport the sampling equipment. It should be obvious at the time of the pre-test plant survey whether help will be necessary for equipment transportation, and that is the time to mention the problem to the plant contact. Given some advance notice, the plant can probably provide the personnel and equipment, such as winches and forklifts, that could make a very difficult job a lot easier.

Once at the sampling location the equipment can be unpacked and assembled for sampling. This includes installation of the rail(s) on which the sample box will ride. If the duct is circular (and vertical), the use of two rails will simplify the sampling. If the duct is rectangular, provisions should be made for the simple movement of the rails as the traverse lines are changed. If these preparations are not done carefully and with some thought, the

Table 10.1

Routine Maintenance and Calibration of the Method 5 Particulate Train

Train System	*System Components*	*Maintenance or Calibration*	*Standard*
Vacuum system	1. pump	1. vacuum check (27-30 in. Hg)	1. vacuum gauge
	2. oilers	2. clean and refill	2. lightweight oil
	3. valves	3. vacuum check (29 in. Hg)	3. vacuum gauge
Gas metering system	1. dry gas meter	1. calibrate (±2%)	1. wet test meter
	2. thermometers	2. calibrate (±5°F)	2. Hg thermometer
	3. orifice meter	3. calibrate, $\Delta m \pm 1.5$-2.0 (±.15)	3. wet test meter
	4. manometer	4. clean and refill	
	5. barometer	5. calibrate (±0.1 in. Hg)	5. Hg-barometer
Pitot system	1. pitot S-type	1. calibrate (±5% of range)	1. standard pitot
	2. pitot lines	2. clear and leak check	2. vacuum gauge
	3. manometer	3. clean and refill	
Probe	1. probe	1. clean and leak check (27-30 in. Hg)	1. vacuum gauge
	2. probe heater	2. check continuity and temperature (250°F)	2. ohmeter; T-C
	3. nozzles	3. measure ID	3. micrometer and calipers
	4. thermocouple	4. calibrate (±5°F)	4. Hg thermometer
Sampling liners and glassware	1. filter holder	1. clean and leak check (27-30 in. Hg)	1. vacuum gauge
	2. filter box heater	2. check continuity and temperature (250°)	2. ohmeter; T-C
	3. impingers & connectors	3. clean	
	4. ice bath	4. water leak check	
	5. umbilical cord	5. continuity and leak check	5. ohmeter, vac. gauge

support rail(s) will become a point of difficulty and will extend the sampling time significantly. During the equipment set up period, one member of the sampling team should make the necessary electrical connections, locate an ice supply, and make explanations to the process operating personnel. Provisions for the recording of important process data should have been made at the time of the pre-test plant visit and they should be reemphasized at this point. Throughout this time the other member of the sampling team should be unpacking the rest of the sampling equipment and assembling the train for sampling. This includes adding the required distilled water and silica gel and plugging in the probe and filter box heater to let them warm up to the sampling temperature (250°F). If a second train is to be used to start another sample immediately after the first, it can be prepared during the first test. It should be noted that Method 5 calls for three repetitions of the particulate sample to constitute a source test.

Preliminary Measurements

Duct measurements and traverse points from the pre-survey should be rechecked. Entrance port measurements should be taken so that the probe or rail can be marked for proper location in the duct. Experience has shown that heat-bonding fiberglass tape is an excellent way to mark probes when T_s is $< 700°F$. The number and location of the traversing points should be checked. Figure 9.1 shows the minimum number of traverse points required by Method 1.[1] Figure 9.2 shows the location of the sampling points in the stack. It should be noted the cross-sectional area of the stack is divided into incremental equal areas with the sampling point located at the centroid of each. Table 9.3 provides the location of traverse points in a circular duct.

An initial pitot traverse is performed. This is not a measurement of stack flow but rather a determination of average, minimum and maximum velocity head, stack temperature, and stack pressure so that the selection of the proper nozzle size can be made. This may be done with a separate pitot tube or with the probe pitot if the sample nozzle has been covered.

The dry molecular weight of the gas stream must be determined. In the case of combustion sources, Method 3 requires the use of an Orsat gas analyzer,[1] the use of which was discussed in Chapter 7. In the case of incinerator testing, integrated Orsat analyses are necessary since the results must be corrected to 12% CO_2. In the case of other tests, an experienced estimate, the use of the nomograph or a simpler commercial method can be used. The error in dry molecular weight is usually a relatively small source of error as discussed in Chapter 8.

The final preliminary information required is the estimation or determination of the moisture content in the stack gases. This may be determined by one of the methods discussed in Chapter 7. EPA Method 4,[1] specified by the *Federal Register*, is the condensation method discussed in that chapter. It should be noted that the moisture content used in final calculations will be determined during the particulate sample. If similar tests have been conducted before, then a reasonable preliminary estimate can be used.

This preliminary information is important for the setting of the nomograph used in adjusting the sampling velocity at the nozzle to isokinetic conditions during sampling. The nomograph and its procedures are described in detail in Chapter 15. All data sheets used during the actual test should be prepared while the preliminary measurements are being taken.

Sampling

The sampling should proceed with the probe nozzle positioned at the first traverse point with the tip pointing directly into the gas stream. The pump is immediately started and flow is adjusted to isokinetic conditions. Samples should be taken at 5-minute intervals at each point. Data should be taken at least at 5-minute intervals and recorded on a data sheet similar to that in Table 10.2 When changes occur in stack conditions, the nomograph will facilitate the quick adjustment of the meter box flow conditions to isokinetic conditions. The flow should be adjusted by opening initially both the main and by-pass valves with the flow being maintained by closing the by-pass valve. This protects the vacuum pump by allowing continual flow through it. An extra filter and holder should be heated so that when the sample filter becomes clogged a minimum amount of time is lost in the filter change. Once the traverse is completed, the sample box is then moved to the next traverse location. A note should be added to record both real and clock time during the sampling steps.

Table 10.2

Particulate Test Data Sheet

General ____	Equipment ____	Conditions ____
Firm ____	Meter No ____	Orifice Δm ____
Source ____	Sample Box No ____	Ass. % H_2O ____ T_s ____
Sample Location ____	Probe No ____ Lt ____ Cp ____	Ass. Δp avg ____ P_s/P_m ____
Tester ____	Nozzle Dia ____	Ass. T_m ____ A_s ____
Date ____	Filter Nos. Silica Gel Nos	Bar. Pressure ____
Test No. ____	____ ____	C Factor ____
Observed by ____	____ ____	Probe Heater Set. ____
		Filter Oven Set. ____

Traverse PT.	Time (min)	Dry gas Meter (cf)	Pitot ΔP ("H O)	Orifice Δm ("H_2O)	Meter inlet (°F)	Temp. outlet (°F)	Stack temp. (°F)	Pump vac ("Hg)	Stack pressure	Remarks*
									Impinger outlet max temp	
									Condensate collected (ml)	
									#1 #2	
									#3	

*Real time, leak check, filter change, observation, process information.

Applicable process information such as control device operating data (*e.g.*, precipitator current) and observations by the sampling team such as Ringlemann numbers and weather conditions should be routinely recorded throughout the test period.

Sample Recovery

At the end of the sampling period care must be taken in shutting down the sampling equipment. No part of the flow train should be disassembled until the probe has been removed from the stack. Disconnecting an umbilical cord or an impinger when the probe side of the filter is under a negative duct pressure will cause undesirable back flow through the filter and in the case of the larger 4- and 5-inch filters rupturing of the filter media may result. The correct shut down procedure would be:

1. close line valve on vacuum pump
2. shut pump off
3. disconnect electricity
4. remove probe from stack
5. break sampling train down.

In particulate sampling the mass collected on the filter and in the probe may be on the order of a few milligrams. Thus the sample recovery should be made in a clean and protected area. Ordinarily a platform, roof, or scaffolding is not an acceptable sample recovery area.

While the probe and filter holder are cooling the volume of water condensed in the impinger train should be measured to the nearest ml, the value recorded on the data sheet, and the water discarded. The silica gel should then be transferred to its original container using a rubber policeman or clean utensil (a long thin screwdriver works well). Since the silica gel is preweighed in 200 gm lots and placed in a container before sampling, returning the next silica gel to its original container for weighing eliminates some material transfer errors. Wide-mouthed 500-ml polyethylene sample bottles work well.

After the filter holder has cooled to an acceptable handling temperature, the filter should be carefully removed using tweezers. The removal should take place in a clean area with little air movement, and it is also good practice to remove the filter over a large sheet of nonporous glossy paper. In the event of any spillage, most of the sample will be recoverable. When removed, the filter media should be placed immediately in a suitable sample container. Clear plastic petri dishes serve well for this purpose when taped securely closed. Paper and cardboard containers should be avoided at all costs. The porosity of such containers makes them unacceptable. It must be remembered that the filter may catch only a few milligrams of loosely packed dust.

Finally the particulate must be recovered quantitatively from all sampling train surfaces that were exposed to the sample stream prior to the fiberglass filter media. This usually includes the nozzle, probe, connecting glassware, and upstream side of the filter holder. The particulate catch from these train components should be placed in the same sample bottle. The particulate catch on the filter holder, connecting glassware, and probe nozzle can usually be removed and collected using a razor blade (or clean penknife) and liberal acetone washings. The length of the sampling line in the probe however calls for another procedure. The authors have had success cleaning probes by first rinsing with acetone to recover all large particulate. The probe can then be thoroughly cleaned with preweighed acetone-soaked cotton gun cleaning patches and a gun cleaning rod. Three or four patches are usually sufficient. When using acetone to recover particulate samples quantitively three important things to remember are:

1. acetone is extremely flammable, and the vapors are toxic and explosive.
2. acetone may contain dissolved impurities and reagent blanks should be taken.
3. acetone is a common organic solvent and may dissolve some equipment made from synthetic material, *i.e.*, sòme plastic sample bottles.

All acetone used should be analytical grade and reagent blanks taken of the actual acetone used in the field. A reagent blank is a sample (100 ml is usually sufficient) taken to determine if the reagent itself has any impurities that will affect test results adversely (*e.g.*, dissolved solids). Acetone blanks are evaporated and the mg of particulate/ml determined so that the particulate mass collected in the acetone washings may be corrected for any acetone impurity.

All sample bottles and petri dishes should be labeled with the source, date, time, test number, sampling procedure and contents. This procedure will prove invaluable both in the lab and in the field. If the sample bottles are identified by number only, considerable time can be lost in the field, counting and smelling samples to determine whether all samples taken can be accounted for.

ANALYTICAL PROCEDURES

Silica Gel

The spent silica gel is weighed to the nearest gram. The sum of the grams of water collected by the silica gel and the milliliters of water condensed in the impinger train (1 ml H_2O = 1 gram) is the total water fraction of the sampled gas.

$$V_{W_{STD}} = (0.0474 \frac{cu.ft.}{ml})\ V_{l_c} \qquad (10\text{-}1)$$

where

$V_{W_{STD}}$ is volume of water vapor (std. cu. ft.) in the gas sample at standard conditions T = 530°R P = 29.92 in. Hg

V_{l_c} is total volume of water collected in impingers and silica gel.

Two pan balances accurate to 0.5 gram and a 250-ml graduated cylinder are sufficient for this moisture determination.

Particulate Mass

The filter and any loose particulate matter are transferred from the sample container to a tared glass weighing dish. The sample is then desiccated and dried to a constant weight ($\pm$ 0.2 mg), and the results are reported to the nearest 0.5 mg of particulate.

The acetone washings are thoroughly mixed then transferred to a tared beaker and evaporated to dryness at ambient temperature and pressure. After drying they are desiccated to a constant weight ($\pm$ 0.2 mg) and weighed to the nearest 0.5 mg. The acetone blank is handled in the same manner. The blank correction is made by ratioing the volumes of the sample and the blank. Particulate concentration is calculated as follows:

$$C_s' = \frac{m_n}{V_{m_{std}}} \qquad (10\text{-}2)$$

where

C'_s is concentration of particulate matter in stack gas. grain/std. cu. ft. dry

m_n is total amount of particulate matter collected, mg, filter media and acetone washings

$V_{m_{std}}$ is volume of gas sample through train at standard conditions, cu. ft., dry.

A desiccator and an analytical balance with an accuracy of $\pm$ 0.2 mg are needed to conduct this analysis properly.

RESULTS

Data Handling

To collect the quantity of data needed during a particulate source test is a rather extensive task. The parameters of velocity pressure, orifice Δm, inlet and outlet meter temperature, meter volume, sample time, and usually stack temperature are needed for every traverse point. Tests involving 48 traverse points are not uncommon. Auxiliary data such as probe and sample box temperature, system vacuum, nozzle diameter, pitot factor, and date and time of test are also needed in the determination of particulate source emissions. Well-developed data sheets are needed so that the quantities of data taken can be clearly understood, and it goes without saying that it would be a difficult task to conduct an acceptable particulate source test using the back of an envelope for a data sheet. A typical data sheet is shown in Table 10.2. Computer coded field data sheets can also be used. In this case the data can be key-punched directly from the field data sheets. The use of a computer program to reduce stack sampling data and perform calculations has several advantages over manual calculation methods:

1. speed
2. consistency because everybody does it the same way
3. cost because printed tables can be used thus cutting typing time appreciably.

Computations

The computations required are presented in the order in which they should be handled. The initial calculation is that of the dry sample volume shown in Equation 10-3:

$$V_{m_{std}} = V_m \frac{T_{std}}{T_m} \frac{P_{bar} + \frac{\Delta m}{13.6}}{P_{std}}$$

$$= 17.71 \frac{°R}{in. Hg} (V_m) \frac{P_{bar} + \frac{\Delta m}{13.6}}{T_m} \quad (10\text{-}3)$$

where

- $V_{m_{std}}$ is the volume of gas sample through the dry gas meter (standard conditions), cu ft
- V_m is the volume of gas sample through the dry gas meter (meter conditions), cu ft
- T_{std} is the absolute temperature at standard conditions, 530°R
- T_m is the average dry gas meter temperature, °R
- P_{bar} is the barometric pressure at the orifice meter, inches Hg
- Δm is the average pressure drop across the orifice meter, inches H_2O
- 13.6 is the specific gravity of mercury
- P_{std} is the absolute pressure at standard conditions, 29.92 inches Hg.

The second calculation is the volume of water vapor collected during the sample.

$$V_{W_{std}} = V_{1_c} \frac{\rho\ H_2O}{M\ H_2O} \frac{RT_{std}}{P_{std}} = 0.0474 \frac{cu\ ft}{ml} (V_{1_c}) \quad (10\text{-}4)$$

where

- $V_{W_{std}}$ is the volume of water vapor in the gas sample (standard conditions), cu ft
- V_1 is the total volume of liquid collected in impingers and silica gel, ml
- $\rho\ H_2O$ is the density of water, 1 g/ml
- $M\ H_2O$ is the molecular weight of water, 18 lb/lb_m-mole
- R is the ideal gas constant, 21.83 in. Hg-cu ft/lb_m-mole-R.

The moisture content of the stack gas may then be calculated:

$$B_{wo} = \frac{V_{w\ std}}{V_{m\ std} + V_{w\ std}} \quad (10\text{-}5)$$

where

- B_{wo} is the proportion by volume of water vapor in the gas stream.

The next parameter to be calculated is the dry molecular weight of the stack gases. This follows from Equation 6-14.

$$M_d = 0.44\ (\%\ CO_2) + 0.32\ (\%\ O_2) + 0.28\ (\%\ N_2 + \%\ CO) \qquad (10\text{-}6)$$

where

- M_d is the dry molecular weight, lb_m/lb_m-mole
- $\%\ CO_2$ is the per cent carbon dioxide by volume, dry basis
- $\%\ O_2$ is the per cent oxygen by volume, dry basis
- $\%\ N_2$ is the per cent nitrogen by volume, dry basis
- $\%\ CO$ is the per cent carbon monoxide by volume, dry basis
- 0.44 is the molecular weight of carbon dioxide divided by 100
- 0.32 is the molecular weight of oxygen divided by 100
- 0.28 is the molecular weight of nitrogen and carbon monoxide divided by 100.

The stack gas molecular weight is then calculated by the following formula:

$$M_s = M_d\ (1 - B_{wo}) + 18\ B_{wo} \qquad (10\text{-}7)$$

where

- M_s is the molecular weight of stack gas (wet basis), lb_m/lb_m-mole.

These calculations lead to the calculation of the average stack velocity. This follows from Equation 7-1.

$$(V_s)_{avg.} = K_p C_p\ (\Delta P)^{1/2}_{avg} \left(\frac{(T_s)_{avg}}{P_s\ M_s}\right)^{1/2} \qquad (10\text{-}8)$$

where

- $(V_s)_{avg}$ is the stack gas velocity, ft per second (fps)
- K_p is the 85.48 $\frac{ft}{sec}\left(\frac{lb_m}{lb_m\text{-mole-}°R}\right)^{1/2}$
- C_p is the pitot tube coefficient
- $(T_s)_{avg}$ is the average absolute stack gas temperature, °R

$(\Delta P)^{1/2}_{avg}$ is the average square root of velocity head of stack gas, in. H_2O

P_S is the absolute stack gas pressure, in. Hg.

The stack flow rate is then calculated by the following equation:

$$Q_S = 3600\ (1-B_{wo})\ (V_S)_{avg}\ A \frac{T_{std}}{(T_S)_{avg}}\ \frac{P_S}{P_{std}} \qquad (10\text{-}9)$$

where

Q_S is the stack volumetric flow rate, dry basis, standard conditions, cu ft/hr

A is the cross-sectional area of stack, ft^2.

The particulate concentration is then calculated

$$\overline{C}_S = \frac{\left(\frac{1}{453{,}600}\ \frac{lb_m}{mg}\right) m_n}{V_{m\ std}} = 2.205 \times 10^{-6}\ \frac{m_n}{V_{m\ std}} \qquad (10\text{-}10)$$

where

$\overline{C}_S$ is the concentration of particulate matter in stack gas, lb_m/scf, dry basis

453,600 is the mg/lb_m

m_n is the total amount of particulate matter collected, mg.

The emission rate may be calculated.

$$\overline{PMR}_{s_c} = Q_S\overline{C}_S \qquad (10\text{-}11)$$

where

$\overline{PMR}_{s_c}$ is the particulate mass emission rate, lb_m/hr.

This number would then be used to compare with the emission standard or regulations. In the case of the new federal emission standards for steam generation, the emission rate, PMR, would be divided by the heat input in 10^6 BTU/hr.

The average isokinetic variation of each sample can be calculated by the following equation:

$$I_{avg} = \frac{(T_S)_{avg}\left[\frac{V_{l_c}(\rho_{H_2O})R}{M_{H_2O}} + \frac{V_m}{T_m}\left(P_{bar} + \frac{\Delta m}{13.6}\right)\right]}{\theta\,(V_S)_{avg}\,P_S\,A_n} \times 100$$

$$= \frac{1.667\ \frac{min}{sec}\ (T_S)_{avg}}{} \left[\frac{\left(0.00267\frac{in\ Hg - cu\ ft}{ml - {}^\circ R}\right) V_{l_c}}{\theta\,(V_S)_{avg}\,P_S\,A_n}\right] +$$

$$\left[\frac{\frac{V_m}{T_m}\left(P_{bar} + \frac{\Delta m}{13.6}\right)}{\theta\,(V_S)_{avg}\,P_S\,A_n}\right] \qquad (10\text{-}12)$$

where

- I_{avg} is the per cent of isokinetic sampling
- θ is the total sampling time, min
- A_n is the cross-sectional area of nozzle, ft^2.

As discussed in Chapter 8, the isokinetic variation of each point within the traverse is more meaningful than I_{avg}. It can be calculated in the following manner. The isokinetic variation will be calculated as a ratio of nozzle velocity to stack gas velocity at each point. The stack gas velocity is calculated as follows:

$$(V_S)_{pt} = K_p C_p \left(\frac{T_S\ \Delta P}{P_S\ M_S}\right)^{1/2} \qquad (10\text{-}13)$$

where

- $(V_S)_{pt}$ is the stack gas velocity at traverse point, f.p.s.
- T_S is the absolute stack gas temperature at traverse point, °R
- ΔP is the velocity pressure at traverse point, in H_2O.

The nozzle velocity is calculated:

$$(V_n)_{pt} = \frac{\Delta V}{\Delta \theta} \frac{T_S}{T_{m_{pt}}} \frac{(P_{bar} + \frac{\Delta m_{pt}}{13.6})}{P_s A_n (1-B_{wo})} \qquad (10\text{-}14)$$

where

$(V_n)_{pt}$ is the nozzle velocity at traverse point, fps

ΔV is the volume sampled at traverse point, dry basis, meter conditions, cu ft

$\Delta\theta$ is the sampling time at traverse point, sec

$T_{m_{pt}}$ is the absolute meter temperature while at traverse point

Δm_{pt} is the pressure drop across the orifice meter while at traverse point, in H_2O.

The ratio then becomes:

$$I_{pt} = ((V_n)_{pt}/(V_s)_{pt}) \times 100 \qquad (10\text{-}15)$$

where

I_{pt} is the per cent of isokinetic sampling at traverse point.

These computations have been presented in the order in which they must be performed to achieve the needed information. The only exception is the concentration, which may be calculated after the sample volume calculation is complete. It should be noted that the calculations are quite extensive, leading to the conclusion that considerable care must be taken in making them. If computer facilities are available, they should be used. They will add both speed and accuracy once a program has been developed.

SUMMARY

Conducting an isokinetic source test for particulate matter is not a simple task. The test cannot be properly performed without adequate preparation. A minimum of two people are required to conduct even the most routine isokinetic particulate tests. Usually an engineer and a technician are required. The team members must be familiar with the process being tested, the operation of the source testing equipment, and the applicable emission regulations. At least one member of the source test team must thoroughly understand the analytical procedures which will be employed after the sample collection.

Many particulate source sampling systems are commercially available. These systems are discussed in detail in Chapter 16 of this text. As discussed in Chapter 8, the pitot tube is the largest source of possible error in determining particulate mass emission rates. When emission regulations are written in terms of normalized emission rates (such as lb of particulate/10^6 BTU and lb of particulate/ton of process input) the determination of process input or output rates is also a large source of error.

REFERENCES

1. "Standards of Performance for New Stationary Sources," Federal Register, Vol 36, No. 247, December 23, 1971.
2. Specifications for Incinerator Testing at Federal Facilities, PHS, NCAPC, 1967.
3. "Proposed Standards of Performance for New Stationary Sources," Federal Register, Vol 36, No. 159, August 17, 1971.
4. Hemeon, W. C. L., and Black, A. W. "Stack Dust Sampling: In-Stack Filter or E.P.A. Train," Journal of the Air Pollution Control Association, Vol 22, No. 7, July, 1972.
5. Govan, F. A., Terracciano, L. A., and Rom, J. "Source Testing of Utility Boilers for Particulate and Gaseous Emissions," Presented at the 65th Annual Meeting of the Air Pollution Control Association, Miami Beach, Florida, June, 1972.
6. Achinger, W. C., and Daniels, L. E. "An Evaluation of Seven Incinerators," Proceedings of the 1970 National Incinerator Conference, ASME, New York, New York, May, 1970.
7. Smith, W. C., et al. "Stack Gas Sampling Improved and Simplified with New Equipment," Presented at the 60th Annual Meeting of the Air Pollution Control Association, Cleveland, Ohio, June, 1967.
8. Smith, W. S., Shigehera, R. T., and Todd, W. F. "A Method of Interpreting Stack Sampling Data," Presented at the 63rd Annual Meeting of the Air Pollution Control Association, St. Louis, Missouri, June, 1970.
9. Crandall, W. A. "Determining Concentration and Nature of Particulate Matter in Stack Gases," Publication No. 71-WA/PTC-8, ASME. New York, New York, September, 1972.
10. Air Pollution Source Testing Manual. Los Angeles County Air Pollution Control District, Los Angeles, California, November, 1965.
11. "Determining Dust Concentration in a Gas Stream," Performance Test Code 27-1957 American Society of Mechanical Engineers, New York, New York.

12. Martin, R. M. "Construction Details of Isokinetic Source Sampling Equipment," Publ. No. APTD-0576, Air Pollution Control Office, EPA, Research Triangle Park, North Carolina, 1971.
13. Rom, J. J. "Maintenance, Calibration, and Operation of Isokinetic Source Sampling Equipment," Publ. No. APTD-0576, Office of Air Programs, EPA, Research Triangle Park, North Carolina, 1972.

CHAPTER 11

DETERMINATION OF SULFUR DIOXIDE EMISSIONS

Many industrial processes, such as smelters, sulfuric acid plants and fossil fuel combustion operations, emit sulfur dioxide as an air pollutant. This chapter presents the federal EPA Method 6 sampling and analysis procedure.[1] It is specifically required for testing of new fossil fuel power plants but could be used for most any source which contained sulfur dioxide in the waste gas. Several other references are of interest regarding this test.[2-4]

BASIC METHOD

A gas sample is extracted by proportional sampling techniques. The basic problem is to detect SO_2 in the presence of SO_3 and H_2SO_4. Therefore the latter components first must be removed from the sample. The sulfur dioxide is then absorbed in a hydrogen peroxide solution and eventually is determined quantitatively by titration with barium perchlorate or barium chloride. Thorin is used as a color indicator for this titration. This wet chemical method is similar to that developed by Shell Development Corporation,[2] the major differences being found in the sampling train construction. Metal sulfates, phosphates, sulfuric acid mist, and cations which complex with the thorin indicator or coprecipitate with barium, interfere but can be eliminated by proper use of a filter.

SAMPLING TRAIN

The sampling train is shown in Figure 11.1. Quartz or Pyrex wool is placed in the end of the probe to prevent particulate matter from entering the scrubbers. The particulate matter contains

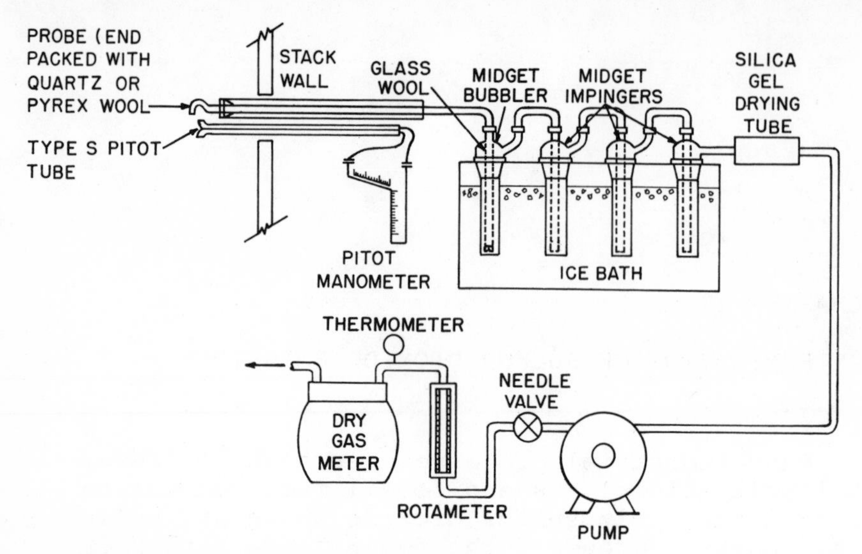

Figure 11.1. Sulfur dioxide sampling train.

sulfates and also other impurities which may precipitate with barium or complex with the thorin indicator.

Acid mist is also removed at this point. The sampling probe, Pyrex or quartz, is heated to prevent condensation prior to sample collection. Matty and Diehl[3] indicate that the probe temperature should be held at 600-700°F. This is above the SO_3 dew point yet seemingly below the temperature at which the glass probe would catalyze the SO_2 oxidation reaction. A thermometer or thermocouple may be placed inside the heater tape to determine the approximate probe temperature.

The sample collector consists of one midget bubbler and three midget impingers placed in an ice bath. The midget bubbler contains an 80% isopropyl alcohol solution. This removes the SO_3 and any carry over H_2SO_4 from the filter. Some glass wool is placed at the top of the midget bubbler to act as a filter and thus preventing any H_2SO_4 mist from carrying over into the following midget impingers. The gas stream becomes saturated with the isopropyl alcohol vapor, which inhibits the oxidation of SO_2 to SO_3 because the alcohol is more readily oxidized.

The SO_2 is removed in the first two midget impingers, which contain hydrogen peroxide solutions. The final impinger is left dry to act as a dry trap and catch any carry over. The SO_2 is absorbed in the inpingers and converted to form H_2SO_4.

Finally is a silica gel drying column to protect the components which follow it. The pump must be of a leakless type because it precedes the rotameter and dry gas meter.

TEST APPARATUS

The following items, categorized into sampling equipment, glassware, and reagents, are necessary for the test. Table 11.1 shows a check list which is helpful in preparing and packing equipment for field tests.

Sampling Equipment

1. Pyrex glass probe (5 to 6 mm ID) with heating system to prevent condensation; a heating tape and a variac work adequately for this
2. Pyrex or quartz wool to act as a filter
3. glass wool for outlet of midget bubbler to act as a filter
4. one midget bubbler
5. three midget impingers
6. silica gel (6 to 16 mesh color indicating type) to dry gas sample
7. drying tube to hold the silica gel, although another impinger may be used
8. needle valve for controlling gas flow rate
9. leakless pump
10. rotameter with 0-10 scfh flow range
11. dry gas meter; the type which indicates 0.01 cf per revolution is preferred but one with 0.1 cf per revolution is acceptable
12. thermometer; the dial gauge type is commonly used
13. pitot tube (S-type) apparatus (same type as for Method 2 in Chapter 7) for determining proportional sampling rates
14. ice bath and ice
15. tubing and fittings to connect various sampling train components
16. mercury manometer for performing leak tests on the sampling train.

Glassware

1. glass wash bottle to hold the deionized water
2. polyethylene storage bottles for storing impinger samples prior to analysis

Table 11.1

Equipment and Chemical Check List
*Gas Sampling Tests**

Client ______________________ Contract No. ________
Test date _______________ Requested by ___________
Date requested ____________ Date needed ___________
Expected return date ____________ Van or plane? ________
Type of tests ________________

Need	Packed	
____	______	Probe, type______ length______ft
____	______	Probe, support, type __________
____	______	Pump, type __________________
____	______	Dry gas meter ___________cf/rev
____	______	Rotameter, range ____________
____	______	Midget bubbler
____	______	Midget impinger, taper tip
____	______	Impinger ice bath
____	______	Impinger transport box
____	______	Impinger connectors and clamps
____	______ ft	Tygon tubing_____" ID _____" OD
____	______ ft	Teflon tubing ____" ID ____" OD
____	______ ft	Polyethylene tubing ____" ID ____" OD
____	______	Silica gel
____	______	Pump oil
		Chemicals:
____	______	____gm(ml)____
____	______	____gm(ml)____
____	______	____gm(ml)____
____	______	____gm(ml)____
____	______	Distilled water
____	______	Squeeze bottles
____	______	Graduated cylinders
____	______	Tool box
____	______	Glass wool
____	______	Duct putty
____	______	Sample bottles, type __________
____	______	Pitot/thermocouple __________ft
____	______	Potentiometer and thermocouple connections
____	______	Manometer and tubing and extra fluid

*Courtesy of the Research Corporation of New England, Wethersfield, Connecticut.

3. transfer pipettes: 5 ml and 10 ml sizes with 0.1 ml divisions and 25 ml size with 0.2 ml divisions
4. volumetric flasks: 50 ml, 100 ml and 1000 ml
5. burettes: 5 ml and 50 ml
6. Erlenmeyer flask: 125 ml
7. dropping bottle for indicator solution.

Reagents

1. deionized distilled water
2. hydrogen peroxide, 3% solution. Prepare by diluting 100 ml of 30% hydrogen peroxide to 1 liter with deionized water. Caution: the 30% H_2O_2 is a strong skin irritant which is not noticeable until a few minutes after the injury occurs. This solution must be prepared daily.
3. isopropyl alcohol, 80% solution. Prepare by mixing 80 ml of isopropyl alcohol with 20 ml of deionized water. This solution is stable.
4. thorin indicator: [1-(0-arsonophenylazo)-2-naphthol-3,6-disulfonic acid, disodium salt (or equivalent)]. Prepare by dissolving 0.2 grams in 100 ml of deionized distilled water. It should be stored in a polyethylene container since it tends to deteriorate if stored in glass.
5. barium perchlorate. 0.01N. Prepare by dissolving 1.95 grams of barium perchlorate, $Ba(ClO_4)_2 \cdot 3H_2O$ in 200 ml of deionized water. Then dilute to 1 liter using isopropyl alcohol. Barium chloride may be used but of course a different weight would have to be added. Dissolve 1.22 grams of barium chloride, $BaCl_2 \cdot 2H_2O$, in 200 ml deionized water and then dilute to 1 liter with the isopropyl alcohol. The normality of this solution should be standardized with sulfuric acid.
6. Sulfuric acid standard, 0.01N. Purchase or standardize to $\pm$ 0.0002 N against 0.01 NaOH which has previously been standardized against primary standard grade potassium acid phthalate.

LABORATORY PROCEDURES

All components must meet the requirements presented in Chapter 4. Key components should be calibrated to eliminate systematic errors, and the sampling train should be assembled and leak tested. This is done by plugging the probe inlet and pulling 10 inches Hg vacuum. Observing the dry gas meter, the leakage rate should not exceed 1% of the desired sampling rate. If the leakage is severe, each joint of the train should be inspected and the leak test performed again. A thin film of silicone grease often helps to make the joints leak tight.

FIELD PROCEDURES

The sampling site for determination of SO_2 must be the same as for the volumetric flow rate determination. The sample point should be the centroid of the cross section of the stack if the area is less than 50 ft^2 or no closer than 3 feet to the wall if the area is greater than 50 ft^2. Minimum sample time is 20 minutes, and the minimum volume collected 0.75 ft^3, corrected to standard conditions. Two samples make up each repetition, taken 1.0 hour apart, and there should be 3.0 repetitions for each stack.

The train should be assembled as in Figure 11.1. Place 15 ml of 80% isopropanol in the midget bubbler and 15 ml of 3% hydrogen peroxide in the first two midget impingers. The remaining impinger will remain dry. Place these in the ice bath with water to keep the temperature of the gases coming out of the last impinger at or below 70°F. The system should be leak checked prior to performing the test.

Once the probe has been heated to the desired temperature, the sample flow rate should be adjusted to be proportional to the stack gas velocity. The method outlined in Chapter 7 for determining moisture content also applies here. Sampling begins by positioning the probe at the desired sampling point and starting the pump. If the probe has a stainless steel sheath, it may be clamped into place. Proportional sampling should be carried out throughout the run. Record readings at least every five minutes and make necessary flow rate adjustments. Table 11.2 shows the data sheet which should be used. Although not specified, a reasonable minimum total gas sample volume is 2 cubic feet.

Table 11.2

SO_2 Test Data Sheet

General	Equipment	Conditions
Firm ______	Meter no. ______	Barometric press ______
Plant location ______	Sample box no. ______	Probe temp ______
Source ______	Probe no. ______ Lt ______	Ambient temp ______
Sample location ______	Filter used ______	Stack area ______
Tester ______	Cont. no. ______	Ave $\sqrt{\Delta P}$ ______
Date ______	Pitot no. ______ Cp ______	T_s Ave ______
Test no. ______		% H_2O ______
Observed by ______		P_s Ave ______

Sample No.	Time Min.	Meter Vol. cf.	ΔP in. H_2O	$\sqrt{\Delta P}$	Rotameter Flow cfh	Meter Temp °F	Sample Bottle No.	Remarks
1								Start
								End
2								Start
								End

At the conclusion of the run make final readings and then shut down the pump. Disconnect the train from the probe and purge the system with clean, ambient air for at least 15 minutes. The purging with SO_2-free ambient air will degassify the water portion of the 80% isopropanol, sending any dissolved SO_2 onto the impingers containing the hydrogen peroxide solution.

Disconnect the impingers after purging and discard the contents of the bubbler. It should be noted here that the SO_3 content of the gas sample could be determined by analyzing the solution from the midget bubbler. This analysis would of course assume that all of the acid mist was removed by the quartz wool filter and that no SO_3 was lost on the wool or the glass probe.

The contents of the two impingers are placed in a polyethylene sample bottle. The three impingers and the connecting tubing are washed with deionized, distilled water and the washings added to the sample bottle. Do not use excessive quantities of wash water; the total volume should be less than 50 ml. Label the sample bottle properly.

ANALYTICAL PROCEDURES

The SO_2 has been oxidized to SO_3 with H_2O_2 and hydrolized to form sulfuric acid in the impingers. The sulfuric acid is titrated against standardized barium usually as the perchlorate or chloride. Thorin is used, which acts to complex, and excess barium is added, thus giving a color indicator.

Transfer the sample to a 50-ml volumetric flask and dilute to the mark with distilled, deionized water. Pipette a 10 ml portion (after mixing) to a 125-ml Erlenmeyer flask and add 40 ml of isopropanol and 2 to 4 drops of the Thorin indicator solution. The solution will then have a yellow-orange color. Titrate to a pink end point with the 0.01 N barium perchlorate and record results.

During the titration the barium is consumed by the sulfate and forms, in the presence of isopropyl alcohol, a gelatinous type of precipitate which equilibrates rapidly. At the first presence of any excess barium, a pink barium-thorin complex forms indicating that all of the sulfate has been consumed. The determination of the end point by the color change may take some practice on the part of the analyst; it is less vivid than some other colorimetric indicators but nonetheless proves to be quite adequate.

The standard laboratory procedure of using a blank of deionized water should be used for all sets of samples, allowing any systematic error due to the reagents or procedures to be accounted for.

CALCULATIONS

The following expression (ideal gas law) is used to calculate the volume of dry gas collected under standard conditions:

$$V_{m_{std}} = V_m \left(\frac{T_{std}}{T_m}\right) \left(\frac{P_{bar}}{P_{std}}\right) \qquad (11\text{-}1)$$

where

- $V_{m_{std}}$ is the volume of gas sample passed through the dry gas meter under standard condition, ft^3
- V_m is the volume of gas sample passed through the dry gas meter under meter conditions, ft^3
- T_{std} is the absolute gas temperature at standard conditions, 530°R
- T_m is the average dry gas meter temperature, °R
- P_{bar} is the absolute pressure at the gas meter
- P_{std} is the absolute pressure at standard conditions, 29.92 in. Hg.

The average concentration of SO is determined using the following expression:

$$C_{SO_2} = [(7.05 \times 10^{-5}\ \frac{lb\text{-}l}{g\text{-}ml})]\ \frac{(V_t - V_{tb})\ N \left(\frac{V_{soln}}{V_a}\right)}{V_{m_{std}}} \qquad (11\text{-}2)$$

where

- C_{SO_2} is the concentration of SO_2 at standard conditions, dry basis, lb/ft^3
- 7.05×10^{-5} is a conversion factor including the number of grams per gram equivalent of sulfur dioxide (32g/g-eq), 453.6 g/lb, and 1000 ml/l, lb-l/g-ml.
- V_t is the volume of barium perchlorate titrant used for the sample, ml
- V_{tb} is the volume of barium perchlorate titrant used for the blank, ml
- N is the normality of barium perchlorate titrant, g-eq/l

V_{soln} is the total solution volume of sulfur dioxide sample, 50 ml

V_a is the volume of sample aliquot titrated, ml.

SUMMARY

The sampling procedure for SO_2 requires constant attention and must be done in a proportional fashion or the sample is not representative.

Laboratory analysis is completely wet chemical and several hazardous solutions are used. For example extreme care must be used when working with the 30% hydrogen peroxide solution. This strong oxidant is a harsh skin irritant, the effect of which is unnoticed until about 5 minutes after exposure.

REFERENCES

1. "Standards for Performance of New Stationary Sources," Federal Register, Vol 36, No. 247, December 23, 1971.
2. "Atmospheric Emissions from Sulfuric Acid Manufacturing Processes," PHS Publ. No. 999-AP-13, U.S. Public Health Service, Cincinnati, Ohio, 1965.
3. Matty, R. E., and Diehl, E. K. "Measuring Flue-Gas SO_2 and SO_3," Power 101, 94-97, November, 1957.
4. Corbett, P. F. "The Determination of SO_2 and SO_3 in Flue Gases," Journal of the Institute of Fuel 24, 237-243, 1961.

CHAPTER 12

DETERMINATION OF SULFURIC ACID MIST EMISSIONS

The previous chapter discussed the determination of SO_2 concentrations; this chapter outlines an alternative method which also allows the determination of sulfuric acid mist. This procedure has been specified by EPA as Method 8 for specific use in determining emissions from sulfuric acid plants.[1] References 2, 3 and 4 provide background information concerning this procedure.

BASIC METHOD

A gas sample is taken isokinetically from the stack, and the acid mist including sulfur trioxide is separated from the sulfur dioxide. Both fractions are then analyzed separately by the barium-thorin titration method presented in Chapter 11. This method should not be used on any source which has a high particulate loading which could cause interferences in the wet chemical analysis.

SAMPLING TRAIN

The sampling train is shown in Figure 12.1. The sample is removed from the stack by a stainless steel nozzle and then passed through a heated Pyrex glass probe. The acid mist and SO_3 are removed effectively in the first Greenburg-Smith impinger which contains an 80% isopropyl alcohol solution. The alcohol vapor inhibits SO_2 oxidation, and the filter removes any acid mist carry-over to the following impingers. The second and third impingers contain hydrogen peroxide solutions which absorb the SO_2 and convert it to sulfuric acid. The fourth impinger is filled with silica gel to dry the gas prior to passing through the remaining sampling

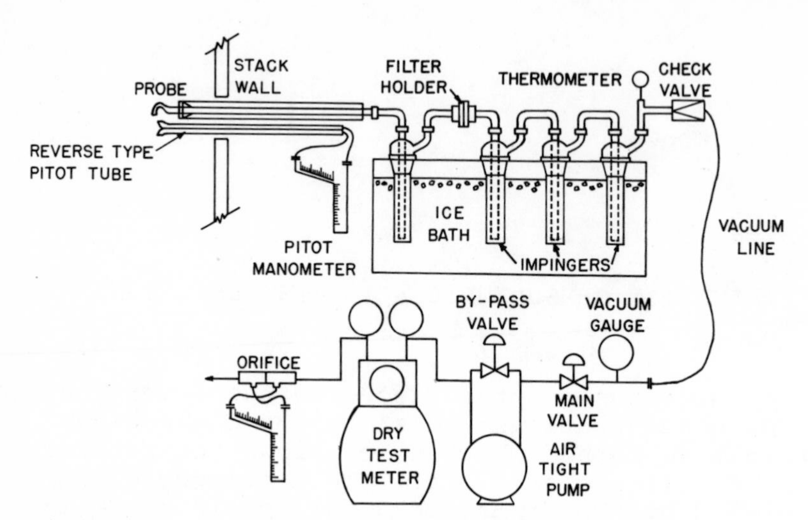

Figure 12.1. Sulfuric acid mist sampling train.

train components. The remaining components are identical to those used for particulate sampling and are presented in detail in Chapter 10.

TEST APPARATUS

The following items are necessary to conduct the test, and are categorized into sampling equipment, glassware, and reagents.

Sampling Equipment

1. type 316 stainless steel nozzle
2. Pyrex glass probe with heating system to prevent condensation during sampling
3. pitot tube apparatus (S-type). Used to determine isokinetic conditions; same as used for particulate sampling
4. Pyrex glass filter holder
5. two Greenburg-Smith impingers
6. two modified Greenburg-Smith impingers; standard tip replaced with 1/2-inch ID glass tube extending to 1/2 inch from the bottom of the impinger flask
7. fiberglass filters: MSA type 1106BH or equivalent, same size filter holder
8. silica gel: 6-16 mesh indicating type (blue to pink)

9. crushed ice and ice bath
10. The remaining components shown in Figure 12.1 are identical to those discussed in Chapter 10.

Glassware

1. clean bottles for holding the deionized water and isopropyl alcohol reagents
2. graduated cylinders: 100 ml, 250 ml, and 500 ml
3. glass sample storage containers
4. pipettes: 25 ml and 100 ml
5. burette: 50 ml
6. Erlenmeyer flask: 250 ml
7. dropping bottle for indicator solution.

Reagents

1. deionized distilled water
2. hydrogen peroxide, 3% solution. Prepare by diluting 100 ml of 30% hydrogen peroxide to 1 liter with deionized water. Caution: the 30% H_2O is a strong skin irritant which is not noticeable until a few minutes after the injury occurs. This solution must be prepared daily.
3. isopropyl alcohol, 80% solution. Prepare by mixing 80 ml of isopropyl alcohol with 20 ml of deionized water. This solution is stable.
4. thorin indicator: [1-(0-arsonophenylazo)-2-naphthol-3,6-disulfonic acid, disodium salt (or equipvalent)]. Prepare by dissolving 0.2 grams in 100 ml of deionized distilled water. It should be stored in a polyethylene container since it tends to deteriorate if stored in glass.
5. barium perchlorate, 0.01N. Prepare by dissolving 1.95 grams of barium perchlorate, $Ba(ClO_4)_2 \cdot 3H_2O$ in 200 ml of deionized water. Then dilute to 1 liter using isopropyl alcohol. Barium chloride may be used but of course a different weight would have to be added. Dissolve 1.22 grams of barium chloride, $BaCl_2 \cdot 2H_2O$, in 200 ml deionized water and then dilute to 1 liter with the isopropyl alcohol. The normality of this solution should be standardized with sulfuric acid.

6. sulfuric acid standard, 0.01N. Purchase or standardize to ± 0.0002 N against 0.01 NaOH which has previously been standardized against primary standard grade potassium acid phthalate.

LABORATORY PROCEDURES

In order to eliminate unnecessary systematic errors, all sampling train components must be calibrated according to the procedures presented in Chapter 4. The sampling train should be leak-tested prior to being taken to the field: assemble the train and plug the inlet to the first impinger, pull a 15 inch Hg vacuum. A leakage rate not to exceed 0.02 cfm is acceptable. If leakage is a problem the fittings should be seated by using a thin film of silicone grease.

FIELD PROCEDURES

After selecting the sampling site and the minimum number of sampling points, the stack pressure, temperature, moisture and range of velocity head must be determined. These determinations should be made according to the methods outlined in Chapters 7 and 9.

Place 100 ml of 80% isopropyl alcohol in the first impinger, 100 ml of 3% hydrogen peroxide in both the second and third impingers and about 200 grams of silica gel in the fourth impinger. Assemble the sampling train without its probe and perform a leak test. If there is no leak, then assemble the train and put ice in the bath around the impingers. Heat up the sampling probes and start the pump to begin sampling. Adjust to isokinetic conditions. Take readings at each sampling point at least every 5 minutes and make necessary adjustments to maintain isokinetic conditions. The nomograph technique presented in Chapter 14 is most expedient and should be used. The minimum total sampling time is 2 hours and the minimum sampling volume is 40 ft^3 corrected to standard conditions. Record all readings on a data sheet such as shown in Table 12.1

At the conclusion of each run turn off the pump and take the final readings. Remove the probe from the stack and disconnect it from the train. Drain the ice bath and purge the remaining part of the train by drawing clean ambient air through the system for 15 minutes. This step is done to strip any

Table 12.1

*Field Data Sheet for Sulfuric Acid Mist Test**

General	Equipment	Conditions
Firm ______	Meter no. ______	Orifice Δm ______
Source ______	Sample box no. ______	Ass. % H_2O ______ T_s ______
Sample location ______	Probe no. ____ Lt ____ Cp ______	Ass. Δp avg ______ P_s/P_m ______
Tester ______	Nozzle dia. ______	Ass. T_m ______ A_s ______
Date ______	Filter nos. ______ Silica gel nos. ______	Bar. pressure ______
Test no. ______	______ ______ ______	C Factor ______
Observed by ______	______ ______ ______	Probe heater set. ______
	Probe wash sample no. ______	Filter oven set. ______

Tra-verse pt.	Time (min)	Dry gas meter (cf)	Pitot ΔP ("H_2O)	Orifice Δm ("H_2O)	Meter inlet (°F)	Temp. Outlet (°F)	Stack temp. (°F)	Pump Vac ("Hg)	Stack pressure	Remarks
									Impinger outlet max temp	
									Sample nos. for impingers	
									#1	
									#2	
									#3	

*Real time, leak check, filter change, observation, process information.

dissolved SO_2 out of the first impinger and collect it in the hydrogen peroxide solution.

Transfer the isopropyl alcohol from the first impinger to a 250 ml graduated cylinder. Use the wash bottle to rinse the probe, first impinger, and all connecting glassware before the filter with 80% isopropyl alcohol. Add the rinse solution to the graduated cylinder and dilute to 250 ml with 80% isopropyl alcohol. Add the filter to the solution, mix, and transfer to a storage container.

Transfer the solution from the second and third impingers to a 500 ml graduated cylinder. Rinse all glassware between the filter and the silica gel impinger with deionized, distilled water and add this rinse water to the cylinder. Dilute to a volume of 500 ml with deionized, distilled water and then put into storage containers.

ANALYTICAL PROCEDURES

The basic analytical procedure is identical to that presented in Chapter 11. Shake the sample container holding the isopropyl alcohol solution and the filter. If the filter breaks up allow fragments to settle for a few minutes before removing a sample. Pipette a 100-ml aliquot of sample into a 250-ml Erlenmeyer flask and add 2 to 4 drops of thorin indicator. Titrate with the barium perchlorate solution to a pink end point. Record titration volume and repeat with a second aliquot of sample.

Shake the container holding the contents of the second and third impingers. Pipette a 25-ml aliquot of sample into a 250-ml Erlenmeyer flask. Add 100 ml of isopropyl alcohol and titrate as before. Repeat process for a second aliquot of sample. Record all results.

Similar titration processes should be carried out on blank samples. This will account for any systematic errors in reagents or procedures.

CALCULATIONS

The following expression is derived from the ideal gas law to determine the volume of dry gas collected under standard conditions:

$$V_{m_{std}} = V_m \frac{T_{std}}{T_m} \left(\frac{P_{bar} + \frac{\Delta m}{13.6}}{P_{std}} \right) \qquad (12\text{-}1)$$

$$V_{m_{std}} = 17.71 \frac{°R}{\text{in Hg}} V_m \left(\frac{P_{bar} + \frac{\Delta m}{13.6}}{T_m} \right) \qquad (12\text{-}2)$$

where

- $V_{m_{std}}$ is the volume of gas sample through the dry gas meter (standard conditions) ft^3
- V_m is the volume of gas sample through the dry gas meter (meter conditions), ft^3
- T_{std} is the absolute temperature at standard conditions, 530°R
- T_m is the average dry gas meter temperature, °R
- P_{bar} is the barometric pressure at the orifice meter, in. Hg
- P_{std} is the absolute pressure at standard conditions, 29.92 in. Hg.
- Δm is the pressure drop across the orifice meter, in. H_2O
- 13.6 is the specific gravity of mercury.

The average concentration of sulfur dioxide is calculated using Equation 12-3.

$$C_{SO_2} = 7.05 \times 10^{-5} \frac{\text{lb-l}}{\text{g-ml}} \frac{(V_t - V_{tb}) N \frac{V_{soln}}{V_a}}{V_{m_{std}}} \qquad (12\text{-}3)$$

where

- C_{SO_2} is the concentration of sulfur dioxide at standard conditions, dry basis, lb/ft^3
- 7.05×10^{-5} is a conversion factor including the number of grams per gram equivalent of sulfur dioxide (32 g/g eq) 453.6 g/lb, and 1000 ml/l, lb-l/g-ml
- V_t is the volume of barium perchlorate titrant used for the sample, ml
- V_{tb} is the volume of barium perchlorate titrant used for the blank, ml
- N is the normality of barium perchlorate titrant, g-eq/l

V_{soln} is the total solution volume of sulfur dioxide, 50 ml
V_a is the volume of sample aliquot titrated, ml
$V_{m_{std}}$ is the volume of gas sample through the dry gas meter (standard conditions) ft^3.

The H_2SO_4 concentration is determined from the following expression

$$C_{H_2SO_4} = [1.08 \times 10^{-4} \frac{lb\text{-}l}{g\text{-}ml}] \frac{(V_t - V_{tb}) N \frac{V_{soln}}{V_a}}{V_{m_{std}}} \quad (12\text{-}4)$$

SUMMARY

This test incorporates all of the complexities of both the particulate matter test and the SO_2 test. The sample must be obtained isokinetically and the wet chemical technique still must be used in the laboratory.

REFERENCES

1. "Standards of Performance for New Stationary Sources," Federal Register, Vol 36, No. 247, December 23, 1971.
2. "Atmospheric Emissions from Sulfuric Acid Manufacturing Processes," PHS Publ. No. 999-AP-13, U.S. Public Health Service, Cincinnati, Ohio. 1965.
3. Corbett, D. F. "The Determination of SO and SO in Flue Gases," Journal of the Institute of Fuel 24, 237-243. 1961.
4. Patton, W. F., and Brink, J. A. Jr. "New Equipment and Techniques for Sampling Chemical Process Gases," J. Air Poll. Cont. Assoc. 13, 162. 1963.

CHAPTER 13

DETERMINATION OF NITROGEN OXIDE EMISSIONS

This chapter presents the federal EPA Method 7 for measuring NO_x emissions, a method intended for the determination of the concentration of NO_x from fossil-fuel fired steam generators and nitric acid plants.[1] Several other references are of interest regarding this test.[2-5]

BASIC METHOD

The sample is removed by a grab sampling technique. It is captured in a 2-liter flask which has been evacuated and which contains an absorbing solution of hydrogen peroxide and sulfuric acid which converts the NO_x (except N_2O) in the captured gas to HNO_3 in solution. The amount of nitrate in solution is determined by the phenoldisulfonic acid method. The amount of nitrate (NO_x) is then determined by colorimetric comparison to standard solutions of potassium nitrate.

SAMPLING TRAIN

The sampling train is shown in Figure 13.1. Quartz or Pyrex wool is packed in the heated glass probe to prevent particulate matter from entering the flask. The stopcocks used are T-bore to insure the proper sequence for evacuating, purging and sampling. There is no required flow rate measurement, the only requirement being that the vacuum pump is capable of achieving 3 in. Hg absolute in the sample flasks.

TEST APPARATUS

The following items are needed for the test. They are categorized into sampling equipment, analysis and reagents.

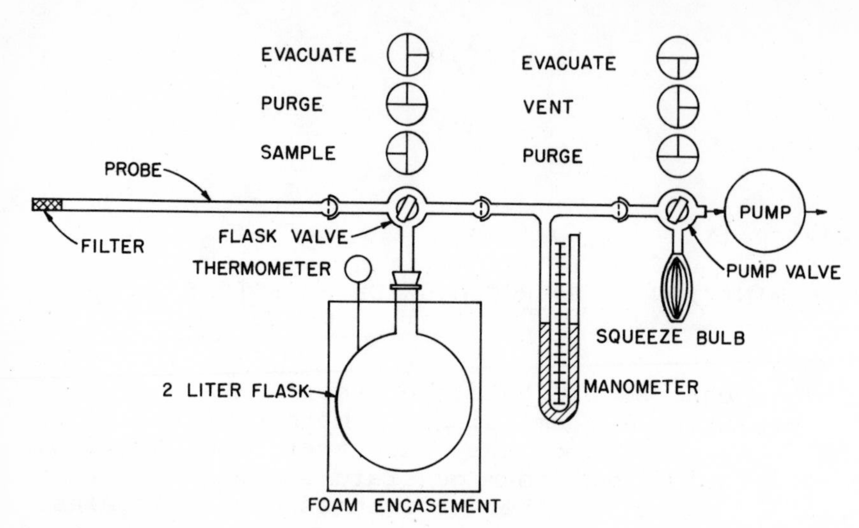

Figure 13.1. Sampling train for oxides of nitrogen (EPA Method 7).

Sampling Equipment

1. Pyrex glass probe with a heating system (necessary only if condensation occurs)
2. Pyrex or quartz wool to act as a filter
3. 2-liter Pyrex round-bottom collection flask with short neck and 24/40 standard taper opening. This should be encased in foam to protect against implosion or breakage.
4. 2 T-bore stopcocks with 12/5 ground-glass sockets for sample lines. One has 24/40 for connection to the flask.
5. 36 in. Hg manometer to measure flask pressure
6. squeeze bulb to purge sampling line (one way)
7. vacuum pump (capable of producing 3 in. Hg absolute)
8. thermometer (gauge type) to measure flask temperature (25°F to 125°F)
9. dropper or pipette, 25 ml
10. vacuum line
11. glass storage containers, protected for shipping
12. glass wash bottle

Analysis

1. steam bath
2. beakers or casseroles, 250 ml

3. volumetric pipettes, 1, 2, and 10 ml
4. transfer pipette, 10 ml with 0.1 ml divisions
5. volumetric flasks: 100 ml for each sample; 1000 ml for standard
6. spectrophotometer, measurement at 420 nm
7. graduated cylinder, 100 ml with 1.0 ml divisions
8. analytical balance, measure to 0.1 mg.

Reagents

1. concentrated H_2SO_4
2. 3% H_2O_2
3. distilled water
4. 1 N NaOH solutions (add 40 g NaOH in distilled water and dilute to one liter)
5. red litmus paper
6. deionized, distilled water
7. 15 to 18% fuming sulfuric acid (by wt free SO_3)
8. phenol (white solid reagent grade)
9. potassium nitrate
10. phenoldisulfonic acid solution: dissolve 25 g of pure white phenol in 150 ml concentrated sulfuric acid on a steam bath; cool, add 75 ml of fuming sulfuric acid, and heat at 212°F for 2 hours. Store in a dark, stoppered bottle.

LABORATORY PROCEDURES

The volume of the flasks must be determined. This is done by filling the assembled flask and valve with water and then measuring the volume of water with a graduated cylinder. Record the volume on each flask.

The absorbing solution is prepared by adding 2.8 ml of concentrated H_2SO_4 to 1 liter of distilled water. To this mixed solution, add 6 ml of 3% hydrogen peroxide. Fresh solution should be prepared weekly and protected from heat or sunlight.

FIELD PROCEDURES

This method determines only the concentration in the gas stream. The volumetric flow rate of the stack must be determined according to EPA Methods 1 and 2 if the pollutant mass emission rate, PMR_s, is to be calculated.

Three repetitions are required per test and four grab samples are required per repetition. These four samples should be taken over a two-hour interval. If the process characteristics are not known these time requirements should be used. If the process is known to be steady (see Chapter 2), then the time can be shortened. Clear justification of this modification should be included in the final report.

Pipette 25 ml of absorbing solution into a sample flask. Insert the flask valve stopper into the flask with the valve in the "purge" position. Assemble the sampling train as shown in Figure 13.1 and place the probe at the sampling point. Turn the flask valve and the pump valve to their "evacuate" positions. Evacuate the flask to at least 3 inches Hg absolute pressure. Turn the pump valve to its "vent" position and turn off the pump. Check the manometer for any fluctuation in the mercury level. If there is a visible change over the span of one minute, check for leaks. Record the initial volume, temperature, and barometric pressure. A field data sheet for this test is shown in Table 13.1. Turn the flask valve to its "purge" position, and then do the same with the pump valve. Purge the probe and the vacuum tube using the squeeze bulb. If condensation occurs in the probe and flask valve area, heat the probe and purge until the condensation disappears. Then turn the pump valve to its "vent" position. Turn the flask valve to its "sample" position and allow sample to enter the flask for about 15 seconds. After collecting the sample, turn the flask valve to its "purge" position and disconnect the flask from the sampling train. Shake the flask for 5 minutes.

This method specifies a minimum sample absorption time of 16 hours. Margolis and Driscolls theoretically predicted a 97% recovery would require 28.7 hrs. Hence the absorption process is the slowest step in the NO_x procedure. If the laboratory is nearby then the samples may be returned to the laboratory in the flask. Otherwise, after absorption they may be transferred to the laboratory in collection flasks.

ANALYTICAL PROCEDURES

Recovery

After the absorption period shake the contents of the flask for 2 minutes. Then connect the flask to a Hg manometer so that the flask pressure can be measured. The barometric pressure is also required at this time. The contents of the

Table 13.1

Data Sheet for EPA Method #7, NO_x Sampling

General	Equipment	Conditions
Firm ________	Meter No. ________	Barometric Press. ________
Location ________	Sample Box No. ________	Ambient Temp. ________
Source ________	Probe No. ____ Lt. ____	Probe Temp. ________
Sample location ________	Vac. Pump No. ________	Average $\sqrt{\Delta P}$ ________
Tester ________		T_s Ave. ________
Date ________		% H_2O ________
Test No. ________		P_s Ave. ________
Observed by ________		Stack Area ________

Sample No.	Flask No.	Flask Vol. ml.	Abs. Soln. Vol. ml.	Initial Cond.		Final Cond.		Bottle No.	Remarks
				P_i "Hg	T_i °R	P_g "Hg	T_f °R		
1									
2									
3									
4									

flask are then transferred to either the shipment container or a 250-ml beaker. The flask is rinsed with two small portions of distilled water (10 ml). An accompanying blank of absorbing solution, along with an equal amount of rinse, is also processed. Then 1N NaOH is added to the absorbing solutions until they are alkaline to litmus paper.

Analysis

If the sample has been shipped in a container, transfer the contents to a 250-ml beaker using a small amount of distilled water. Evaporate the solution to dryness on a steam bath and then cool. Add 2 ml phenoldisulfonic acid solution to the dried residue and titrate thoroughly using a glass rod. Make sure the solution contacts all the residue. Add 1 ml distilled water and 4 drops of concentrated sulfuric acid. Heat the solution on a steam bath for 3 minutes with occasional stirring. Cool, add 20 ml distilled water, mix well by stirring, and add concentrated ammonium hydroxide dropwise with constant stirring until alkaline to litmus paper. Transfer the solution to a 100-ml volumetric flask and wash the beaker three times with 4 to 5 ml portions of distilled water. Dilute to the mark and mix thoroughly. If the sample contains solids, transfer a portion of the solution to a clean, dry centrifuge tube, and centrifuge, or filter a portion of the solution. Measure the absorbance of each sample at 420 nm using the blank solution as a zero. Dilute the sample and the blank with a suitable amount of distilled water if absorbance falls outside the range of calibration.

Calibration

The standard solution is prepared by dissolving 0.5495 g of KNO_3 in distilled water and diluting to 1 liter. The working solution is prepared by diluting a 10-ml portion of the standard solution to 100 ml; 1 ml of this solution then is equivalent to 25 μg of NO_2. The calibration curve is prepared by adding 0.0 to 16.0 ml of standard solution to a series of beakers. To each beaker add 25 ml of absorbing solution and 1 N NaOH dropwise until alkaline to litmus paper. This will take about 25 to 35 drops. The analysis procedure is then followed. A calibration curve is then drawn of concentration in μg NO_2 per sample versus absorbance at 420 nm.

CALCULATIONS

The sample volume at standard conditions is calculated by the following expression.

$$V_{sc} = (17.71 \frac{°R}{in.Hg}) (V_f - V_a) \left(\frac{P_f}{T_f} - \frac{P_i}{T_i}\right) \qquad (13\text{-}1)$$

where

- V_{sc} is dry sample volume at standard conditions, ml
- V_f is volume of flask and valve, ml
- V_a is volume of absorbing solution, 25 ml
- P_f is final absolute pressure of flask, in. Hg
- P_i is initial absolute pressure of flask, in. Hg
- T_f is final absolute temperature of flask, °R
- T_i is initial absolute temperature in flask, °R.

The calibration curve is used in conjunction with the following expression to determine stack gas concentration.

$$C_{NO_x} = 6.2 \times 10^{-5} \frac{lb/scf}{\mu g/ml} (\frac{m}{V_{sc}}) \qquad (13\text{-}2)$$

where

- C_{NO_x} is concentration of NO_x as NO_2 (dry basis), lb/scf
- m is mass of NO_2 in gas sample, µg
- V_{sc} is sample volume.

SUMMARY

The NO_x field test can be conducted easily by one person. In fact this field test can be run at the same time as other pollutant tests. Thus if a team of 2 or 3 people is conducting an isokinetic particulate matter test, the NO_x samples can be run concurrently.

The NO_x laboratory procedure is lengthy, requires a number of steps and analysis must be performed by a qualified chemist or laboratory technician. Errors may arise due to improer procedures and deterioration of reagents.

The velocity traverse must be run concurrently with this NO_x test. Also the pertinent process information for the sampling period is essential. Remember, the NO_x concentration data may only be part of what is needed to determine compliance. Regulations are sometimes stated on the basis of pounds per million BTU's input or pounds per ton of process weight.

REFERENCES

1. "Standards of Performance for New Stationary Sources," _Federal Register_, Vol 36, No. 247, December 23, 1971.
2. _Standard Methods of Chemical Analysis_, Vol. I, 6th ed., D. Van Nostrand Co., Inc., New York. 1962.
3. "Standard Method of Test for Oxides of Nitrogen in Gaseous Combustion Products (Phenoldisulfonic Acid Procedure)," In _1968 Book of ASTM Standards_, Part 23, ASTM Designation D-1608-60. Philadelphia, Pa. 1968. pp. 725-729.
4. Jacob, M. B. The Chemical Analysis of Air Pollutants. Vol. 10, Interscience Publishers, Inc., New York. 1960. pp. 351-356.

CHAPTER 14

PROPERTIES OF AEROSOLS

An aerosol is a suspension of liquid droplets or solid particles in a gas. In this chapter, however, we shall discuss only two major ones: particle size and light attenuation. Light attenuation is important regarding plume opacity; in this respect EPA Method 9 is presented. Aerosol size affects plume opacity and is also important in the selection and design of air pollution control equipment. In addition, the size of an aerosol determines whether it might be lodged in the respiratory system.

PLUME OPACITY

When process gases are discharged into the atmosphere, they frequently are visible to the human eye. The plume is seen because it contrasts with its background, being either considerably lighter (greater luminance) or darker (less luminance). Professor Maximillian Ringelmann used this contrasting luminance feature as the basis of rating black plumes.[1] He developed a scale, shown in Figure 14.1, composed of different shades of black and white.[2] The Ringelmann numbers are related to opacity, the degree to which the transmitted light is obscured by the plume. If a light source, as shown in Figure 14.2a, interacts with a plume, the light intensity is reduced due to absorption and scattering. In 14.2b the contrast occurs because the luminance of the plume, B_p, is different from the luminance of the background, B_b.

The transmittance is defined as

$$\%TR = \left(\frac{I}{I_o}\right) 100 \qquad (14\text{-}1)$$

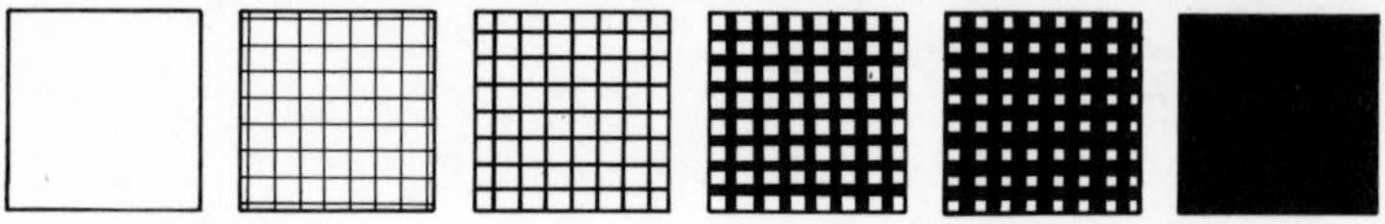

% BLACK AREA	0	20	40	60	80	100
RINGELMAN SCALE	0	1	2	3	4	5
% OAPACITY	0	20	40	60	80	100
% TRANSMITTANCE	100	80	60	40	20	100

Figure 14.1. *The relationship of Ringelmann scale to plume opacity and transmittance.*

(a) Top view showing light source behind the plume.

(b) Top view showing light source behind observer.

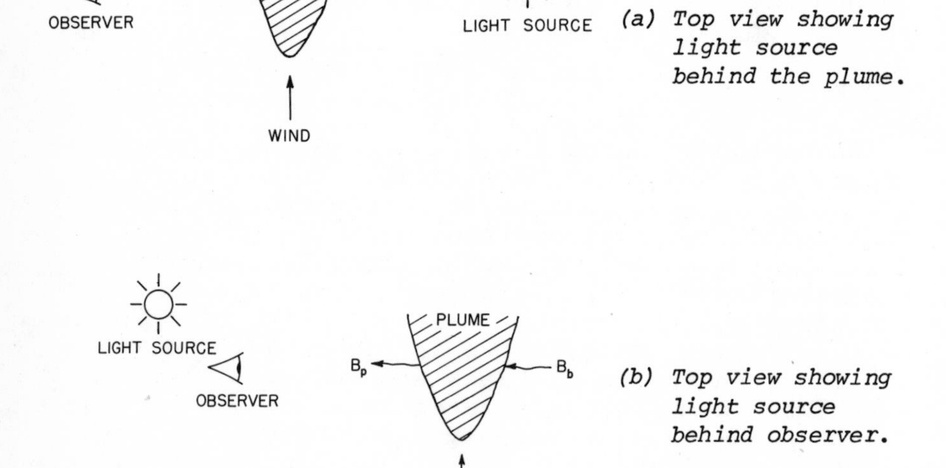

Figure 14.2. *Light obscured by smoke stack plume.*

where

I is the intensity of the light passing through the plume

I_o is the intensity of the light entering the plume

Opacity is defined mathematically as

$$\%OP = 100 - \%TR \tag{14-2}$$

Thus if a plume is perfectly transparent, it has 0% opacity. If a plume obscures all of the light, it has an opacity of 100%.

The Ringelmann chart was incorporated into the Boston Smoke Ordinance in 1910. Since that time many different governmental agencies have used it as a basis for air pollution control.[3] For example, many regulations state that it is illegal to emit smoke of a darker shade than Ringelmann No. 2 for more than 2 or 3 minutes in an hour. This short grace period is to allow for starting up and/or soot blowing operations.

The usefulness of this approach was extended to other types of plumes by defining equivalent opacity, a term referring to the extension of the Ringelmann Chart approach to judge the degree to which a visible plume of any color obscures the light. Thus the observer assigns an opacity rating to the amount of light obscured through *any color* plume, not just black or white. Hence, the plume opacity approach has received considerable use in the past. It will continue to be used in the future because of its ease of application. Basically this approach constitutes the EPA Method 9 which was recently promulgated for new sources.[4] This method is presented in detail in following sections.

Visual Effects

A plume is visible because it contrasts with the background. The luminance contrast is defined

$$CP = \frac{B_p - B_b}{B_b} \tag{14-3}$$

where

B_p is the luminance of the plume

B_b is the luminance of the background.

But B_p is related to the background luminance as follows

$$B_p = B_a + B_b \text{ (TR)} \tag{14-4}$$

where

B_a is the luminance caused by light being scattered by the plume into the observer's sight path (called plume air-light)

TR is the fraction of background luminance transmitted through the plume.

Substituting Equation 14-4 into Equation 14-3, we obtain

$$CP = \frac{B_a}{B_b} + (TR-1) \tag{14-5}$$

If the plume does not scatter light, $B_a = 0$ and CP = TR - 1. The particle size, shape and composition govern the amount of visible light scattered and therefore the value of B_a.

The transmittance of light energy through a plume follows Bougher's (Beer-Lambert) Law:

$$I = I_o \text{ EXP } (-bx) \tag{14-6}$$

where

b is the extinction coefficient due to gases and aerosols in the plume

x is the thickness of the plume.

The extinction coefficient is actually made up of several components

$$b = b_1 + b_2 + b_3 + b_4 \tag{14-7}$$

where

b_1 is the scattering due to aerosols

b_2 is the scattering due to gas molecules

b_3 is the absorbance due to gases and vapors

b_4 is the absorbance due to aerosols.

But from Equations 14-1 and 14-6

$$TR = \frac{I}{I_o} - \text{EXP } (-bx) \tag{14-8}$$

and hence Equation 14-5 becomes

$$CP = \text{EXP } (-bx) - 1 \tag{14-9}$$

when $B_a \simeq 0$. Thus we see that the contrast is a function of plume thickness, x, and the extinction coefficient due to composition and concentration, b. If the plume is transparent, b = 0 and the contrast is also zero. In any given plume all of these components of b exist. The actual predominant component is often different depending upon the type of source and hence the characteristics of the plume. Factors affecting light attenuation by aerosols are particle size and shape, particle composition, wave length of the light, angle of incidence, and particulate concentration. Of all the gases, nitrogen dioxide absorbs the strongest in the visible region of light. Green and Lane,[5] Cadle,[6] Davies,[7] and Conner and Hodkinson[8] all present the theoretical considerations for describing light attenuation by aerosols. Robinson[9] has extended the light attenuation theory for calculating plume opacity. More recently Pilat and Ensor[10,12] have calculated plume opacities using light attenuation theory.

The visible nature of plumes varies with both aerosol size and chemical composition. The plumes from fuel oil and coal combustion operations are usually a dark shade due to the unburned carbon (fly ash). However, if the fuel contains a high percentage of sulfur, and a good particulate control is used, then the plume appears bluish due to the sulfur trioxide. These plumes may also have a brownish tint due to the nitrogen oxides formed during combustion. Incinerator plumes are typified by blackness under poor operating conditions. However, when properly fired, their plume is most obvious due to its high moisture content. In Portland cement operations the rotary kiln emits large quantities of dust; the plume also has a high moisture content. Often the dust is still visible after the water has evaporated. The tail gases from a nitric acid plant contains acid mist, nitric oxide, nitrogen dioxide, and oxygen and nitrogen. The acid mist and nitrogen dioxide cause the plume to be brownish. The tail gas from a sulfuric acid plant is bluish due to the sulfur trioxide; this plume also contains an acid mist.

Plume Evaluation

The opacity of plumes may be evaluated by trained observers or in-stack equipment. The trained observer technique known as EPA Method 9 "Visual Determination of the Opacity of Emissions from Stationary Sources" is presented here. The in-stack photoelectric method is presented in Chapter 17.

Procedure. The trained observer stands at approximately two stack heights, but not more than a quarter of a mile from the base of the stack with the sun to his back. From a vantage point perpendicular to the plume (see Figure 14-2b) the observer studies the point of greatest opacity in the plume. This point is not necessarily at the very top of the stack. The data required in Table 14.1 are recorded every 15 to 30 seconds to the nearest 5% opacity. A minimum of 25 readings is taken. The average opacity is then computed to determine if the source is within compliance. This procedure applies to new sources as given in Table 14.2.

Table 14.1

Field Data Sheet for Plume Opacity Tests

sec / min	0	15	30	45	sec / min	0	15	30	45
0					30				
1					31				
2					32				
3					33				
4					34				
5					35				
6					36				
7					37				
8					38				
9					39				
10					40				
11					41				
12					42				
13					43				
14					44				
15					45				
16					46				
17					47				
18					48				
19					49				
20					50				
21					51				
22					52				
23					53				
24					54				
25					55				
26					56				
27					57				
28					58				
29					59				

Observation data

Plant ____________

Stack location ____________

Observer ____________

Date ____________

Time ____________

Distance to stack ____________

Wind direction ____________

Wind speed ____________

Sum of nos. recorded ________

Total no. of readings ________

Opacity:

$$\frac{\text{Sum of nos recorded}}{\text{Total no. readings}}$$

____

Table 14.2

Opacity Requirements for New Priority I Sources[4]

Process	*Opacity Regulation*
Fossil-fuel steam generator	Not greater than 20% except that 40% shall be permissable for not more than 2 minutes in any hour.
Incinerators	Not applicable.
Portland Cement Plant (applicable areas): kiln, clinker cooler, raw mill system, finish mill system, raw material storage, clinker storage, finished product storage, conveyor transfer point, bagging and bulk loading and unloading systems.	Not greater than 10% except where the presence of uncombined water is the only reason for failure to meet the regulation.
Nitric Acid Plant	Not equal to or greater than 10%.
Sulfuric Acid Plant	Not equal to or greater than 10%.

Smoke Reading Aids. A number of smoke reading devices have been developed to aid in determining plume opacity. These include smoke tintometer, umbrascope, smokescope smoke comparator charts, and photoelectric devices.[3] However most of these only apply to black-gray plumes and are usually not worth hauling around.

Trained Observers. To perform the EPA Method 9 an observer must have successfully completed a smoke reading course, which must be an EPA course or an equivalent approved course. In order to certify, the candidate must assign opacity readings in 5% increments to 25 different black and 25 different white plumes with an error not to exceed 15% on any one reading and an average error not to exceed 7.5% in each category. The smoke generator used to qualify the observer must be equipped with a calibrated smoke indicator or light transmission meter located in the source stack. In this way the observer is trained to read plumes based upon the actual light transmission (opacity) characteristics of the plume. All qualified observers must pass this test every six months in order to remain certified.

Smoke Generating Equipment. The training and testing of smoke observers requires the use of a device which manufactures black and white smoke. The production of shades of black smoke can be accomplished by creating incomplete combustion. The white smoke is produced by heating a distillate-type oil so that it vaporizes into a gas and then cooling it so that the vapor condenses into an aerosol cloud. This cloud is white and its opacity varies with the amount of oil vaporized. The control of the visual densities of the smoke plumes is accomplished by measuring the opacity before the plume is emitted and altering the flow of combustible material (black) or vaporizing liquid (white) until the desired opacity is achieved. The plume opacity is measured in-stack with a photoelectric device which is calibrated using neutral density filters. Figure 14.3 shows the equipment for producing black and white smoke.

Training Procedures. Training on the smoke generator begins with the familiarization of the personnel with known densities (opacities) of black and white smoke. The instructor calls out the opacity meter reading for various plumes. The candidate is thus allowed to observe a wide range of plume opacities. After about an hour of such practice the testing period begins. Readings are taken at the sound of a horn (or other such signal) and recorded on a form similar to that shown in Table 14.3. Between plume readings the candidate should not stare at the plume; this causes fatigue and thus errors. Dark or tinted glasses should not be worn during the test unless they will always be worn during time when reading plumes. At the end of the run the instructor will read off the transmissometer (instrument opacity) readings and the candidate enters these values in the "Transmissometer Reading" column. Ringelmann numbers are converted to opacity by using the relationship shown in Figure 14.1. Then the deviations are computed and recorded in the appropriate plus or minus column. The

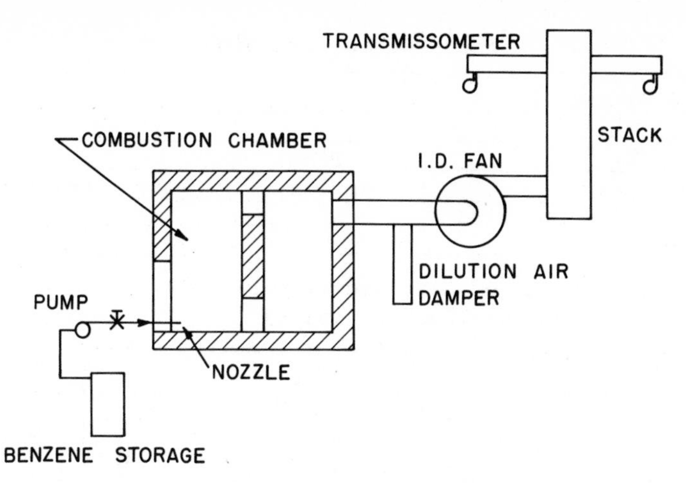

(a) Black smoke generating equipment.

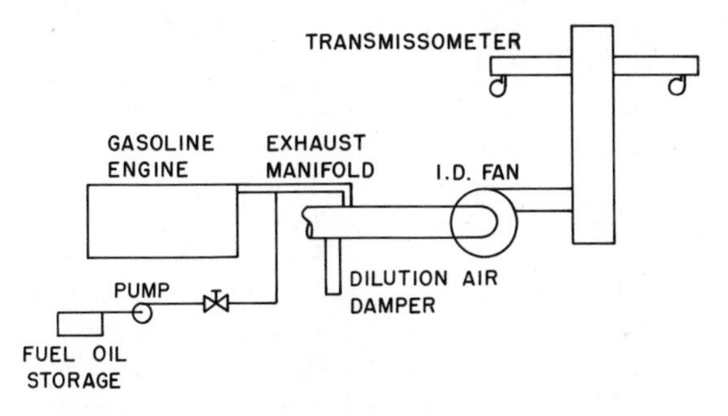

(b) White smoke generating equipment.

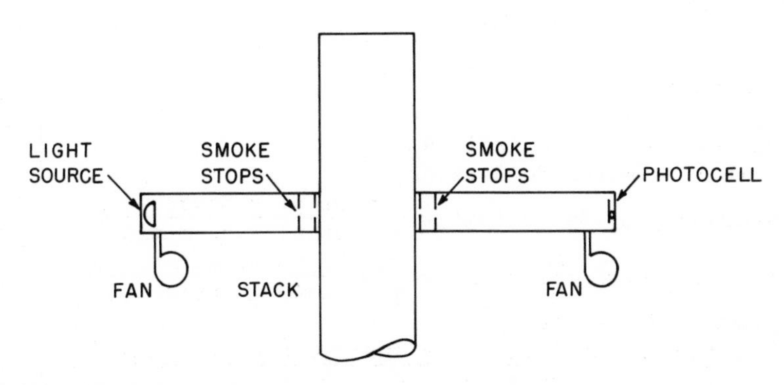

(c) Transmissometer

<u>Figure</u> <u>14.3</u>. *Smoke generating equipment for training course.*

Table 14.3

Smoke School Training Form

Name of observer ________________
Affiliation ________________
Date ________ Time ________
Wind speed ______ Direction ______ Sky condition ______
Observer's position ________________
Corrected by ________________

(Record Black or Gray smoke in Ringelmann No. - 1/4 Unit Smallest Division)
(Record All other smoke in % opacity - 5% smallest division)

Run No. 1B (black)

Reading no.	Observer reading	Transmissometer reading	+ Deviation	- Deviation	Reading no.	Observer reading	Transmissometer reading	+ Deviation	- Deviation
1					13				
2					14				
3					15				
4					16				
5					17				
6					18				
7					19				
8					20				
9					21				
10					22				
11					23				
12					24				
					25				

Run No. 2W (white)

Reading no.	Observer reading	Transmissometer reading	+ Deviation	- Deviation	Reading no.	Observer reading	Transmissometer reading	+ Deviation	- Deviation
1					13				
2					14				
3					15				
4					16				
5					17				
6					18				
7					19				
8					20				
9					21				
10					22				
11					23				
12					24				
					25				

Run number		
Number correct		
Number of plus deviations		
Number of minus deviations		
Average plus deviations = $\frac{\text{sum of plus deviations}}{\text{no. of plus deviations}}$		
Average minus deviations = $\frac{\text{sum of minus deviations}}{\text{no. of minus deviations}}$		
Average deviation = $\frac{\text{(sum of plus deviations) + (sum of minus deviations)}}{\text{total no. of readings}}$		
Number of readings 15% deviation and over		

candidate is certified if no single observation deviates by more than 15% and if the average error for black and white plumes does not exceed 7.5% for each category. After successfully passing this training procedure, the observer is expected to retain by memory the various shades of opacity for a period of up to six months.

Certification tests should not be conducted during windy and/or rainy weather. The rain can make operation of the smoke generator somewhat hazardous due to possible electrical shocks. The wind rapidly disperses the plume and makes it almost impossible for the candidate to certify.

Comments on Opacity Measurements

Opacity methods are used because they are economical, do not require highly trained personnel and are easily applied. The equipment required to operate a smoke school costs about $5000; its operation cost is only a few dollars an hour. Once trained, an inspector can make many observations in one day.

A number of errors and difficulties are involved with this method. The plume reading depends on (1) the geometry with respect to the observer, the plume and background light, (2) the ability of the observer to judge opacity, and (3) the chemical and physical nature of the plume. The errors caused by the first two hopefully are reduced by the training school. But it should be noted that the inspectors (observers) are trained using only black and white plumes although the regulation applies to all color shades of opacity. Also presumably the regulations apply 24 hours a day but it is difficult to read plumes at night.

The nature of the plume can pose problems. For example, if the plume contains large quantities of water vapor and mist, it may be impossible to read the plume at all. The inspector may have to read the plume at a point where the water has evaporated; this may occur some distance from the stack after the plume has had an opportunity to disperse. In some cases the inspector may have to wait for a hot dry day, in which case the steam will dissipate rapidly.

The opacity of a plume can be reduced by adding dilution air which reduces the concentration. For a given volume flow rate the plume opacity can also be reduced if a smaller diameter stack is used. In this latter case the light passes through a thinner plume and hence less light is obscured. Also the amount of light obscured by a plume depends upon the particle size distribution. The submicron particles have a great light scattering capability and tend to make the plume opaque; in this case a plume may meet the particulate regulations for mass emission rate and still fail due to its opacity. Conversely if a plume contains only very large particles it may pass the opacity regulation and not meet the mass emission rate regulation. Depending upon particle size, a given mass concentration can provide a wide range of plume opacities. There is no easy way an inspector can take these latter situations into account. The dilemma of the plume opacity method reveals that a plume could be condemned when viewed on one day and accepted another, or condemned when viewed from one direction and accepted from another, even when its composition has not changed.

PARTICLE SIZING

The size of a particle is an important parameter. As discussed in the previous sections it affects plume opacity. It is also an important factor for the selection and design of air pollution control equipment. Therefore the following sections cover the common sampling techniques, analysis methods and data handling procedures available for determining particle size distribution.

Particle Size Statistics

Since the individual particles in any particulate sample are of many different sizes, statistical methods offer an excellent way to present particle size data. Particle size data are usually presented as frequency or cumulative distributions in terms of number or mass. The distributions are generated by grouping observed size data in classes or intervals of size (such as all particles between 1.0-1.2 microns) and determining the number of particles or amount of mass in each interval. The results of such a data analysis are then displayed using a histogram as shown in Figure 14.4, plotting the number of particles or mass of each interval against each interval. If cumulative frequency distributions are desired the number of particles or mass of particles in each interval can be successively added and plotted against

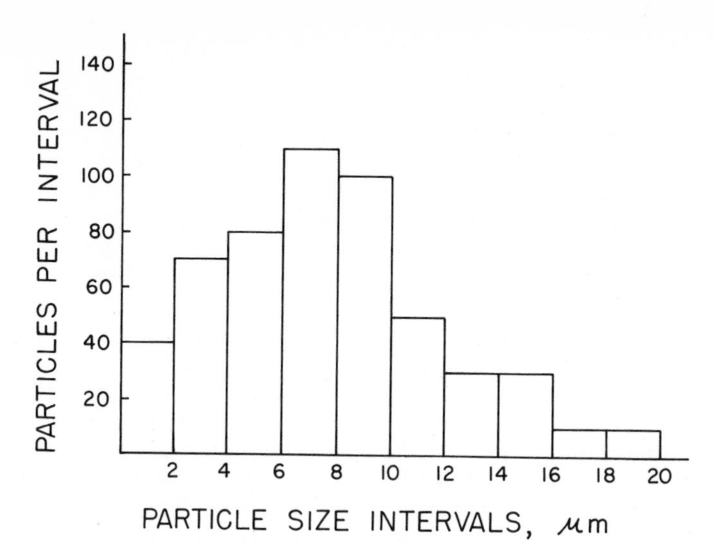

Figure 14.4. *Histogram of particle size data.*

the interval numbers as shown in Figure 14.5. Presenting particle size data in this way is concise and easily interpreted. Reference 14 discusses the statistics of particle size data in considerably more detail.

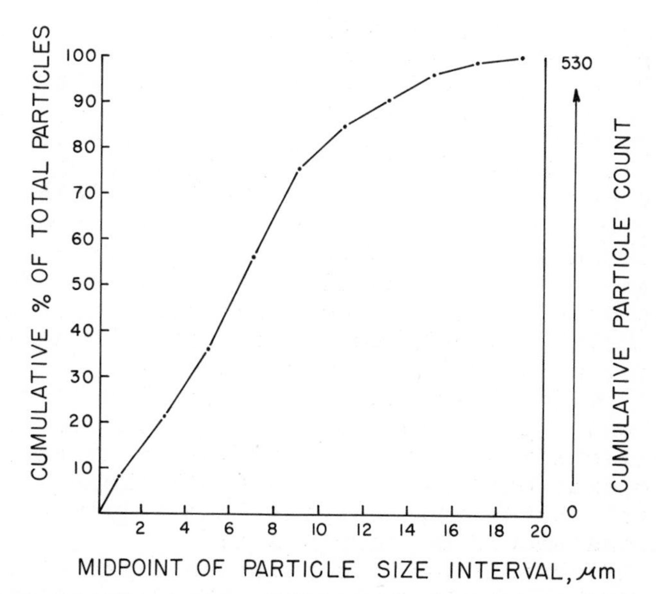

Figure 14.5. *Cumulative number size distribution of particle data.*

Sample Collection Errors

Determining the size distribution of particulate air pollution emissions is a complex problem. Even if an analytical method exists that will yield accurate size and size-mass distribution data, it is virtually impossible to collect a particulate sample without altering its size distribution. Furthermore, as discussed in Chapters 8 and 10, in any finite duct or stack (a nonpoint source) the particle size distributions at individual points within the dust are unique. Therefore, a distribution of individual point particle size distributions exists. Since it is usually impossible to sample the full flow of a source, an integration of many point particle size distributions must be used. Approximate integrations can be accomplished by traversing the cross-sectional area and sampling at several points as discussed in Chapter 10. This approximate integration method works reasonably well in determining unbiased particulate concentrations, but the possible errors encountered are of even greater concern when trying to determine a size distribution.

Ideally to reduce the sampling error incurred, the ultimate collection device should itself be directly in the gas flow as close to the location of interest (*e.g.*, control device, etc.). However, this is not usually possible or even desirable. The usual *in situ* stack or duct environment of air pollution sources is too extreme to allow direct insertion of the ultimate collection device. Furthermore, the physical size of most collection devices is usually sufficient to disturb the local velocity profile and thus the local particle size distribution. Therefore, some sort of a sampling probe is usually used. Usually these probes are fabricated of glass or stainless steel and are as small as possible to reduce local flow disturbances without being so small as to discriminate against the larger particles in the gas sample.

Properly designed and used (within the design limitations), probes could deliver reliable samples to a collection device. However, when using any probe to extract a particulate sample for subsequent collection and sizing analysis, the following limitations should be recognized:

1. If nonisokinetic conditions exist at the probe nozzle entrance, the particulate size distribution may be altered.

2. Particulate impaction may occur at all bends in nozzle and probe lines, resulting in particulate losses.
3. Particulate may settle out in horizontal probes.
4. The gas velocity in vertical probes may be insufficient to support larger particles.
5. Water and other condensible fractions in the original gas stream may condense or chemical reactions may occur that will change the nature of the sampled particulate.
6. Small particles may diffuse to the probe and sample collecting device walls due to thermal or electrostatic potential gradients and brownian diffusion.
7. Agglomeration or fracturing of particles in the probe lines or collection device may alter the size distribution.

Size Analysis Methods

If a representative sample of an aerosol can be reliably collected, several analytical methods exist that can be used to quantify the size distribution. However, the size analysis of a given particulate sample as determined by a specific method of analysis is usually unique and cannot be duplicated using another analysis method. Therefore:

1. Before choosing a sizing method, determine which method will best generate the information needed with respect to the test purpose.
2. When reporting particle size distribution data, indicate clearly how the sample was extracted and what specific analysis method was used.

Due to the size of the particles of interest, microscopy is the only method commercially available for direct counting and/or sizing of individual particles. Generally, a particle size distribution can be determined in only three ways:

1. collecting an integrated sample of particulate with subsequent microscopic counting and sizing of the individual particles.
2. collecting a particulate sample that has been separated into size intervals by some indirect physical means with subsequent gravimetric weighing of the intervals.
3. collecting a particulate sample that has been separated into size interval by some

indirect physical means with subsequent microscopic counting of particles within the intervals.

Microscopic Comparison. Optical and electron microscopies can be used to measure particle size by comparison to known standards. However since the images are two-dimensional, only linear or area measurements of size can be made. Since most particles encountered are not spherical, measurements are usually expressed in projected or statistical diameters. The projected diameter is the diameter of a standard circle whose area is estimated to be the same as the particle of concern. Sizing particles on the basis of statistical diameters assumes that the particles sized are randomly oriented. Therefore, if the number of particles sized is large enough, the projected diameters measured will be statistically representative of the true size. The length of the longest horizontal dimension across a particle as viewed and the length of the horizontal line that divides the projected particle area in half are also often used to estimate particle size.

Linear distances can be measured very accurately with a microscope. The lower sensitivity limit of light microscopes is about 0.5 μm. Electron microscope sensitivity is on the order of 0.001 μm.[15] Since microscopic methods entail two-dimensional optical comparison, samples for analysis must be collected in such a way as to prevent particle layering, aggregation and fracturing. The samples are usually collected by filtration using a membrane filter or by impaction on a glass slide. Collecting too large a particulate sample may make two-dimensional size and count analysis difficult. Thus, many particulate air pollution sources emit particulate concentrations high enough to severely limit the total sample collection time and hence the reliability of the measurement. With membrane filter collection, the sampling time for collecting a particulate sample from a coal-fired boiler for microscopic size and count analysis is on the order of a minute.

Aerodynamic Separation. Particulate samples can be indirectly separated according to size in many ways. The basic methods routinely used for air pollution sources are aerodynamic in nature:[6]

1. sedimentation
2. elutriation
3. inertial impaction
4. centrifugal separation.

Sedimentation particle size separation is a method that relies on the fact that aerodynamic drag can cause particles to settle out of slowly moving gas streams. The drag force exerted on a particle is a function of its diameter, indicating that larger particles will settle faster. Thus, if everything else is held constant, particles can be separated with respect to their size alone. All sedimentation separators operate in a horizontal flow mode. A schematic of a sedimentation separator is shown in Figure 14.6. The overall applicable size range for sedimentation separators is 1-50 μm.[16]

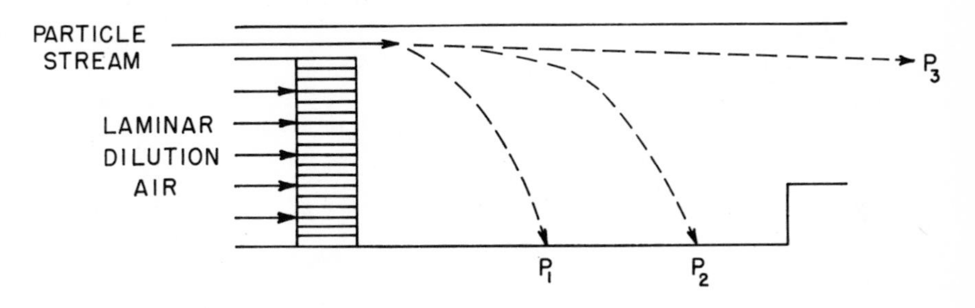

Figure 14.6. Schematic of sedimentation particle separator.

Elutriation separators differ from sedimentation separators only in flow direction. As shown in Figure 14.7 the gas flow through elutriation devices is in vertical direction against the acceleration of gravity. The separation size cutoff is adjusted by varying the velocity of the gas stream. As in the case of sedimentation the drag force that causes particles to settle is a function of particle size. The applicable size range of gravitational elutriation separators is 1-100 μm.[16]

Inertial impaction devices rely on the inertial properties of particles for separation. As indicated in the schematic in Figure 14.8 the inertia characteristics, which are a function of particle size, cause particles to impact on obstruction surfaces located directly at the exit of nozzles. The gas stream molecules and small particles are able to negotiate the sharp bends around the obstruction surface. The particle size that will impact can be regulated by varying the gas velocity and the

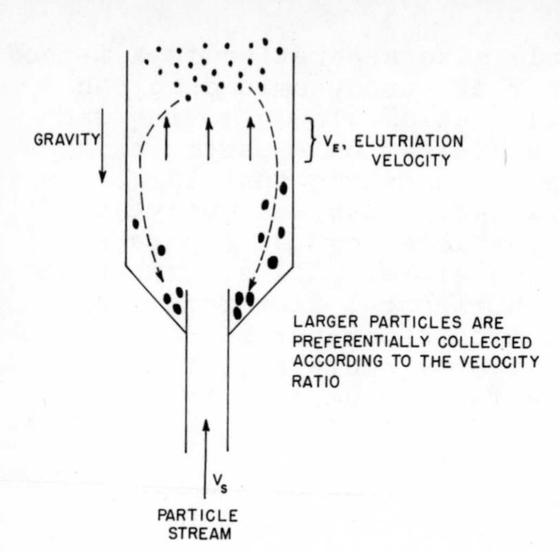

Figure 14.7. Schematic of elutriation particle separator.

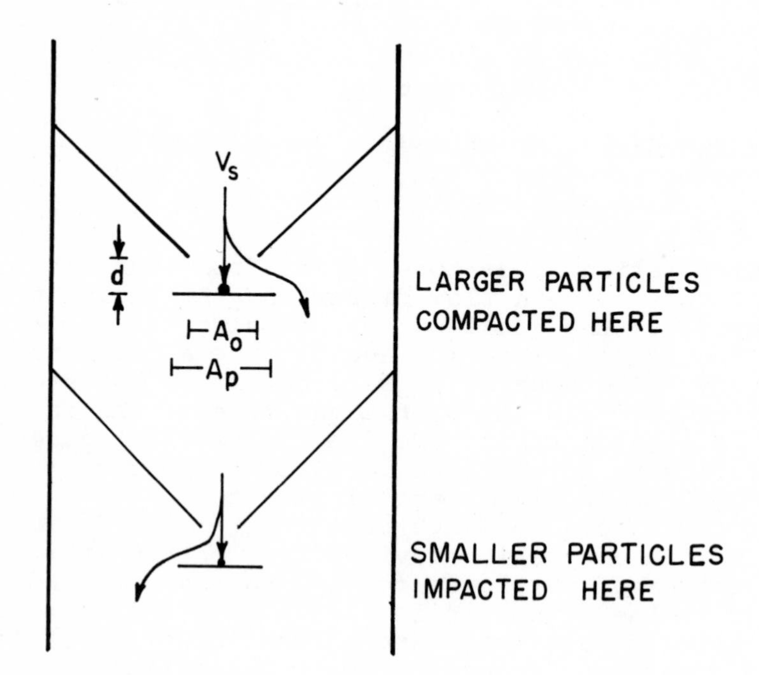

Figure 14.8. Schematic of multiple inertial impactor particle separator.

distance from the gas nozzle exit to the impactor plate. Multiple stage impactors that will generate particle size distribution information are commercially available. The applicable particle size range for separation is 0.2-30 μm.[16]

Centrifugal separators also rely on particle inertia for separation. The gas stream is caused to rotate by introducing it to a stationary cyclone device which rotates because of the entering gas velocity or a mechanically rotated surface. Particles are impacted and thus collected according to their size. The familiar Bahco particle classifier is an example of such a device. The applicable particle size range for the Bahco classifier is 1-60 μm. Other types of centrifugal separators, applicable down to 0.03 μm,[16] are available.

Comments on Particle Sizing Methods. All particle sizing methods require that unbiased particle samples be taken from the sampled air stream. As indicated in this section it may not be possible to determine the particulate size distribution from common air pollution sources without bias. All particle sizing theory is based on spherical particles, but particulate air pollution sources rarely emit spherical particles. The common aerodynamic separators used have several disadvantages. The size cutoffs between sizing intervals are usually not precise. It is difficult to mechanically handle particulate matter without causing some agglomeration or fracturing of particles. Since the particles are not spherical, aerodynamic separation is dependent on the size and shape of the particles.

The density of individual particles is also a factor in aerodynamic separation. Aerodynamic methods cannot be calibrated theoretically, and laboratory calibration of the instruments is necessary. Thus, the instruments should be calibrated with well-defined particulate (of known size and weight distribution) that is similar in all ways to the particulate to be separated. This is very difficult. Also, it is the nature of all aerodynamic separators that particle size cutoffs are a function of flow rate through the sampler, which practically negates the possibility of direct isokinetic sampling of emitted particulate.

A recent development in particle sizing is the use of piezoelectric crystals in conjunction with one of the methods presented in this section. The theory of piezoelectricity is presented in Chapter 18. Figure 14.9 shows a particle sizing device

Figure 14.9. Piezoelectric device for sizing aerosols.[17]

designed and developed by Stiles.[17] A gas sample enters the settling chamber. It is then allowed to remain static under constant temperature and pressure conditions. The particles settle to the bottom of the chamber and are collected on the piezoelectric microbalance. Assuming that Stokes' settling conditions exist, the rate of mass increase information can be used to calculate the size distribution. Stiles has applied this method for sizing of aerosols from combustion processes. The device is self-contained and battery-operated. Possibly such devices will eventually be used to determine particle size distributions from many different types of sources.

REFERENCES

1. Ringelmann, M. "Method of Estimating Smoke Produced by Industrial Installations," Rev. Technique, 268, June, 1898.
2. "Ringelmann Smoke Chart," Information Circular 8333, U.S. Department of the Interior. 1967.
3. "Air Pollution Control Field Operations Manual," Publ. No. 937, U.S. Public Health Service, Washington, D.C. 1962.
4. "Standards of Performance of New Stationary Sources," Federal Register, Vol 36, No. 247, December 23, 1971.
5. Green, H. L. and Lane, W. R. Particle Clouds: Dusts, Smokes and Mists, E. & F. N. Spon Ltd., London, England. 1964.
6. Cadle, R. D. Particle Size Theory and Industrial Applications, Reinhold Publishing Corp., New York, New York. 1965.
7. Davies, C. N., editor. Aerosol Science, Academic Press, Inc., New York, New York. 1966.
8. Conner, W. D. and Hodkinson, J. R. "Optical Properties and Visual Effects of Smoke Stack Plumes," Publ. No. 999-AP-30, U.S. Public Health Service, Washington, D.C. 1967.
9. Robinson, E. Air Pollution, Vol I, Stern, A. C. editor. Academic Press, Inc., New York, New York. 1968.
10. Pilat, M. J. and Ensor, D. S. Atmospheric Environment Vol. 4, pp. 163-173. 1970.
11. Ensor, D. S. and Pilat, N. J. "Calculation of Smoke Plume Opacity From Particulate Air Pollutant Properties," Paper presented at 63rd Air Pollution Control Association Meeting, St. Louis, Missouri, June, 1970.
12. Ensor, D. S. and Pilat, M. J. "The Relationship Between the Visibility and Aerosol Properties of Smoke Stack Plumes," Paper presented at the Second International Union of Air Pollution Prevention Associations, Washington, D.C. December, 1970.

13. "Air Quality Criteria for Particulate Matter," Publ. No. AP-49, U.S. Public Health Service, Washington, D.C., January, 1969.
14. Air Sampling Instruments, American Conference of Governmental Industrial Hygenists, Cincinnati, Ohio, 1972.
15. McCrone, W. C., Draftz, R. G., and Delly, J. G. The Particle Atlas, Ann Arbor Science Publishers, Ann Arbor, Michigan, 1967.
16. Sem, G. J., Borgor, J. A., Whitby, K. T., and Liu B. Y. B. State of the Art: 1971 Instrumentation for Measurement of Particulate Emissions from Combustion Sources, Vol. III: Particle Size, Contract No. 70-23, U.S. Environmental Protection Agency, July, 1972.
17. Stiles, A. R. School of Civil Engineering, Purdue University, West Lafayette, Indiana.

CHAPTER 15

COMPUTATIONAL METHODS

The purpose of this chapter is to acquaint the reader with the mechanics involved in computing the source test parameters and results as discussed in the preceding chapters. This chapter contains examples of most of the calculations needed in source testing. A method concerning the use of significant digits in computations and a discussion of the nomographs most widely used to calculate isokinetic conditions are also discussed.

COMPUTATIONAL TECHNIQUES

Even though source sampling methods have been established[1] there are many possibilities for experimental error as discussed in Chapter 8 of this text. This error, or uncertainty, should be reflected in the final results obtained. Results should be expressed in such a way as to make the uncertainty of the results obvious. This can be accomplished rigorously by using the statistical analysis of errors outlined in Chapter 8. However, when this is not practical, other less rigorous methods can be used to approximate the order of uncertainty.

The order of magnitude of the uncertainty of any measurement is expressed approximately by the number of significant digits in the numerical measurement. When making a measurement it is good practice to record all digits from the graduated scale used, plus one estimated digit. A measurement taken in this way, estimating the final digit, contains the maximum number of significant digits attainable. Usually the last digit is assumed to be uncertain by $\pm$ 1. However, the authors feel that the instruments used in source sampling, as they are used in the field and in the laboratory, have the uncertainties listed in Table 15.1. Thus, the inclined oil monometers used in source sampling should be read to the nearest $\pm$ 0.005 in H_2O. A velocity pressure, ΔP, of 0.050 in H_2O has an absolute uncertainty of 0.005 in H_2O and a relative uncertainty of 1 part in 10. A velocity pressure of 1.000 in H_2O has an absolute uncertainty of 0.005 in H_2O and a relative uncertainty of 1 part in 200. Furthermore, a ΔP of 1.50 in H_2O has an absolute uncertainty of 0.02 in H_2O and a relative uncertainty of 1 part in 75.

Table 15.1

The Uncertainty of Source Sampling Instrument Measurements

Instrument	*Applicable range*	*Authors' estimate of uncertainty*
Field Instruments		
Hg-Manometer	0.0-29.9 in. Hg	0.1 in. Hg
Inclined oil manometer	0.000-1.000 in. H_2O	0.005 in. H_2O
Vertical oil manometer	1.00-10.00 in. H_2O	0.02 in. H_2O
Dial thermometer	0-250°F	2°F
Thermocouple	0-2000°F*	1-5°F*
Dry gas meter (1CF)	**	0.01 CF
Dry gas meter (0.1CF)	**	0.001 CF
Orifice meter	0.00-1.50	0.02 CFM
Laboratory Instruments		
Analytical balance	0-100 g	0.0002 g
Two pan balance	0-500 g	0.1 g
Volumetric burrette	0-50 ml	0.01 ml
Graduated cylinder	0-100 ml	0.5 ml

*Depending on the thermocouple and read out device used.
**Typical particulate tests require a sample volume of 40-100 CF. Typical SO_x tests require 0.75-5 CF.

The degree of uncertainty of any computation involving the manipulation of experimental measurements should be commensurate with the data involved. That is, the final result is at least as uncertain as the most uncertain measurement used in computing a result. The final result should show the order of this uncertainty. Significant digits are used to indicate the certainty to which a measurement or calculation can be made.

Rules and conventions have been adopted to facilitate the expression of results with the correct number of significant digits. These computation rules govern only how many significant digits should be retained in a computed result. They do not estimate the error of a computed result. Chapter 8 contains a discussion of errors in source sampling. The rules covering the use of significant digits are as follows:

1. All digits after and including the first nonzero digit in a number are significant. If a zero appears at the end of a number, it is significant only if it represents a reliable measured value.

Example: 0.012 has 2 significant digits, 1.02 has 3 significant digits, 2400 has 2, 3, or 4 significant digits, 2.4×10^3 has 2 significant digits, 2.40×10^3 has 3 significant digits, and 2.400×10^3 has 4 significant digits.

2. When expressing an experimental quantity, read all digits from the instrument scale and estimate the last digit. This last digit is the first uncertain digit and the last significant one.

Example: If the smallest scale division on an automobile speedometer is 10 mph, the speed of the car can be estimated to the nearest 1 mph (*e.g.*, 51 mph has 2 significant digits, 102 mph has 3 significant digits).

3. When eliminating nonsignificant digits from a number (rounding off):
   a. increase the last retained digit by 1 if the eliminated digit is greater than 5.
   b. retain the last significant digit unchanged if the first nonsignificant digit is less than 5.
   c. if the first nonsignificant digit is exactly 5, either retain the last significant digit unchanged if it is even, or increase the last significant digit by 1 if it is odd.

Example: When rounded to three digits, the numbers 2.0361, 2.03489, 2.025, 2.035, become respectively, 2.04, 2.03, 2.03, and 2.04.

4. The resultant from addition or subtraction should contain only as many decimal places as the component with the least number of decimal places (greatest uncertainty).

Example: addition / subtraction

$$\begin{array}{r} 51.2 \\ 2.35 \\ +\ \underline{1.376} \\ 54.9 \end{array} \qquad \begin{array}{r} 1.376 \\ -\underline{1.02} \\ 0.36 \end{array}$$

5. The resultant of a multiplication or division computation must not contain any more significant digits than the component in the computation that has the least number of significant digits (the most uncertain component).

Example: multiplication

$15 \times 2.57 \times 301 = 11{,}603.55$ or 1.2×10^4

↑ 2 significant digits ↑

division

$$\frac{79}{3.175} = 24.881 \quad \text{or} \quad 25$$

↳ 2 significant digits ↑

Example calculations for source sampling according to the federal EPA performance standards[1] will be illustrated in the next section of this chapter. Although no estimate of relative probable error will be made, results will be obtained using the rules outlined for the use of significant digits.

Table 15.2 presents the relative uncertainty (with respect to significant digits) allowed by the federal performance standards. These standards may be interpreted to say that compliance test results only need be reported with the equivalent number of significant digits as the standard that applies. If this is the case, many questions can be raised concerning the standards and the reporting of test data. For instance the operator of an affected steam generating facility might argue that even though the plant's three boilers (coal, oil, and gas respectively) emit 0.7049, 0.3049, and 0.2049 lb of $NO_x/10^6$ BTU, the three boilers are in compliance with the performance standards. In this case the measured emissions are approximately 0.71%, 1.7%, and 2.5% in excess of the allowable emissions before the measured values are rounded off to the number of significant digits of the applicable standard. There is some question if this is the proper procedure.

Table 15.2

Relative Uncertainty Allowed in Applying the Federal Performance Standards for Stationary Sources

Affected process	*Emission standards (note significant digits)*	*Allowable emissions (considering significant digits of standards)*	*Approximate relative uncertainty allowed*
STEAM GENERATORS (with heat inputs greater than 250 x 10^6 BTU/hr)	(1) 0.10 lb part./10^6 BTU	(1) 0.1049	(1) 1 part in 20
	(2) 0.80 lb $SO_2/10^6$ BTU (liquid fuel)	(2) 0.8049	(2) 1:16
	(3) 1.2 lb $SO_2/10^6$ BTU (solid fuel)	(3) 1.249	(3) 1:24
	(4) 0.20 lb $NO_x/10^6$ BTU (gaseous fuel)	(4) 0.2049	(4) 1:40
	(5) 0.30 lb $NO_x/10^6$ BTU (liquid fuel)	(5) 0.3049	(5) 1:60
	(6) 0.70 lb $NO_x/10^6$ BTU (solid fuel)	(6) 0.7049	(6) 1:140
	(7) 20% opacity	(7) 20.49	(7) 1:40
INCINERATORS (burning > 50 ton/day of at least 50% municipal waste)	(1) 0.08 gr/dscf or particulate (corrected to 12% CO_2)	(1) 0.0849	(1) 1:16
PORTLAND CEMENT	(1) 0.30 lb part/ton of cement (kiln)	(1) 0.3049	(1) 1:60
	(2) 0.10 lb part/ton of cement (klinker cooler)	(2) 0.1049	(2) 1:20
	(3) 10% opacity	(3) 10.49	(3) 1:20
HNO_3 PRODUCTION	(1) 3 lb NO_x/ton of acid (100%)	(1) 3.49	(1) 1:6
	(2) 10% opacity	(2) 10.49	(2) 1:20
H_2SO_4 PRODUCTION	(1) 4 lb SO_2/ton of acid (100%)	(1) 4.49	(1) 1:8
	(2) 0.15 lb H_2SO_4 mist/ton H_2SO_4	(2) 0.1549	(2) 1:30
	(3) 10% opacity	(3) 10.49	(3) 1:20

The standards applying to nitric acid production are even less precise. The standard is 3 lb of NO_x/ton of acid produced. Since the standard contains only 1 significant digit, does an HNO_3 plant that has emissions of 3.49 lb of NO_x/ton HNO_3 (actually over 16% above the standard) comply with the performance standards?

SAMPLE CALCULATIONS

For purposes of illustration, a hypothetical source test (*one replicate only*) on an oil-fired boiler will be discussed. Steam boiler #3 (10^9 BTU/hr) at the Dawson plant of BT&Y utilities was installed after August 17, 1971. Therefore, the unit is required to comply with the federal performance standards for new stationary sources. According to these standards for new oil-fired steam boilers of greater than 250 x 10^6 BTU/hr heat input, the facility must not emit air pollutants in excess of:

(a) 0.10 lb particulate/10^6 BTU input, as measured by EPA Method 5
(b) 20% opacity (40% allowed 2 min/hr), as determined by EPA Method 9
(c) 0.80 lb SO /10^6 BTU input, as determined by EPA Method 6
(d) 0.30 lb NO_x/10^6 BTU input, as determined by EPA Method 7.

Specifications for the boiler tested are given in Table 15.3. The boiler operating conditions during the hypothetical tests are given in Tables 15.4 and 15.5. The BT&Y boiler discussed was hypothetically tested for the following emissions:

(a) particulate per EPA Method 5
(b) SO_2 per EPA Method 6
(c) NO_x per EPA Method 7
(d) SO_3 and H_2SO_4 mist per EPA Method 8.

The following presentation is meant to be an illustration of the calculations involved. The hypothetical source test data collected is shown on the data sheets in Examples 15.1, 15.2, 15.3, and 15.4.

Table 15.3

Boiler Specification Sheet

Unit No. 3 - Dawson, U.S.A.
BT&Y Utility Company

Date Installed: September 1, 1971

Fuel: Oil

Fuel Burning Equipment: Tilting Tangential Burners

Furnace: Heating Surface - 19,220 ft^2
Volume - 68,500 ft^3

Boiler: Manufacturer: Dawson Engineering
Pressure: Design - 1650 psi
Operating - 1500 psi

Additional Equipment: Superheater - Two stage
with Spray Type Desuperheater
Reheater
Economizer
Air Heater

Performance: lb steam per hour - 750,000 Rated
800,000 Max 12 hr Guaranteed
850,000 Max 4 hr Guaranteed

Feedwater Temp. - Economizer 450°F
Boiler 510°F

Steam Temp. 1000°F

Heat Release BTU/cu ft/hr 15,000

Temp. Gas from Air Heater - 260°F

Temp. Air from Air Heater - 510°F

Overall Efficiency - 88%

Table 15.4

Boiler Operation

Unit No. 3 - Dawson, U.S.A.
BT&Y Utility Company

Source Test #1	
Generation Load MW	112
Average Steam Load 10^3 lb/hr	790
Percent of Rated Steam Load	105%
Steam Temperature °F	990°F
Boiler Drum Pressure PSIG	1600 PSIG
Average Oil Flow 10^3 gal/hr	7.50
Excess Air at Particulate Samp. Loc.	36%
Stack Flow Rate dscfh	1.23×10^7

Table 15.5

Fuel Oil Analysis - Test #1

Unit No. 3 - BT&Y Utilities 6/12/72

No. 2 Low-Sulfur distillate oil

Carbon	85.73% (by weight)
Hydrogen	12.35%
Sulfur	0.97%
Nitrogen	0.25%
Oxygen	0.58%
Ash	0.12%
BTU/lb	19,063
lb/gal	7.20

Example 15.1
Particulate Source Test Data

General
Firm *BT&Y Utilities*
Source *Oil Boiler #3*
Sample location *Stack*
Tester *R.Y., D.L.B. & C.D.T.*
Date *6/12/72*
Test no. *1*
Observed by *P.R.W.*

Equipment
Meter no. *17 (.1cf)*
Sample box no. *2*
Probe no. *5* Lt *12'* Cp. *.850*
Nozzle dia. *.255 in.*
Filter nos. *#145* Silica gel nos. *SG-8*
Probe wash sample no. *PW-14*

Conditions
Orifice Δm *1.85*
Ass. % H_2O *10%* T_s *318°*
Ass. Δp avg *.75* P_s/P_m *1.0*
Ass. T_m *75°F* A_s *94.9 ft^2*
Bar. pressure *29.8*
C Factor *.850*
Probe heater set. *250°F*
Filter oven set. *250°F*

Traverse pt.	Time (min)	Dry gas meter (cf)	Pitot ΔP ("H_2O)	Orifice Δm ("H_2O)	Meter inlet (°F)	Temp. outlet (°F)	Stack temp. (°F)	Pump vac ("Hg)	Stack Pressure	Remarks
1	10.0	1.75	.530	1.49	73	70	310	5	-.50" H_2O	Test start 10:00 am
2	10.0	8.50	.580	1.63	73	70	315	5		Test end 11:15 am
3	10.0	15.57	.630	1.77	75	70	320	5	Impinger outlet max temp 50°F	
4	10.0	22.93	.240	2.05	75	70	320	7		
5	10.0	30.92	.820	2.29	75	72	320	7		7.50 x 10^3 GPH of oil
6	10.0	39.34	.870	2.42	77	72	320	7		
7	10.0	47.99	.950	2.62	77	72	320	8	Condensate collected (ml)	CO_2 / O_2
8	10.0	57.05	.940	2.60	77	75	320	8		11.5% / 5.8%
9	10.0	66.05	.830	2.30	77	75	320	7		12.0% / 5.6%
10	10.0	74.51	.630	1.77	77	75	320	7	#1 130	11.0% / 6.0%
11	10.0	81.87	.600	1.68	77	75	320	7	#2 65	
12	10.0	89.05	.580	1.63	77	75	310	7	#3 5	Test end 11:15 am
		96.13							m_{ave} = 2.11" H_2O	T_m ave = 74°F

Example 15.2

Laboratory Analysis Report

Client *BT&Y Utilities*
Sent by *R.F.Y.*
Report to *R.F.Y.*

Laboratory no. *001026*
Date received *6/14/72*
Date completed *6/21/72*

Identification No.			Sample source	Material sampled	Sampling & analysis	Results	Remarks
Test	Sample	Bottle					
1	1	#145	Boiler #3	Part. filter	EPA #5	38.0 mg	-
1	1	PN-14	(oil fired)	Probewash	EPA #5	94.5 mg	blank - considered
	1	SG-8	↓	Silic gel	EPA #5	46 grams	
1	1	SO_2-11		SO_2	EPA #6	78.5 mg	as SO_2
	2	SO_2-19		SO_2	EPA #6	105.0 mg	as SO_2
1	1	N-2		NO_x	EPA #7	390 g	as NO_2
	2	N-7		NO_x	EPA #7	398 g	as NO_2
	3	N-9		NO_x	EPA #7	405 g	as NO_2
	4	N-31		NO_x	EPA #7	387 g	as NO_2
1	1	H-23		H_2SO_4	EPA #8	51.5 mg	as H_2SO_4
		B-10		Acetone Blank	EPA #5	2.0 mg	Vol = 100 ml
		B-15		DI H_2O Blank	EPA #6,7,&8	neg.	

Example 15.3

SO_2 Test Data

General
Firm *BT&Y Utilities*
PLT Location *Dawson, U.S.A.*
Source *Oil Boiler #3*
Sample location *Stack*
Tester *R.Y., D.L.B. & C.D.T.*
Date *7/12/72*
Test no. *#1, Samples #1 & #2*
Observed by *P.R.W.*

Equipment
Meter no. *11 (.1 cf)*
Sample box no. *SO_2 -1*
Probe no. *7* Lt *5'*
Filter used *no*
Cont. no. ______
Pitot no. ______ Cp. ______

Conditions
Barometric press. *29.8*
Probe temp. *350°F*
Ambient temp. *75°F*
Stack area ______
Ave $\sqrt{\Delta P}$ *from part. test*
T_s Ave. ______
% H_2O ______
P_s Ave. ______

Sample no.	Time min.	Meter vol. cf.	ΔP in. H_2O	$\sqrt{\Delta P}$	Rotameter flow cfh	Meter temp °F	Sample bottle no.	Remarks
1	*30*	*10.765*	*.630*	*.795*	*4.0*	*75°*	*SO_2-11*	*Start 10:00 am*
		12.932	*.615*					*End 10:30 am*
2	*40*	*13.572*	*.620*	*.789*	*5.0*	*75°*	*SO_2-19*	*Start 11:00 am*
		16.622	*.620*					*End 11:40 am*

Example 15.4

NO_x Source Test Data

General

Firm: *BT&Y Utilities*
Location: *Dawson, U.S.A.*
Source: *Boiler #3*
Sample location: *Stack*
Tester: *R.Y., D.L.B. & C.D.T.*
Date: *6/12/72*
Test no.: *1*
Observed by: *P.R.W.*

Equipment

Meter no.:
Sample box no.: *2*
Probe no.: *2* Lt.: *4'*
Vac. pump no.: *10*

Conditions

Barometric press.: *29.8*
Ambient temp.: *75°F*
Probe temp.: *300°F*
Ave. $\sqrt{\Delta P}$: *see Ex. 15.1*
T_s ave.: *318°F*
% H_2O: *10%*
P_s ave.: *-.50 in. H_2O*
Stack area: *94.9 ft^2*

Sample no.	Flask no.	Flask vol ml	Abs. soln. vol ml	Initial cond.		Final cond.		Bottle no.	Remarks
				P_i "Hg	T_i °R	P_g "Hg	T_f °R		
1	NO_x-17	2010	25	1.8	535°	29.8	530°	N-2	
2	NO_x-19	2005	25	1.8	535°	29.8	530°	N-7	
3	NO_x-13	1990	25	1.8	535°	29.8	530°	N-9	
4	NO_x-6	1990	25	1.8	535°	29.8	530°	N-31	

Particulate Emission Calculations

From the particulate source test data in Example 15.1:

Δm ave = 2.11 in. H_2O
T_m ave = 75°F = 534°R.

Volume of Gas Sampled

$$V_{m_{std}} = V_m \frac{T_{std}}{T_m} \frac{P_{bar} + \frac{\Delta m_{ave}}{13.6}}{P_{std}} \quad (15\text{-}1)$$

$$= 95.38 \frac{530}{534} \frac{29.8 + \frac{2.11}{13.6}}{29.92} = 94.9\text{dscf}$$

where

$V_{m_{std}}$	is the volume of gas sample through the dry gas meter at standard conditions (70°F, 29.9 in. Hg), cf
V_m	is the volume of gas sample through the dry gas meter at meter conditions, cf
T_{std}	is 70°F, 530°F
T_m	is the average dry gas meter temperature, 74°F, 534°R
P_{bar}	is the uncorrected barometric pressure at the outlet of the orifice meter, 29.8 in. Hg
Δm_{ave}	is the average pressure drop across the orifice meter, 2.11 in. H_2O.
13.6	is the specific gravity of mercury
P_{std}	is the absolute pressure at standard conditions, 29.92 in. Hg.

Volume of H_2O Vapor in Stack Gas: (from Examples 15.1 and 15.2)

water condensed in impingers = 200 ml
water absorbed on silica gel = 46 grams

$$V_{w_{std}} = V_{l_c} \frac{\rho_{H_2O}}{M_{H_2O}} \frac{R\,T_{std}}{P_{std}} \quad (15\text{-}2)$$

$$= 246 \frac{1.00}{18.0} \frac{(21.83)(530)}{29.92}$$

$$= 12.1 \text{ cf.}$$

where

$V_{w_{std}}$ is the total volume of water in the sampled gas at standard conditions (70°F, 29.92 in. Hg), cf

V_{l_c} is the total volume of liquid H O collected in impingers and on silica gel, 246, ml

$\rho\ H_2O$ is the density of water at standard conditions, 1.00 g/ml

R is the ideal gas constant, 21.83 in. Hg cf/lb-mole-°R

T_{std} is the standard temperature, 530°R

P_{std} is the standard pressure, 29.92 in. Hg.

Moisture Content of Stack Gas:

$$B_{w_o} = \frac{V_{w_{std}}}{V_{m_{std}} + V_{w_{std}}} \tag{15-3}$$

$$= \frac{12.1}{94.9 + 12.1} = 0.113 \text{ or } 11.3\%$$

where

B_{w_o} is the mole fraction or proportion by volume of H_2O vapor in the stack gas

$V_{w_{std}}$ is 12.1 cf H_2O vapor at standard conditions

$V_{m_{std}}$ is 94.9 cf of dry sampled gas at standard conditions.

Molecule Weight of Stack Gas:

From orsat analysis per EPA Method 3:

CO_2 avg = 11.5%

O_2 avg = 5.8% by volume (dry basis)

CO avg = 0%

From moisture analysis per EPA Method 5:

H_2O avg = 11.3%

From EPA Method 3:

$$\%\ N_2 \text{ (dry basis)} = 100\% - (\%\ CO_2 + \%\ O_2 + \%\ CO) \text{ (dry basis)}$$

$$= 100\% - (27.3\%) = 82.7\%$$

$$M_s = (1-B_{H_2O})(B_{CO_2}M_{CO_2} + B_{O_2}M_{O_2} + B_{CO}\,M_{CO} + B_{N_2}M_{N_2}) + B_{H_2O}\,M_{H_2O} \qquad (15\text{-}4)$$

$$= (1-.113)[(0.115)(44.0) + (0.048)(32.0) + (0.0)(28.0) + (0.827)(28.0)] + (0.113)(18.0)$$

$$= 27.6 \text{ lb/lb-mole}$$

where

- M_s is the molecular weight of the stack gas, lb/lb-mole
- B_i is the mole fraction of the component gas
- M_i is the molecular weight of the component gas
- B_{H_2O} is the mole fraction of the water vapor in the stack gas.

Excess Combustion Air (per EPA Method 3):

$$\%\ EA = \frac{(\%O_2) - 0.5(\%CO)}{0.264(\%N_2) - (\%O_2) + 0.5(\%CO)} \times 100 \qquad (15\text{-}5)$$

$$= \frac{(5.8) - 0.5(0)}{0.264(82.7) - (5.8) - 0.5(0)}$$

$$= 36\%$$

where

- %EA is the per cent excess combustion air
- $\%O_2$ is the per cent O_2 by volume on a dry basis
- $\%N_2$ is the per cent N_2 by volume on a dry basis
- %CO is the per cent CO by volume on a dry basis
- 0.264 is the ratio of O_2 to N_2 in air by volume.

Particulate Emissions Concentration

$$C'_s = (0.0154 \frac{\text{grain}}{\text{mg}}) \frac{M_n}{V_{m_{std}}} \qquad (15\text{-}6)$$

$$= (0.0154) \frac{132.5}{94.9} = 0.0214 \text{ gr/dscf}$$

where

C'_s is the concentration of particulate matter in the stack gas, gr/dscf

M_n is the total mass of particulate collected on filter media, 132.5 mg from laboratory analysis report, Example 15.2

$V_{m_{std}}$ is the volume of stack gas sampled (volume through dry gas meter), dscf.

also,

$$C_s = 2.205 \times 10^{-6} \frac{\text{lb}}{\text{mg}} \frac{M_n}{V_{m_{std}}} \qquad (15\text{-}7)$$

$$= 2.205 \times 10^{-6} \frac{132.5}{94.9} = 3.08 \times 10^{-6} \text{ lb/dscf.}$$

where

C_s is the concentration of particulate matter in sampled stack gas, lb/dscf.

Stack Gas Volumetric Flow Rate (per EPA Method 2)

$$(V_s)\text{avg} = K_p\ C_p\ (\sqrt{\Delta p})\text{avg} \ \frac{(T_s)\text{avg}}{P_s\ M_s}^{1/2} \qquad (15\text{-}8)$$

where

(V_s)avg is the average stack gas velocity, fps.

K_p is $85.48 \ \frac{\text{ft}}{\text{sec}} \ \frac{\text{lb}}{\text{lb-mole-°R}}^{1/2}$

C_p is the pitot tube coefficient

(T_s)avg is the average stack gas temperature, °R

$(\sqrt{\Delta p})$avg is the average square root of velocity head of stack gas, in H_2O, determined according to EPA Methods 1 and 2

P_s is the absolute stack gas pressure, in. Hg
M_s is the molecular weight of the stack gas, lb/lb-mole (wet basis).

From Example 15.1, the Particulate Test Data Sheet:

$$(\sqrt{\Delta p})\,avg = 0.848$$

$$C_p = 0.850$$

$$(T_s)\ avg = 778°R$$

$$P_s = P_{bar} + (P_s) = 29.8 + \frac{-0.50}{13.6}$$

$$= 29.8 \text{ in. Hg}$$

$$M_s = 26.2 \text{ lb/lb-mole-°R}$$

$$(V_s)\,avg = (85.48)(0.850)(0.848)\left(\frac{778}{(29.8)(27.6)}\right)^{1/2}$$

$$(V_s)\,avg = 59.8 \text{ fps.}$$

Then calculating the average stack gas volumetric flow rate:

$$\overline{Q}_s = 3.600 \times 10^3 \ \frac{sec}{hr}\ (1 - B_{H_2O})\,(V_s)\,avg A_s \frac{T_{std}\ P_s}{(T_s)\,avg\ P_{std}} \qquad (15\text{-}9)$$

where
$\overline{Q}_s$ is the average volumetric flow rate at dry standard conditions, dscfh
B_{H_2O} is the mole fraction of water vapor in the stack gas, dimensionless
A_s is the cross-sectional area of the stack at the sample point, 94.9 ft^2 (from the Particulate Test Data Sheet).

Then,

$$\overline{Q}_s = (3.600 \times 10^3)(.887)(59.8)(94.9)\frac{(530)(2918)}{(778)(29.92)}$$

$$= 1.23 \times 10^7 \text{ dscfh.}$$

Therefore, the pollutant mass emission rate can be calculated by

$$\overline{PMR}_c = \overline{Q}_s\ C_s \qquad (15\text{-}10)$$

where

$\overline{PMR}_c$ is the average pollutant mass emission rate, lb/hr (calculated by the concentration method)

$$\overline{PMR}_c = (1.23 \times 10^7)\ \frac{dscf}{hr}\ (3.08 \times 10^{-6})\ \frac{lb}{dscf}$$

$$= 3.79 \times 10^1 \text{ lb/hr of particulate.}$$

However, since the federal emission standards are written in terms of lb/10^6 BTU, the heat input to the boiler during the test must be calculated. Using the average oil flow as determined by an in-plant oil flow meter, the heat input can be calculated.

From the Particulate Test Data Sheet in Example 15.1:

$$(\dot{B}_{oil})avg = 7.50 \times 10^3 \text{ GPH},$$

and the fuel analysis in Example 15.3, as required by the performance standards,

$$\dot{H} = (7.50 \times 10^3\ \frac{gal}{hr}\ (7.20)\ \frac{lb}{gal}\ (19{,}063)\ \frac{BTU}{lb}$$

$$= 103 \times 10^7 \text{ BTU/hr.}$$

Then, the particulate emission rate in terms of lb of particulate/10^6 BTU can be calculated:

$$\overline{PMR}_{BTU} = \frac{37.9 \text{ lb/hr}}{1030 \times 10^6 \text{ BTU/hr}} = 0.0368 \text{ lb/}10^6 \text{ BTU}$$

The federal performance standards also require that heat input be calculated by performing a material balance over the combustion process. This is discussed in more detail in Chapter 3.

<u>Calculation of the Average Isokinetic Ratio</u>

$$I_{avg} = \frac{(V_n)avg}{(V_s)avg} \times 100 \tag{15-11}$$

where

I_{avg} is the average per cent isokinetic

$(V_n)avg$ is the average gas velocity into the nozzle entrance

$(V_s)avg$ is the average gas velocity of the stack gases.

Writing the equations for the nozzle velocity independent of the velocity pressure measured in the stack:

$$(V_n)avg = \frac{V_m \dfrac{T_s P_m}{T_m P_s M_d}}{A_n \theta} \qquad (15\text{-}12)$$

where

$(V_n)avg$ is the average nozzle velocity

V_m is the volume of gas through the dry gas meter, cf (dry)

T_s is the average absolute stack gas temperature °R

P_s is the average absolute stack pressure, in. Hg

P_m is the average absolute dry gas meter pressure, in. Hg

$$\left(P_m = \frac{\Delta m}{13.6} + P_{bar}\right)$$

T_m is the average absolute meter temperature, °R

M_d is the average mole fraction of the dry stack gas, $(1-B_{H_2O})$

A_n is the area of the nozzle, ft^2

θ is the total sampling time, sec.

Then,

$$(V_n)avg = \frac{(95.38)\ \dfrac{778}{534}\ \dfrac{30.0}{29.8}\ \dfrac{1}{0.887}}{(3.54 \times 10^{-4})ft^2 (120)min(60)\dfrac{sec}{min}}$$

$$= 61.6 \text{ fps.}$$

$$I_{avg} = \frac{(V_n)avg}{(V_s)avg} = \frac{61.6}{59.8} = 1.03 \text{ or } 103\%.$$

To conform with the federal performance standards the average isokinetic ratio, I_{avg}, must be greater than 0.90 and less than 1.10.

Sulfur Dioxide Emissions

Assuming that all the sulfur in the fossil fuel burned is converted to SO_2 according to the equation

$$S + O_2 \rightarrow SO_2,$$

a theoretical SO_2 emission rate can be calculated. This emission rate will always be higher than the actual rate. Some of the sulfur is converted to SO_3. Metallic sulfates and H_2SO_4 are also formed to some degree. Part of the sulfur is also retained in the bottom ash. Typical S to SO_2 converstion yields for fossil fuel combustion are in the range of 95-98%.

Given the fuel burn rate, $\dot{B}$, and the sulfur content of the fuel the theoretical SO_2 emissions can be calculated using the following relationship

$$C'_{SO_2} = (\dot{B})_{avg}(\rho_f)(\%S)\frac{M_{SO_2}}{M_s} \qquad (15\text{-}13)$$

where

C'_{SO_2} is the theoretical mass emission rate of SO_2, lb/hr

$(\dot{B})_{avg}$ is the fuel flow rate, in this case, gal/hr

f is the density of the fuel oil, 7.20 lb/gal

%S is the per cent by weight of sulfur in the fossil fuel

M_{SO_2} is the molecular weight of SO_2, $\frac{64.1 \text{ lb}}{\text{lb-mole}}$

M_S is the molecular weight of sulfur, $\frac{32.0 \text{ lb}}{\text{lb-mole}}$.

Then,

$$C'_{SO_2} = (7.50 \times 10^3)(7.20)(0.97 \times 10^{-2})\frac{64.1}{32.0}$$

$$= 1.05 \times 10^3 \text{ lb } SO_2/\text{hr}.$$

The data from the SO_2 test data sheet, Example 15.3, and the laboratory analysis report, Example 15.2, can be used to calculate the measured mass emission rate of SO_2. The federal performance standards require two samples per SO_2 test. Therefore from the examples above:

Sample 1: 78.5 mg SO_2 in 2.167 dcf (at 75°F, 29.8 in. Hg).

Sample 2: 105.0 mg SO_2 in 3.050 dcf (at 75°F, 29.8 in. Hg).

$$C_{SO_2} = \frac{M_{SO_2} \dfrac{1}{454} \dfrac{lb}{g}}{V_m \dfrac{T_{std}}{T_m} \dfrac{P_{bar}}{P_{std}}} \qquad (15\text{-}14)$$

where

C_{SO_2} is the emitted concentration of SO_2, lb/dscf

M_{SO_2} is the mass of SO_2 collected, g

$$C_{SO_2} \#1 = \frac{\dfrac{78.5 \times 10^{-3}}{454}}{2.167 \dfrac{530}{535} \dfrac{29.8}{29.92}} = 8.08 \times 10^{-5} \text{ lb/scf}$$

$$C_{SO_2} \#2 \quad \frac{\dfrac{105 \times 10^{-3}}{454}}{3.050 \dfrac{530}{535} \dfrac{29.8}{29.92}} = 7.71 \times 10^{-5} \text{ lb/scf.}$$

Now to determine the emission rate of SO in lb/10 BTU, average the two samples,

$$(C_{SO_2})_{avg} = \frac{(8.08 \times 10^{-5} + 7.71 \times 10^{-5})}{2}$$

$$= 7.93 \times 10^{-5} \text{ lb } SO_2\text{/dscf.}$$

then multiply by the volumetric flow rate and divide by the heat input rate in 10^6 BTU/hr. Thus

$$(\overline{PMR}_{SO_2})_{BTU} = \frac{(7.93 \times 10^{-5} \frac{lb}{scf})(1.23 \times 10^{7} \frac{scf}{hr})}{(1.03 \times 10^{3})\ 10^{6}\ \frac{BTU}{hr}}$$

$$= 0.946 \frac{lb\ SO_2}{10^6\ BTU}$$

Nitrogen Oxide Emissions

The nitrogen oxides, NO, NO_2, N_2O_4, and N_2O_5, are the most common NO_x air pollutants from combustion processes. A small amount of the NO_x emitted may be formed by the combustion of nitrogenous compounds in the fuel burned. However most of the nitrogen oxides emitted are formed by the oxidation of N_2 in the combustion air by excess combustion oxygen. The main factors in NO_x production in combustion processes are flame temperature, residence time, and excess combustion oxygen. Therefore, there are no simple calculations that can be made for theoretical NO_x production (as can be made for SO_2 emissions). Fossil-fueled combustion processes, however, typically emit NO_x in concentrations of 100 to 1000 ppm (as NO_2). Most of the nitrogen oxides emitted to the atmosphere are in the form of NO and NO_2.

According to the new federal performance standards, EPA Methods 1, 2, 3, and 7 are used to determine NO_x emission rates. An NO_x source test consists of four individual samples. Using the data from Examples 15.1, 15.2, and 15.4, the NO_x and particulate source test data sheets and the laboratory analysis data sheet, the NO_x emissions can be calculated as follows:

Grab Sample Flask Volume

$$V_{SC} = \frac{T_{std}}{P_{std}}\left(\frac{P_f}{T_f} - \frac{P_i}{T_i}\right)(V_f - V_a) \qquad (15\text{-}15)$$

where

V_{SC}	is the sample volume at standard conditions (dry basis), ml
P_f & P_i	are the final and initial absolute pressures of the sampled gas, in. Hg
T_f & T_i	are the final and initial absolute pressures of the sampled gas, °R
V_f	is the volume of the flask and valving (usually 2000 ml), ml
V_a	is the volume of absorbing solution added (usually 25 ml), ml.

From the preceding data sheets:

Sample	V_f (ml)	V_a (ml)	P_i (in. Hg)	P_f (in. Hg)	T_i (°R)	T_f (°R)
1	2010	25	1.8	29.8	535	530
2	2005	25	1.8	29.8	535	530
3	1990	25	1.8	29.8	535	530
4	1990	25	1.8	29.8	535	530

Then

$$V_{SC_1} = \frac{530}{29.92}\ \frac{29.8}{530}\ \frac{1.8}{535}\ (2010-25)$$

$$= 1.85 \times 10^3 \text{ ml}$$

and

$$V_{SC_2} = 1.85 \times 10^3 \text{ ml}$$

$$V_{SC_3} = 1.83 \times 10^3 \text{ ml}$$

$$V_{SC_4} = 1.83 \times 10^3 \text{ ml.}$$

<u>NO_x Concentration Calculation.</u> Given the sample volume and the mass of NO_x collected, the concentration can be calculated.

$$C_{NO_x} = \frac{M_{NO_x}}{V_{SC}\text{ml}}\ \frac{28.3 \times 10^3 \frac{\text{ml}}{\text{scf}}}{454 \text{ g/lb}} \qquad (15\text{-}16)$$

where

C_{NO_2}	is the concentration of NO_x (as NO_2) (dry basis), lb/dscf
M	is the mass of NO_x collected (as NO_2), grams
V_{SC}	is the volume of the sample at standard conditions (dry basis), ml.

Then using the data from the laboratory analysis,

$$NO_{x_1} = 390\ \mu g \text{ as } NO_2$$

$$NO_{x_2} = 398\ \mu g \text{ as } NO_2$$

$$NO_{x_3} = 405\ \mu g \text{ as } NO_2$$

$$NO_{x_4} = 387\ \mu g \text{ as } NO_2$$

$$C_{NO_{x_1}} = \frac{3.90 \times 10^{-4}}{1.85 \times 10^{3}} \frac{2.83 \times 10^{4}}{4.54 \times 10^{2}}$$

$$= 1.31 \times 10^{-5} \text{ lb/dscf}$$

and

$$C_{NO_{x_2}} = 1.34 \times 10^{-5} \text{ lb/dscf}$$

$$C_{NO_{x_3}} = 1.38 \times 10^{-5} \text{ lb/dscf}$$

$$C_{NO_{x_4}} = 1.32 \times 10^{-5} \text{ lb/dscf.}$$

According to EPA Method 5 for each replicate NO_x source test consisting of four samples:

$$\overline{C}_{NO_x} = \frac{\sum_{i=1}^{4} C_{NO_{x_i}}}{4} \tag{15-17}$$

where

$\overline{C}_{NO_x}$	is the average NO_x concentration for each repetition, lb/scf.

Thus,

$$\bar{C}_{NO_x} = \frac{(1.31 + 1.34 + 1.38 + 1.32)(10^{-5})}{4}$$

$$= 1.34 \times 10^{-5} \text{ lb/scf.}$$

The mass emission rate of NO_x is then calculated:

$$\overline{PMR}_{NO_x} = \bar{C}_{NO_x} \bar{Q}_S \qquad (15\text{-}18)$$

$$= (1.34 \times 10^{-5}) \frac{\text{lb}}{\text{dscf}} (1.23 \times 10^{7}) \frac{\text{dscf}}{\text{hr}}$$

$$= 1.65 \times 10^{2} \text{ lb/hr.}$$

Computing the emission rate standardized according to heat input,

$$(\overline{PMR}_{NO_x})_{BTU} = \frac{\overline{PMR}_{NO_x}}{\dot{H}} = \frac{1.65 \times 10^{2} \frac{\text{lb}}{\text{hr}}}{(1.03 \times 10^{3})\ 10^{6} \frac{\text{BTU}}{\text{hr}}}$$

$$= 0.160 \text{ lb } NO_x/10^{6} \text{ BTU.}$$

Sulfur Trioxide and H_2SO_4 Mist Emissions

A small portion of the sulfur in fossil fuels may be oxidized to sulfur trioxide during the combustion process. At lower temperatures sulfur trioxide hydrolyzes to form sulfuric acid when reacted with water droplets or vapor. Thus SO_3 and H_2SO_4 mist might be somewhat of an air pollution problem from some combustion processes. EPA Method 8 is intended for use in quantifying the SO_3 and H_2SO_4 mist emissions from sulfuric acid production processes. However, here the method is applied to SO_3 and H_2SO_4 emissions from the combustion process.

For purposes of illustration, the velocity traverse and sample volume data from the preceding particulate test data sheet in Example 15.1 will be used.

Thus

$$\overline{Q}_s = 1.23 \times 10 \quad \text{dscfh}$$

$$V_{m_{std}} = 94.9 \text{ dscf}$$

$M_{H_2SO_4} = 51.5$ mg (SO_3 and H_2SO_4 as H_2SO_4) from lab analysis data sheet Example 15.2

Then

$$C_{H_2SO_4} = \frac{M_{H_2SO_4}}{V_{m_{std}}} \; \frac{1 \times 10^{-3} \text{ g/mg}}{4.54 \times 10^{2} \text{ g/lb}} \tag{15-19}$$

$$= \frac{51.5}{94.9} \; \frac{1 \times 10^{-3}}{4.54 \times 10^{2}}$$

$$= 0.120 \times 10^{-5} \text{ lb/dscf}$$

and

$$\overline{PMR}_{H_2SO_4} = (0.120 \times 10^{-5}) \frac{\text{lb}}{\text{dscf}} \; (1.23 \times 10^{7}) \frac{\text{dscf}}{\text{hr}} \tag{15-20}$$

$$= 14.7 \; \frac{\text{lb}}{\text{hr}} \text{ of } SO_3 \text{ and } H_2SO_4 \text{ (as } H_2SO_4\text{)}$$

ISOKINETIC SAMPLING FIELD CALCULATIONS

As discussed in Chapter 10, isokinetic conditions must be met when sampling for particulate matter. Particulate matter is distributed according to size and density with the velocity of the stack gas. Thus, to extract a representative sample, the velocity at the entrance to the sampling nozzle must be equal to the stack gas velocity at the sampling. When using the EPA particulate train,[1] an S-type pitot tube is used to measure stack gas velocity and an orifice meter is used to measure the sample gas velocity at the nozzle entrance. The stack gas velocity at a sampling point as measured with an S-type pitot is expressed as

$$V_s = K\, C_p \left(\frac{\Delta P\, T_s}{P_s\, M_s} \right)^{1/2} \qquad (15\text{-}21)$$

where

- V_s is the stack gas velocity, fps
- K is a constant, 85.48 $\frac{ft}{sec} \left(\frac{lb}{lb\ mole\text{-}°R} \right)^{1/2}$
- C_p is the pitot tube correction factor
- ΔP is the velocity pressure, in. H_2O
- T_s is the absolute stack gas temperature, °R
- P_s is the absolute stack gas pressure, in. Hg
- M_s is the molecular weight of the stack gas, $\frac{lb}{lb\ mole}$.

The flow rate through the calibrated orifice meter is

$$Q_m = C_m\, A_m \left(\frac{T_m\, \Delta m}{P_m\, M_m} \right)^{1/2}$$

where

- Q_m is the volumetric flow rate, cfs
- A_m is the orifice area, ft^2
- T_m is the absolute temperature of the metered gas, °R
- Δm is the orifice pressure differential in. H_2O
- P_m is the absolute pressure of the metered gas, in. Hg
- M_m is the molecular weight of the metered gas, $\frac{lb}{lb\text{-}mole}$
- C_m is the orifice meter coefficient.

The velocity at the nozzle entrance can then be calculated

$$V_n = \frac{\left(Q_m \frac{T_s\, P_m}{T_m\, P_s\, M_d} \right)}{(A_n)} \qquad (15\text{-}22)$$

where

A_n is the area of the nozzle, ft^2

M_d is the mole fraction of the dry gas in the stack.

To maintain isokinetic conditions, the velocity at the nozzle entrance, V_n, must equal the stack gas velocity at the sampling point, V_s. Thus

$$V_n = V_s$$

or

$$\frac{Q_m T_s P_m}{A_n T_m P_s M_d} = K\, C_p \left(\frac{\Delta P\, T_s}{P_s M_s}\right)^{1/2} \qquad (15\text{-}23)$$

Since

$$Q_m = C_m A_m K \left(\frac{T_m \Delta m}{P_m M_m}\right)^{1/2} \qquad (15\text{-}24)$$

Then Equation 15-23 can be reduced to

$$\Delta m = \left(\frac{A_n}{C_m A_m} M_d\, C_p\right)^2 \frac{T_m P_s M_m}{P_m M_s} \frac{\Delta P}{T_s} \qquad (15\text{-}25)$$

or since $A_n = \frac{\pi D^2}{4}$ where D is the nozzle diameter, ft,

$$\Delta m = K_2 D^4 \frac{\Delta P}{T_s}, \qquad (15\text{-}26)$$

where

Δm is the orifice meter pressure differential, in. H_2O

K_2 is $\left(\frac{\pi}{4C_m A_m} M_d C_p\right)^2 \frac{T_m P_s M_m}{P_m M_s}$ (15-27)

D is the diameter of the nozzle, ft

ΔP is the velocity pressure, in. H_2O

T_s is the absolute stack gas temperature at the sampling point, °R.

Thus for any velocity pressure (assuming K_2 and D are constant) encountered in source sampling the orifice meter differential pressure, Δm, which will produce isokinetic sampling conditions, can be calculated, given the stack gas temperature at the sampling point. However, even though K_2 is relatively constant throughout any specific sampling period, it may change from test to test on the same source. The calculation of K_2 is tedious and requires a fair amount of time. Furthermore the possibility of error when doing the calculation under field conditions is great. Nomographs can be developed to reduce the calculation time and the possibility of error. Smith *et al.*[2] have developed such nomographs, shown in Figures 15.1 and 15.2. These nomographs are commercially available (see Chapter 16) and in widespread use. Therefore, the use and limitations of these nomographs will be discussed. The nomograph in Figure 15.1 is a four independent variable (Δm, T_m M_d, and $\frac{P_s}{P_m}$) nomograph used to calculate C, the ratio of the true value of K_2, to some assumed value of K_2.

In thiscase the assumed value of K_2 used is calculated on the following basis.

$$M_{d_{stack}} = 0.95$$

$$P_s = P_m = 29.92 \text{ in. Hg}$$

$$T_m = 70°\text{F}$$

$$\Delta m = f(C_m, A_m) = 1.84 \text{ (}\Delta m\text{ which gives 0.75 scfm)}$$

$$M_{d_{meter}} = 1.00$$

$$M_m = 29 \text{ lb/lb-mole}$$

$$C_p = 0.85$$

and

$$M_s = 0.95\ M_m + 0.05\ (18).$$

The nomograph is in terms of Δm, T_m, M_d, and P_s/P_m; therefore it corrects for encountered values of Δm, T_m, M_d, and P_s/P_m that do not equal the assumed values. However, the nomograph (Figure 15.1) cannot correct for moisture passing through the orifice meter, for gases that have dry molecular weights other than 29, and for pitot factors, C_p,

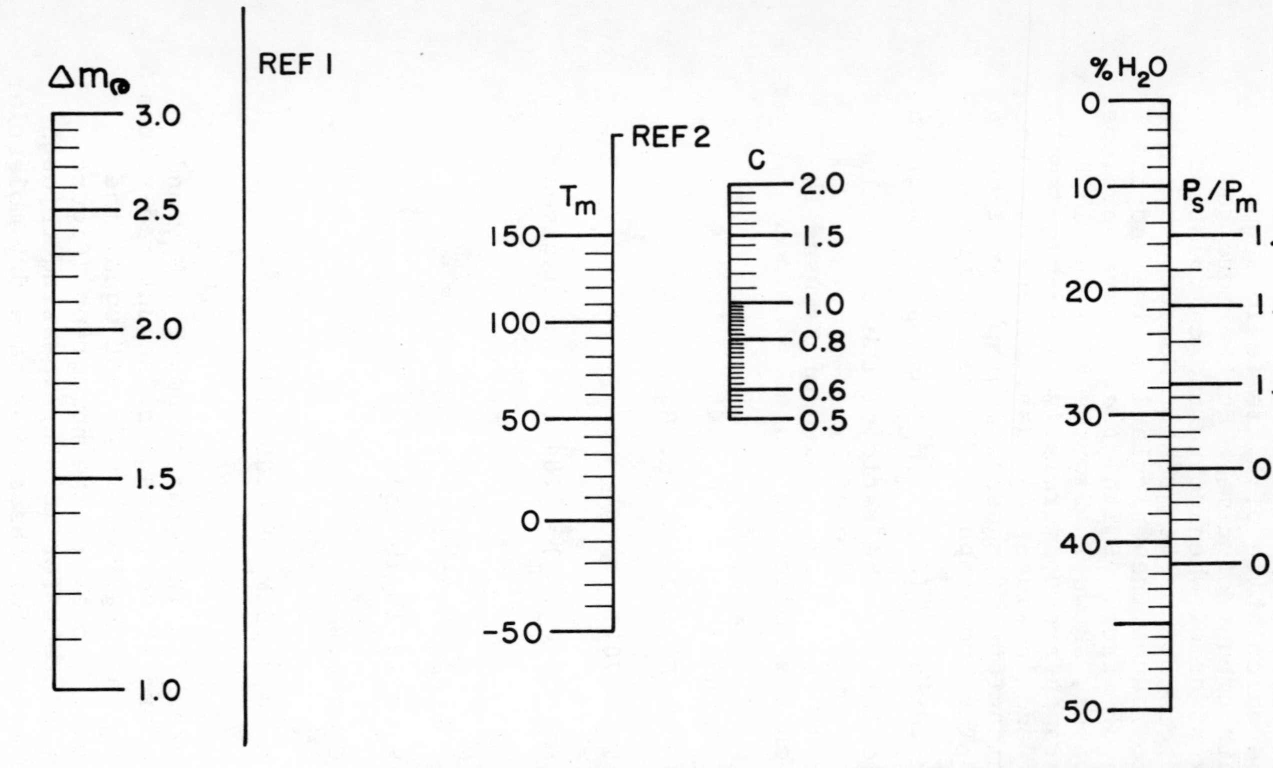

Figure 15.1. *Nomograph for calculating C, a ratio of true to assumed K_2.*[2]

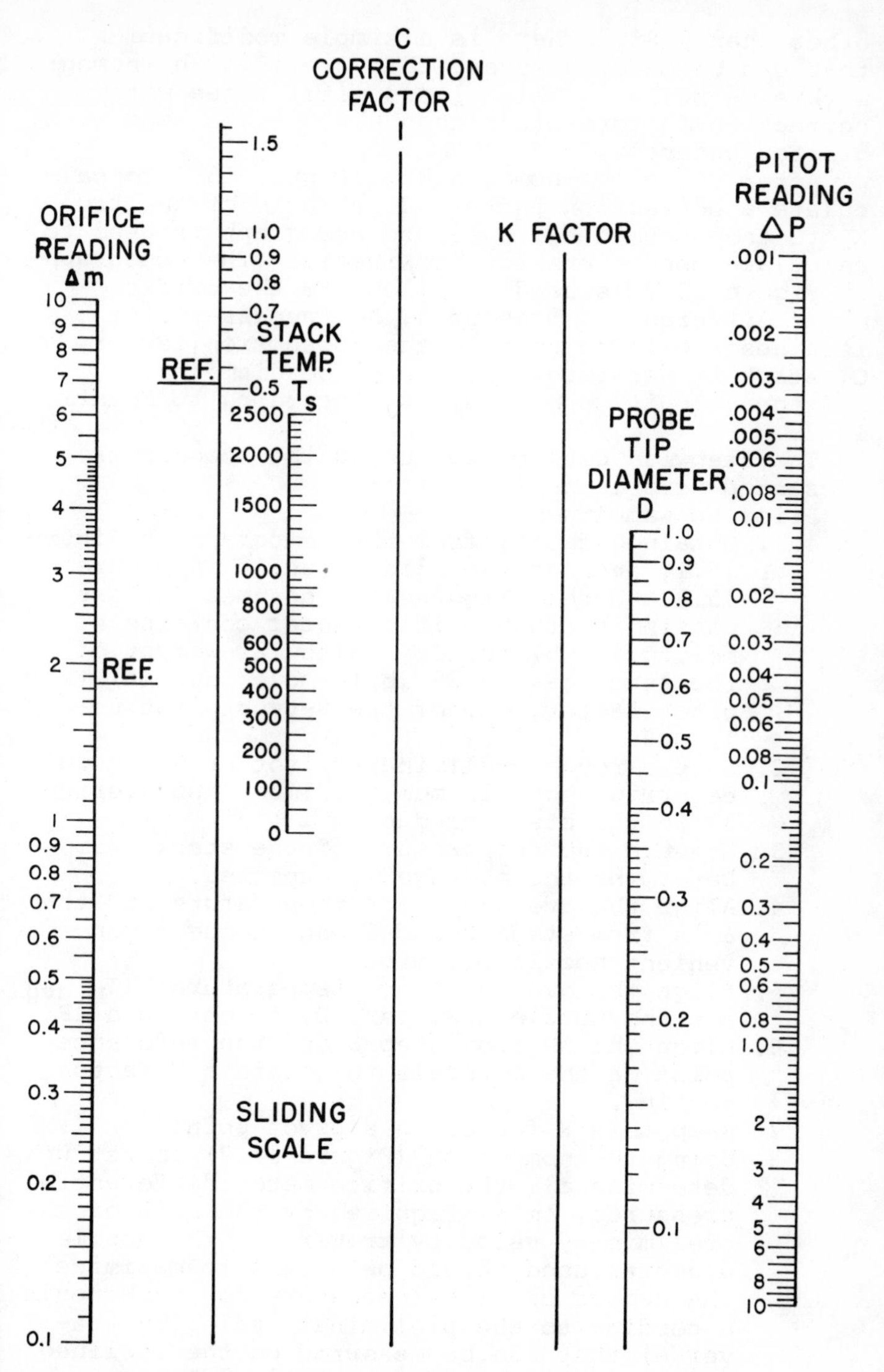

Figure 15.2. *Nomograph for calculating isokinetic source sampling conditions given C (K_2 ratio), T_s, D, and ΔP.*[2]

other than 0.85. There is a simple modification that can be made, however, that provides the nomograph with a C_p scale.[3] This allows pitot tubes with correction factors other than 0.850 to be used with greater accuracy.

After using the nomograph in Figure 15.1 to calculate a correction factor, C, a ratio of the true K_2 to the assumed K_2, a second nomograph is used to calculate conditions for isokinesis. The nomograph in Figure 15.2 is used to calculate the orifice meter differential pressures, Δm, necessary for isokinesis with respect to the correction factor, C, stack temperature, T_s, the probe diameter, D, and the velocity pressure, ΔP, according to Equation 15-25.

The stepwise outline for using the nomographs presented here is as follows:

Prior to sampling:

1. Obtain C factor from the nomograph in Figure 15.1, and set the sliding scale in Figure 15.2. Note: Figure 15.1 assumes (a) gas passing through orifice meter contains no moisture, (b) the dry molecular weight of the stack gas is 29 lb/lb-mole, and (c) the pitot factor, C_p, of the S-type pitot used is 0.85.
2. Make a rough preliminary pitot traverse to determine the minimum, maximum, and average ΔP in the stack or duct.
3. Measure the temperature of the stack gas. Determine the average T_s expected.
4. Align the average stack temperature and the ΔP's from steps 2 and 3 and choose a convenient nozzle diameter.
5. Align the average stack temperature, $(T_s)_{avg}$, and the nozzle diameter, D, to obtain a ΔP.
6. Align the ΔP from step 5 and the reference point on the Δm scale to obtain a K factor setting.
7. Keep this K factor as a pivot point.
8. Using the nomograph (Figure 15.2) as set up, determine all the orifice meter differential pressures, Δm's, required by the ΔP's of the preliminary velocity traverse. The nozzle diameter used should be chosen to maximize the number of Δm's (necessary for isokinesis according to the preliminary velocity traverse) that can be measured on the inclined scale of the manometer. This scale is considerably more accurate than the vertical scale.

During sampling:

9. For each velocity pressure, ΔP, encountered, determine the orifice meter differential pressure, Δm, necessary for isokinetic sampling conditions. This assumes T_s does not change significantly from point to point.
10. If T_s changes significantly (*e.g.*, ΔT_s °R/$T_{s_{avg}}$°R > 0.02), repeat steps 3 through 9.

REFERENCES

1. "Standards for Performance of New Stationary Sources," Federal Register, Vol 36, No. 247, December 23, 1971.
2. Smith, W. S., Martin, R. M., Durst, D. E., Hyland, R. G., Logan, T. J., and Hagar, C. B. "Stack Gas Sampling Improved and Simplified with New Equipment," Paper 67-119 58th Annual Meeting of the Air Pollution Control Association, June, 1967.
3. Campbell, Donald., The Research Corporation of New England, Wethersfield, Connecticut, Personal Communication, June, 1972.

CHAPTER 16

COMMERCIAL SOURCE TESTING EQUIPMENT

This chapter discusses commercially available source testing equipment. Comparisons are made of various trains on a component basis. The authors have not used all of these units and therefore the comments will be restricted to discussion of the components only. All performance information is that provided by manufacturers. Chapter 4 should be used as a reference for the presentation of the theory, operation and merits of various components used in these trains.

The emphasis of the chapter is on the Federal EPA Procedures[1] and the commercial trains available for these methods. Figure 16.1, Method 5 (Chapter 10) and Figure 16.2, Method 8 (Chapter 12), are reproduced here. The gaseous trains, Figure 16.3, Method 6 (Chapter 11) and Figure 16.4, Method 7 (Chapter 13) are also presented here. These are the basic trains which the commercial units follow.

It is the opinion of the authors that the construction of a complete train is ill-advised if the builder has not had previous sampling experience. The authors also advise against the construction of a complete train following the Construction Details.[2] Once sampling experience has been developed, it may be less expensive to build the components necessary. Most commercial trains are designed for general sampling cases. Trains fabricated in-house may be better designed for specific cases, the separation of filter oven and condensor discussed in Chapter 10 being an example of this.

Inherent in the proposition of actually building a train or components is the existence of a shop facility. The building or routine maintenance of source sampling equipment requires a shop working area. This area can also be used for calibration

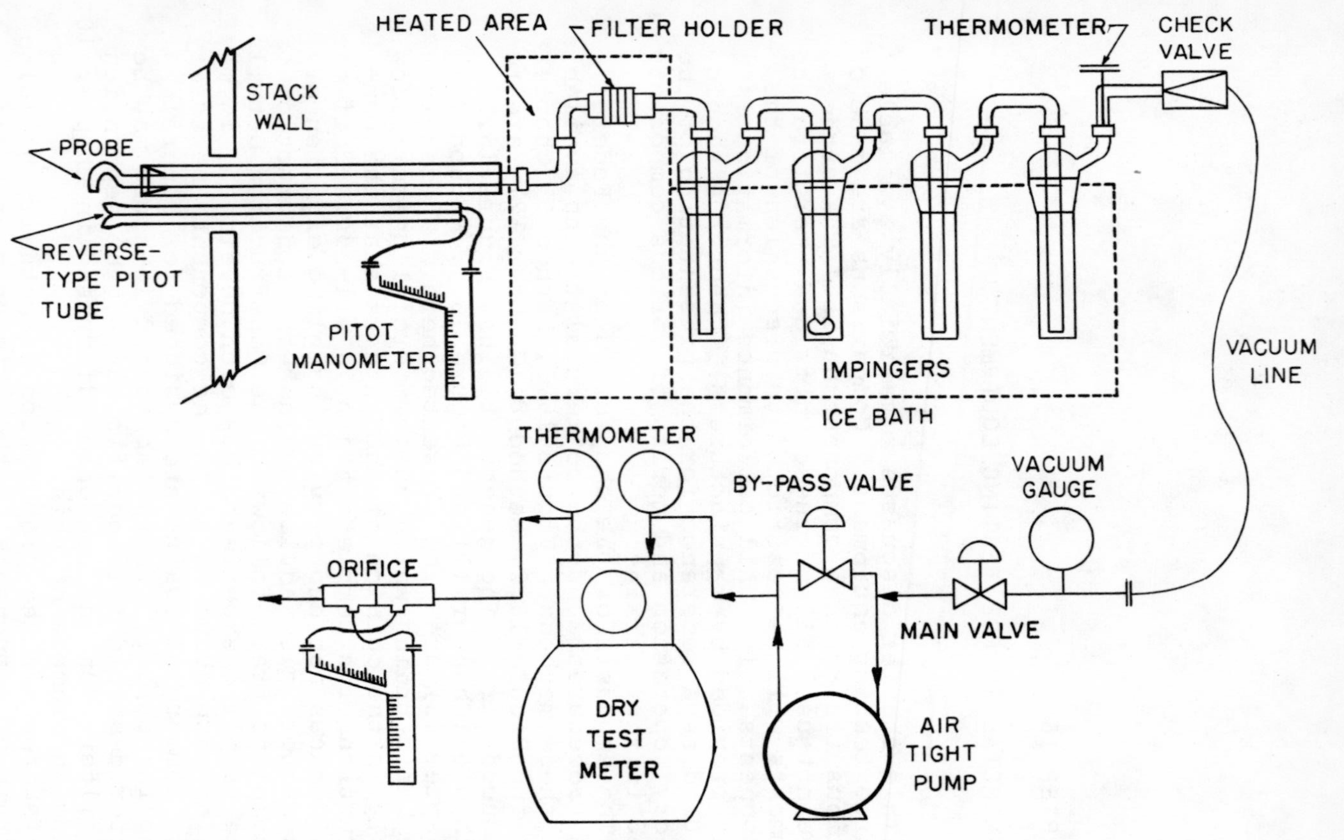

Figure 16.1. *Particulate sampling train (EPA Method 5).*

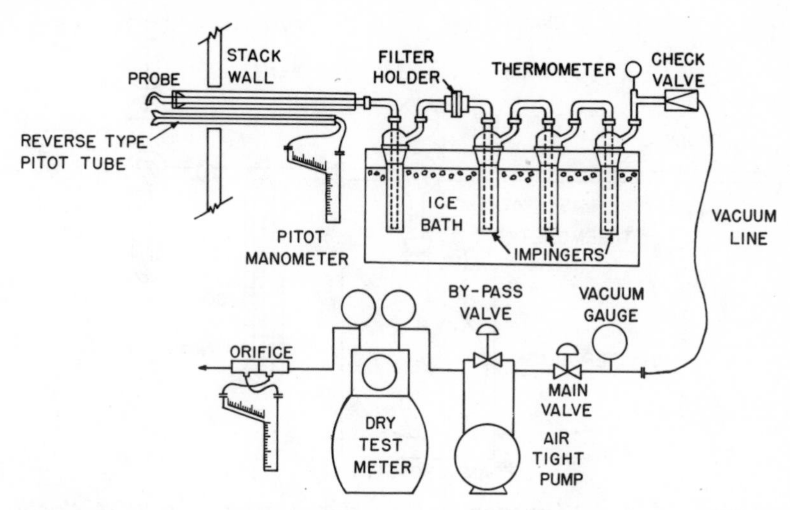

Figure 16.2. Sulfuric acid mist sampling train (EPA Method 8).

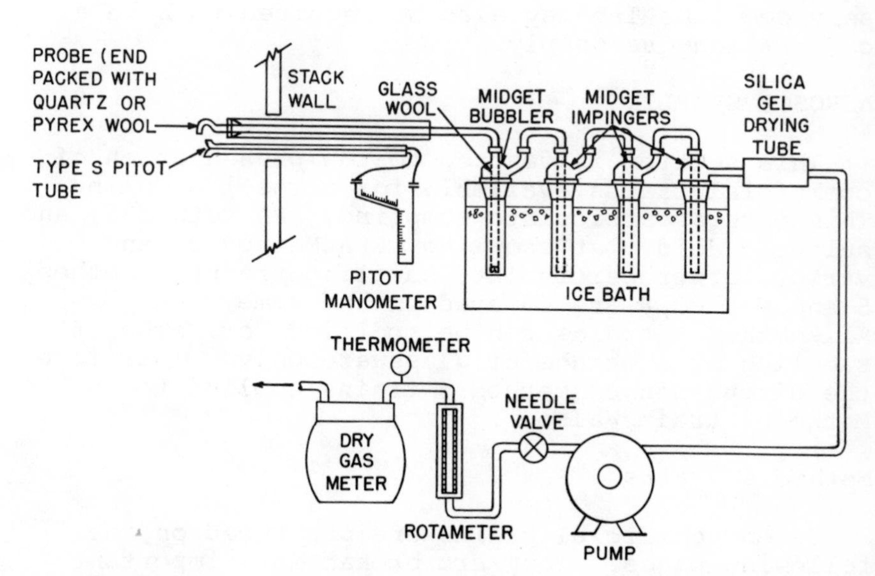

Figure 16.3. Sulfur dioxide sampling train (EPA Method 6).

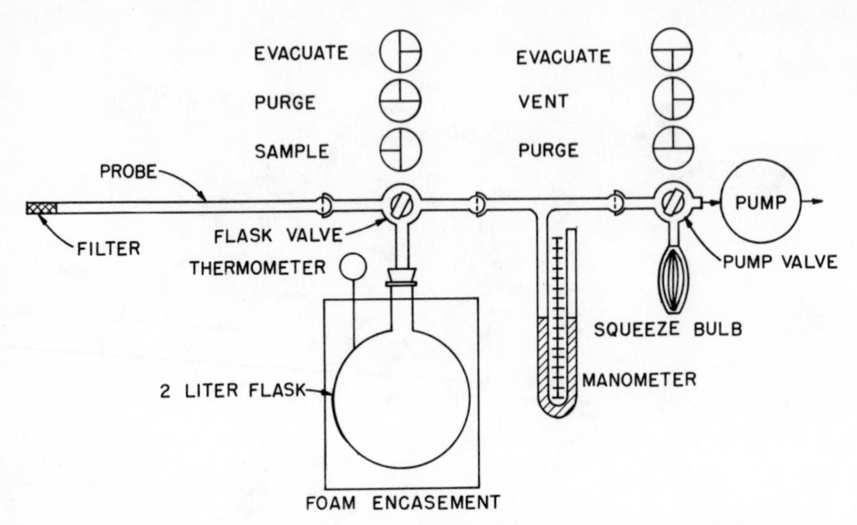

Figure 16.4. Nitrogen oxides sampling train (EPA Method 7).

of the equipment in preparation for testing. A laboratory is also necessary to successfully conduct a source sampling program. It must contain an analytical balance, oven, fume hood, gas supply, chemical supply, and other standard glassware and equipment. Depending on the sampling devices serviced, the lab may also be required to have a calibration gas supply.

AEROSOL SAMPLING

This section is devoted to the presentation of commercial trains available for aerosol sampling. This covers particulate sampling, EPA Method 5, and sulfuric acid mist sampling, EPA Method 8, and various other particulate sampling trains. Method 5 and Method 8 are covered in the same section. All Method 5 trains can be modified for Method 8 sampling by a change of glassware only. Therefore the discussion of Method 5 trains applies to Method 8 trains also.

Method 5 Trains

Seven commercial trains are presented on the following pages. They are broken into important categories which can be used in comparing components between trains. Also listed are the *advertised unit price* and a *second price*. This second price is based on the following suggested sampling equipment supply. It is not purported as an ideal train since each sampling program will have different requirements in design. It is hoped, however, that this train will give a somewhat more realistic comparison price than the manufacturer's advertised price.

Suggested Sampling Equipment Supply

2 sample units (filter box and impinger bath)
2 5-ft probes (heated and with thermocouples and pitots)
100 ft of umbilical cord
1 control module
1 nomograph
4 sets of glassware (with maximum filter size available)
2 sample unit support systems
1 thermocouple temperature read out device.

This example train list was developed based on the following assumptions:

1. All stacks and ducts can be sampled with 5 ft probes.
2. No sampling location requires more than 100 ft of umbilical cord.
3. All sampling is done with the probe in a horizontal position.
4. All particulate sampling must be done with the impingers connected directly to the probe.
5. Sampling will be done at one location at a time (*e.g.*, no efficiency tests will be run).
6. All calibration equipment (*e.g.*, wet test meters, standard pitot, wind tunnel, temperature baths, etc.) is available.
7. That two sample support systems will be used in tandem whenever possible to reduce sampling time.
8. Extra equipment such as vacuum pumps and dry gas meters are available from local distributors.

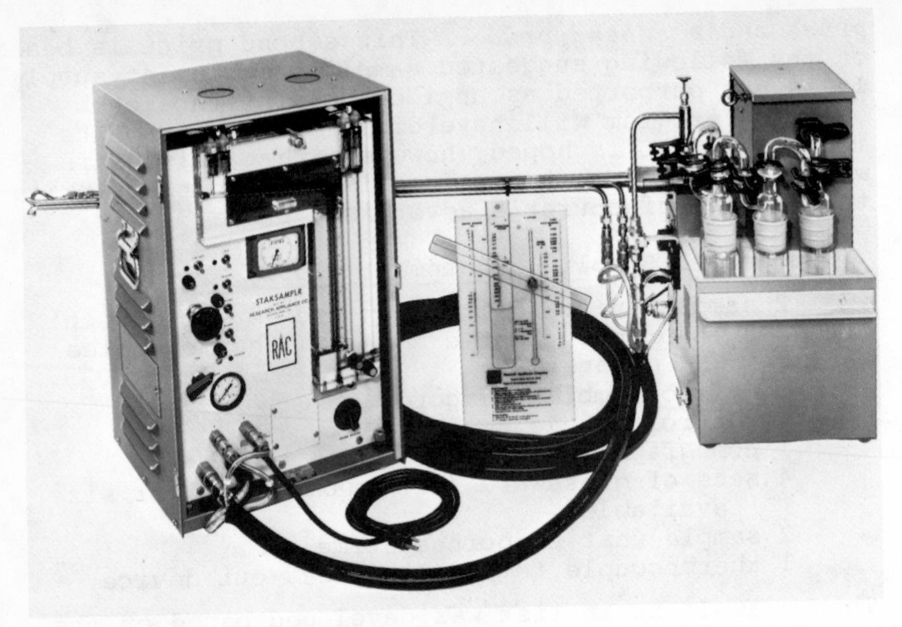

Figure 16.5. *RAC train Staksamplr, Model 2343 (courtesy of Research Appliance Co.).*

Manufacturer	Research Appliance Co. Route 8 Allison Park, Pennsylvania 15107
Basic Design	Probe, sample unit, and control module (follows EPA Construction Details)
Probe	Combination probe and pitot with detachable thermocouple Standard lengths: 3, 5, and 10 ft Outer liners available: stainless steel Inner liners available: stainless steel, Incronel, Pyrex, or quartz Nozzle sizes: 1/4, 3/8, and 1/2 in. Jacketed high temperature probes available Special features: in-stack filters or impactors available

Sample Unit	Single case for filter oven and impinger bath Filter: maximum 4 in. dia, 4 screw retainer Cyclone may be used as precleaner Filter oven temperature control: thermostat Temperature measurements: dial thermometers for filter oven and final impinger Umbilical cords: 25, 50, and 100 ft (can be interconnected) Support system: duo rail Special features: vertical sampling capability.
Control Module	Vacuum pump with fiber vanes. Dry gas meter: 0.1 cf/rev Pressure measurement: 0-10 in. H_2O dual manometer Temperature measurement: dry gas meter uses dial thermometers; thermocouple-separate pyrometer Heater control for probe Nomograph available Special features: timer
Specifications	Control module: 115V, 20 amps, 87 lbs, 24H x 15W x 13½D in. Sample unit: 39½ lbs with glassware, 21H x 12W x 15D in.
Costs	$2661/$5715
Remarks	All the above information was obtained from the manufacturer's bulletins and personal communications.[3]

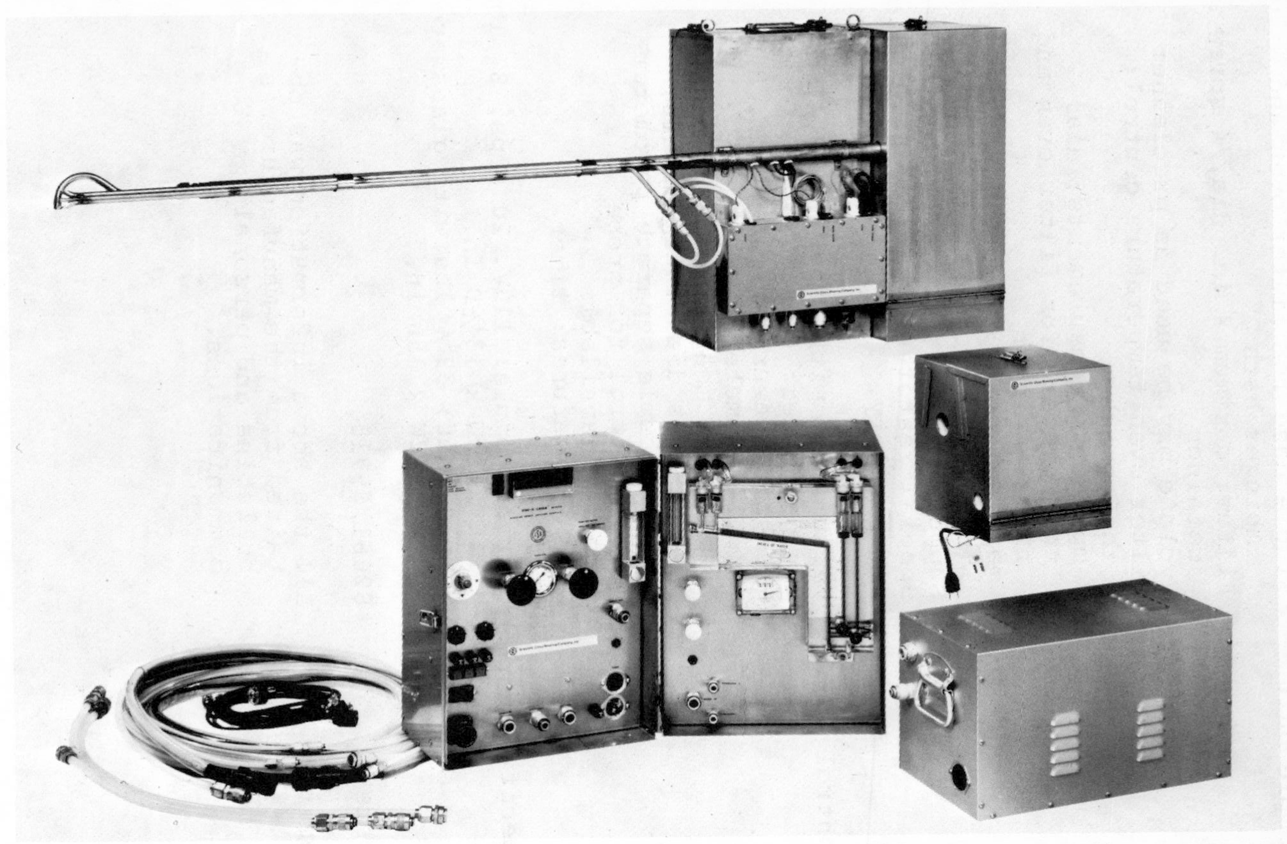

Figure 16.6. *Stac-o-lator Model AP-5000 (courtesy of Scientific Glass Blowing Co., Inc.).*

Manufacturer	Scientific Glass Blowing Co., Inc. P.O. Box 18306 Houston, Texas 77023
Basic Design	Probe, sample unit (separable impinger case, cyclone oven, and filter oven) and control module (separate vacuum pump).
Probe	Probe with detachable pitot and thermocouple combination Standard lengths: 3, 5 and 10 ft Outer liner available: stainless steel Inner liners available: Pyrex, stainless steel, quartz or Inconel Nozzle sizes: 1/4, 3/8, and 1/2 in. Inner liner heater standard Jacketed high temperature probes available Special features: probe liner thermocouple
Sample Unit	Modular sample case, impinger bath attaches to interchangeable filter or cyclone ovens Filter: maximum 4 in. dia, 2 piece compression nut retainer Cyclone may be used as precleaner Filter oven temperature control: input controller Temperature measurements: filter ovens and final impinger uses thermocouples Umbilical cords: 25, 50, 75 and 100 ft (can be interconnected) Support system: monorail Special features: vertical sampling capability; can also be adopted for gas sampling (EPA Methods 4, 6, 7 and 8)
Control Module	Vacuum pump with fiber vanes (separate from control module) Dry gas meter: 0.1 cf/rev Pressure measurement: 0-5 in. H_2O dual manometer

	Temperature measurement: dry gas meter with thermocouple temperature read out, pyrometer calibrated for the 6 train thermocouples Heater control for probe and filter ovens Nomograph available Special features: timer, rotameter, valving system for gas methods
Specifications	Control module: 115v, 15 amps, 40 lbs, 18H x 14W x 18D in. Vacuum pump case: 8 amps, 4 cfm at 0"Hg vacuum, 45 lbs, 18H x 9W x 10D in. Impinger case: 13½ lbs without glassware, 21½H x 12W x 9½D in. Cyclone oven: 12½ lbs without glassware, 21½H x 9½W x 9½D in. Filter oven: 8 lbs without glassware, 10½H x 9½W x 9½D in.
Costs	$5000*/6475 without filter ovens *quoted on individual basis no base limit
Remarks	This unit could be extremely useful to a company which plans its own tests in that, properly outfitted, it can be used for both particulate and gaseous source testing. All of the above information was obtained from manufacturer's bulletins and personal communications.[4]

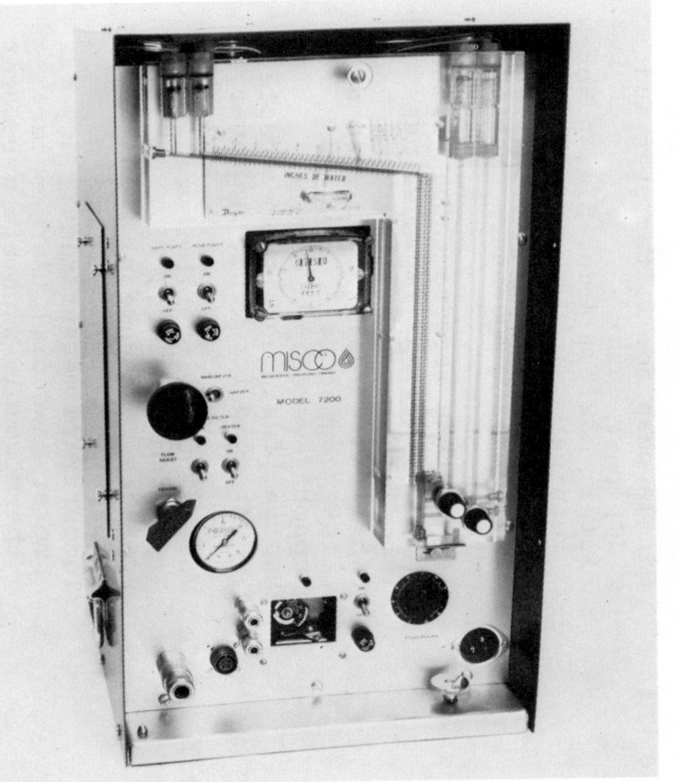

a) meter box

b) sample box

Figure 16.7. MISCO Source Sampler, Model 7200 (courtesy of Microchemicals Specialities Co.).

Manufacturer	Microchemical Specialities Co. 1825 Eastshore Highway Berkeley, California 94710
Basic Design	Probe, sample unit, and control module (follows EPA Construction Details)
Probe	Combination probe and pitot with detachable thermocouple Standard lengths: 3, 5, 7, and 10 ft Outer liners available: stainless steel Inner liners available: Pyrex, stainless steel, or quartz Nozzle sizes: 1/8, 1/4, 3/8 and 1/2 in. (gooseneck or straight) Inner liner heater optional Jacketed high temperature probes available Special features: glass nozzles available; in-stack filters or impactors available
Sample Unit	Single case for filter oven and impinger bath Filter: maximum 4 in. dia, 4 screw retainer Cyclone may be used as precleaner Filter oven temperature control: thermostat Temperature measurements: filter oven and final impinger uses dial thermometer Umbilical cords: 50 ft (can be interconnected) Support system: duo rail Special features: thermocouple available for final impinger
Control Module	Vacuum pump with fiber vanes Dry gas meter: 0.1 cf/rev Pressure measurement: 0-10 in. H_2O dual manometer Temperature measurement: dry gas meter uses dial thermometer; thermocouple-separate pyrometer

	Heater control for probe Nomograph available Special features: high flow forced air cooling; timer
Specifications	Control module: 115v, 14 amps, 70 lbs, 23¾H x 13½W x 13½D in. Sample unit: 115v, 7 amps, 34 lbs, 20H x 12W x 15D in.
Costs	$2290/4810
Remarks	All the above information was obtained from the manufacturer's bulletins and personal communications.[5]

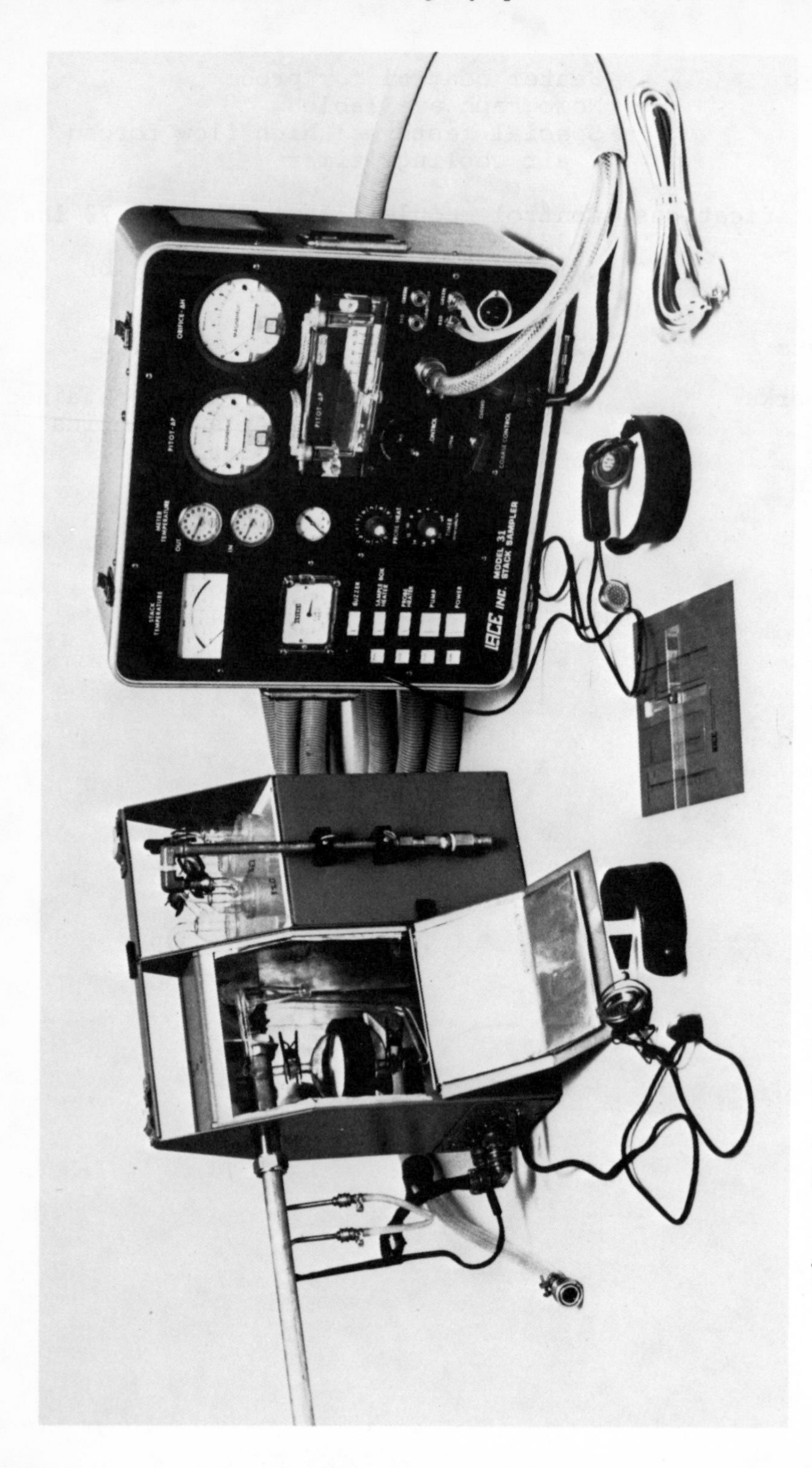

Figure 16.8. Stack Sampler Model 31 (courtesy of LACE Engineering Co.).

Manufacturer	LACE Engineering Co. 8829 North Lamar P.O. Box 9757 Austin, Texas 78766
Basic Design	Probe with sample unit (separable oven and impinger bath) and control module
Probe	Combination pitot and probe with detachable thermocouple Standard lengths: 5 and 10 ft Outer liner available: stainless steel Inner liners available: Pyrex or stainless steel Nozzle sizes: 1/4, 3/8, and 1/2 in. Inner liner heated standard Jacketed high temperature probes available
Sample Unit	Separable impinger bath and filter oven Filter: maximum 4 in. dia, compression nut retainer Cyclone may be used as precleaner Filter oven temperature control: thermostat Temperature measurements: filter oven (none); final impinger uses dial thermometer Umbilical cords: 25, 50, 75, 100 ft (can be interconnected) Support system: monorail Special features: vertical sampling capability
Control Module	Vacuum pump with diaphragm (no surge tank) Dry gas meter: 0.1 cf/rev. Pressure measurements: ΔP, 0-4 in. H_2O Magnehelic and 0-0.5 in. H_2O inclined manometer; Δm, 0-10 in. H_2O Magnehelic gauge. Temperature measurements: dry gas meter uses dial thermometers; pyrometer readout for stack thermocouples

	Heater control for probe Nomograph available Special features: forced air cooling fan; fiberglass case; telephone communication system; timer
Specifications	Control module: 120v, 10 amps, 55 lbs, 20H x 21W x 14D in. Filter oven: 5 lbs without glassware, (7) 19H x 8W x 8D in. Impinger case: 5 lbs without glassware, (7) 19H x 8W x 8D in.
Costs	$3285/6190
Remarks	The use of a low range manometer along with the Magnehelic gauge is an extremely good idea. All the above information was obtained from the manufacturer's bulletins and personal communications.[6]

Figure 16.9. GII-200 Source Sampler (control module shown) (courtesy of Glass Innovations).

Manufacturer	Glass Innovations, Inc. P.O. Box B Addison, New York 14801
Basic Design	Probe, sample unit, and control module
Probe	Probe with detachable pitot and thermocouple combination Standard lengths: 3, 5, and 10 ft Outer liners available: stainless steel or Inconel

	Inner liners available: stainless steel, Inconel, Pyrex, or quartz Nozzle sizes: 1/8 to 3/4 in. Inner liners heated standard Jacketed high temperature probes available Special features: probe liner thermocouple
Sample Unit	Single case for filter oven and impinger bath Filter: maximum 6 in. dia, 2-piece compression nut retainer Cyclone may be used as precleaner Filter oven temperature control: rheostat Temperature measurements: filter oven and final impinger uses thermocouples Umbilical cords: 25 ft (can be interconnected, however thermocouple requires recalibration) Support system: monorail Special features: probe located on center line of sample unit; reversed impinger taper joints to prevent impinger catch contamination; vertical sampling capability; umbilical cord clamp
Control Module	Vacuum pump uses diaphragm (no surge tank) Dry gas meter: 0.1 cf/rev Pressure measurement: 0-10 in. H_2O dual manometer Temperature measurement: dry gas meter uses thermocouple; temperature readout--pyrometer calibrated for the 6 train thermocouples Heater controls for probe and filter oven Nomograph available Special features: three-way valves on manometers for rezeroing during sampling; timer

Specifications	Control module: 110v, 15 amps, 65 lbs, 20½H x 21½W x 13½D in. Sample unit: 45 lbs with glassware, 22H x 10W x 20D in.
Cost	$3284/5632
Remarks	All the above information was obtained from the manufacturer's bulletins and personal communications.[7]

a) control unit

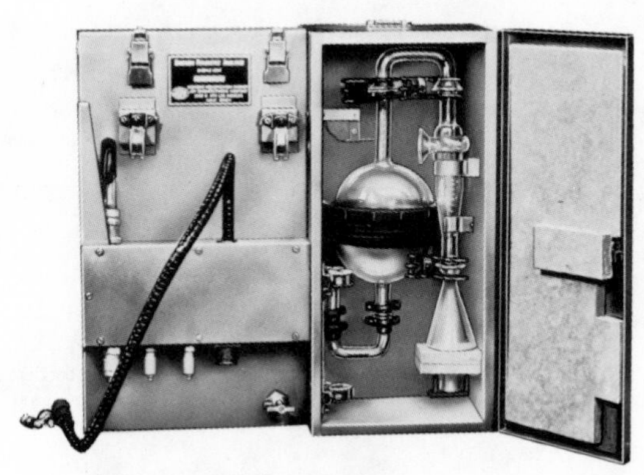

b) sampling unit

Figure 16.10. Emission Parameter Analyzer (courtesy of Western Precipitation Division/Joy Manufacturing Co.).

Manufacturer	Western Precipitation Division/ Joy Manufacturing Co. 1000 West Ninth Street Los Angeles, California 90015
Basic Design	Probe, sample unit, and control module
Probe	Probe with attached pitot and thermocouple combination Standard lengths: 5 and 10 ft Outer liner available: stainless steel

	Inner liner available: Pyrex or stainless steel Nozzle sizes: 1/4, 3/8 and 1/2 in. Inner liner heated standard
Sample Unit	Single case for filter and impinger bath Filter: maximum 5 in. dia, 2-piece compression nut retainer Cyclone may be used as precleaner Filter oven temperature control: rheostat Temperature measurements: final impinger, filter oven, entry impinger Umbilical cord: 25, 50, 75, and 100 ft (cannot be interconnected) Support system: monorail Special features: vertical sampling capability; condensor to replace impingers; portable refrigeration unit
Control Module	Vacuum pump: diaphragm (with surge tank) Dry gas meter (temperature compensated) Pressure measurement: 2 Magnehelic gauges Temperature measurements: dry gas meter temperature readout Heater control for probe and filter oven Nomograph (electronic meter rate computer) available Special features: timer; forced air cooling fan
Specifications	Control module: v, amps, lbs, H x W x D in. Sample unit: lbs glassware, H x W x D in.
Cost	Not available
Remarks	This is a carefully designed and compact unit.

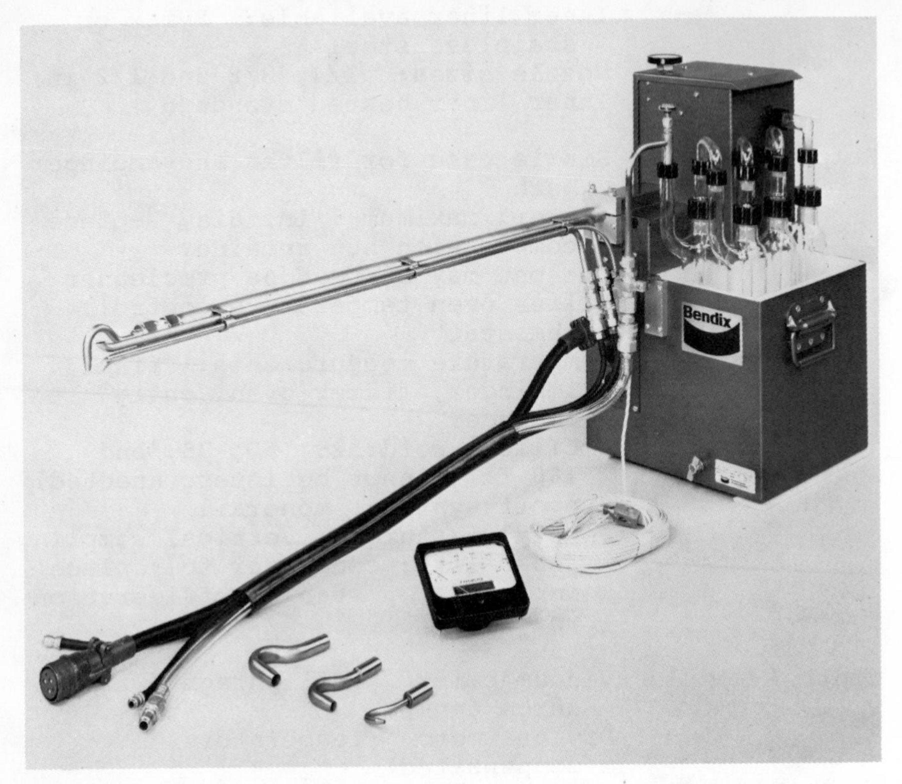

Figure 16.11. Bendix Isokinetic Stack Sampler-4500 Series (sample unit shown) (courtesy of Environmental Science Division of Bendix).

Manufacturer	Environmental Science Division of Bendix 1400 Taylor Avenue Baltimore, Maryland 21204
Basic Design	Probe, sample unit, and control module (follows EPA Construction Details)
Probe	Combination probe and pitot with detachable thermocouple Standard lengths: 3, 5, 10 and 15 ft Outer liners available: stainless steel Inner liners available: Pyrex Nozzle sizes: 1/8, 1/4, 3/8, and 1/2 in. Inner liners heated option

Sampling Unit	Single case for filter oven and impinger bath Filter: maximum 3 in. dia, 2-piece compression nut retainer Cyclone may be used as precleaner Filter oven temperature control: thermostat Temperature measurement, filter oven and final impinger uses dial thermometer Umbilical cords: 25, 50, 75 and 100 ft (can be interconnected) Support system: duo rail Special features: carrying case for glassware available; vertical sampling capability; no ground glass ball joints (these have been eliminated by the use of compression nut joints on all glass joints); auxiliary 110 v power outlet
Control Module	Vacuum pump with diaphragm (no surge tank) Dry gas meter: 0.1 cf/rev Pressure measurement: 0-10 in. H_2O dual manometer Temperature measurement: dry gas meter--dial thermometers Heater control for probe Nomograph available Special features: forced air cooling fan available; two-way walkie-talkie system available
Specifications	Control module: 110v, 75 lbs, 24H x 15W in. Sample unit: 30 lbs with glassware, 22H x 12W
Costs	$2700/7603
Remarks	This system has a unique approach to the glassware used in the sampling train. All the above information was obtained from the manufacturer's bulletins and personal communication.[8]

Other Particulate Trains

This section makes no attempt to present *all* the other commercial trains available. Rather, it singles out certain trains to demonstrate the various approaches or functions available. Each train will be presented in essentially the same format as the Method 5 trains. All information reported was obtained from manufacturers' bulletins.

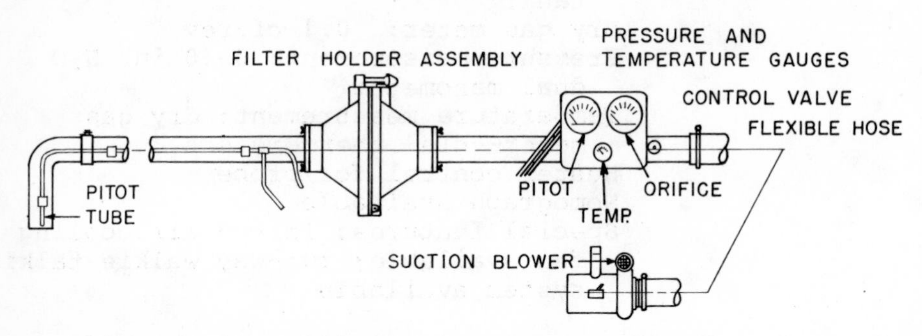

Figure 16.12. Rader Hi-Volume Sampler (courtesy of Rader Pneumatics Inc.).

Manufacturer	Rader Pneumatics, Inc. P.O. Box 20128 Portland, Oregon 97220
Principle	High volume flow rate to reduce sampling time
Basic Design	Probe, filter holder and blower
Probe	Length 30 and 48 in. Nozzle: $1\frac{7}{8}$ in. dia

Standard pitot tube used (0-4 in. H_2O Magnehelic readout)
Material: aluminum
Filter size: 8 x 10 in. fiberglass
Flow measurement: 1¾ orifice used (0-2 in. H_2O Magnehelic readout)
Temperature measurement: dial thermometer
Flow control: butterfly valve downstream from filter
Blower: Gelman Hurricane Blower

Specifications: Sampler: 17½ lbs, 88 or 106 length x 12W x 10D in.
Blower: 9½ lbs

Remarks: This shortened approach has the advantage that many samples can be taken quickly. The lack of any total volume measurement reduces the accuracy of sample volume determination which might reduce sample accuracy. This unit makes sampling for very low concentrations possible. It has been applied to such sources as hog fuel-fired boilers, asphalt batch plants, wood waste incinerators and seed cleaning plants.

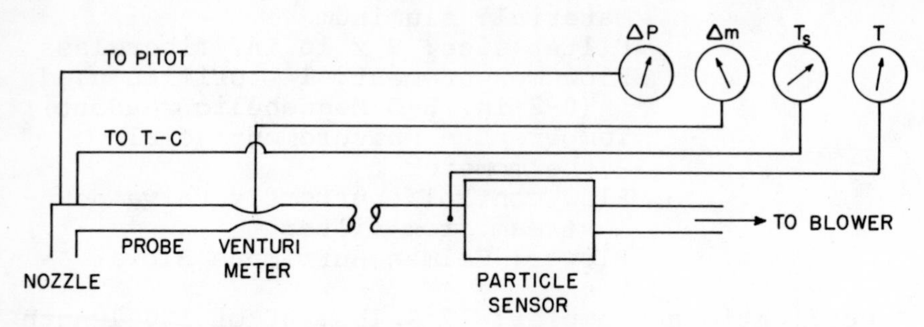

Figure 16.13. IKOR Continuous Particle Monitor.

Manufacturer	IKOR Incorporated Second Avenue Burlington, Massachusetts 01803
Principle	Continuous measurement of the mass flow of particles by measuring the electron transfer of the impinging particle
Basic Design	Probe, sensor unit, control unit
Probe	Contains venturi for sample flow rate measurement Nozzle: 0.93 in. Standard lengths: 5 and 7 ft S-type pitot incorporated Probe heated standard as a preheater Thermocouple available
Sensor Unit	Contains electronic particulate sensor Integral blower and motor Heated umbilical cord
Control Unit	Pressure measurement: 2 Magnehelic gauges Temperature measurement: stack temperature pyrometer; sensor temperature pyrometer Sensor readout: 0-100%, and sensitivity control

	Special features: Unit does not require use of nomograph to maintain isokinetic conditions. Designed for a one-to-one correspondence for velocity head and venturi pressure drop.
Specification	Particulate loading limits: 0.001-100 gr/scf Particulate size limits: 0.1-100 microns Control unit: 110-120v, 60 Hz or 220v, 50 Hz Sensor unit: 28 lbs, 18 x 11 x 17 in. Control unit: 26 lbs, 26 x 11 x 20 in.
Remarks	Permanent installation unit also available. System has some appeal since it gives an instantaneous measurement of particulate concentration.

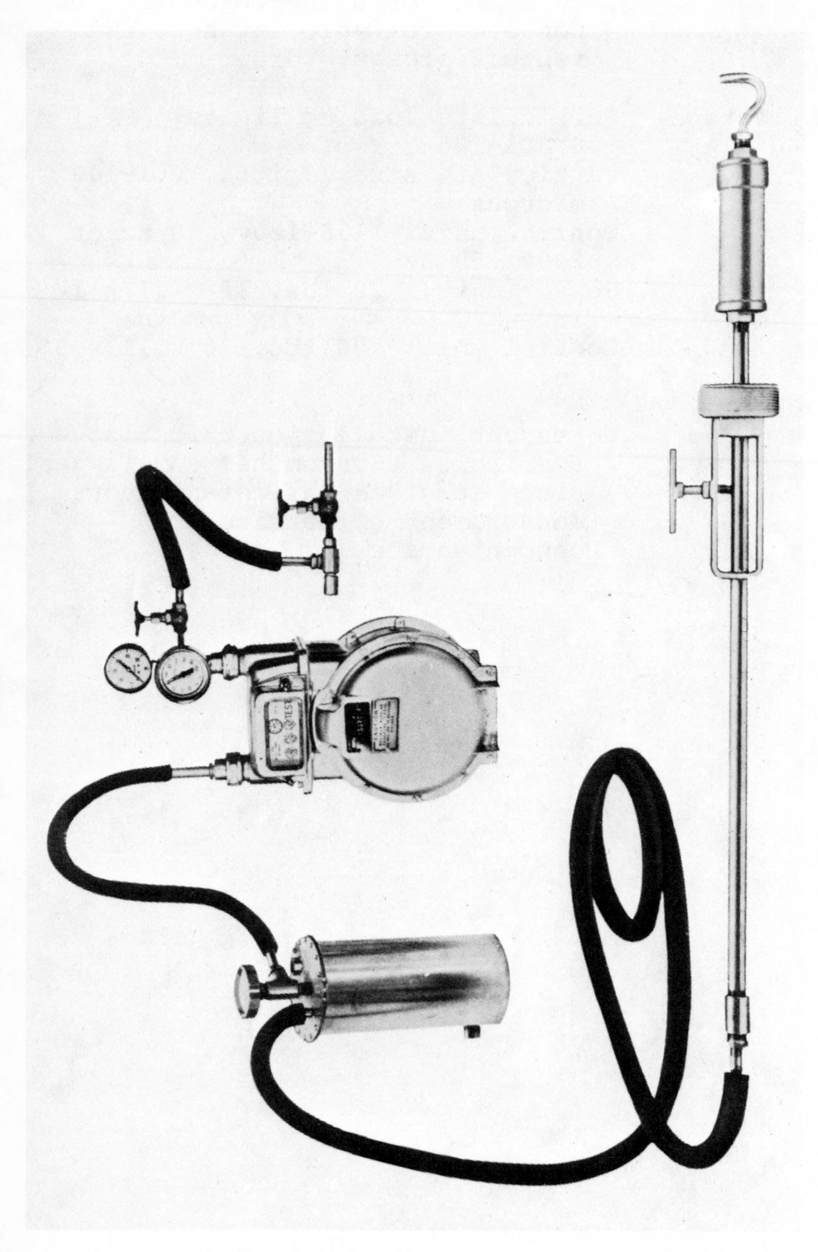

Figure 16.14. Joy Alundum Thimble Assembly (courtesy of Western Precipitation Division/Joy Manufacturing Co.).

Manufacturer:	Western Precipitation Division/ Joy Manufacturing Co. 1000 West Ninth Street Los Angeles, California 90015
Principle	In-stack filtration of particulate
Basic Design	Stack head, probe, condensor, dry gas meter and pump
Probe	Alundum thimble holder Nozzle sizes: 1/8, 1/4, 3/8, and 1/2 in. Probe consists of lengths of stainless steel
Condensor	Standard is fitted for use of cooling water
Dry Gas Meter	1.0 cf/rev
Pump	Any vacuum source
Remarks	Since this type of sampling train allows in-stack collection of particulate samples, many people prefer it to the more complex EPA Method 5 sampling equipment.

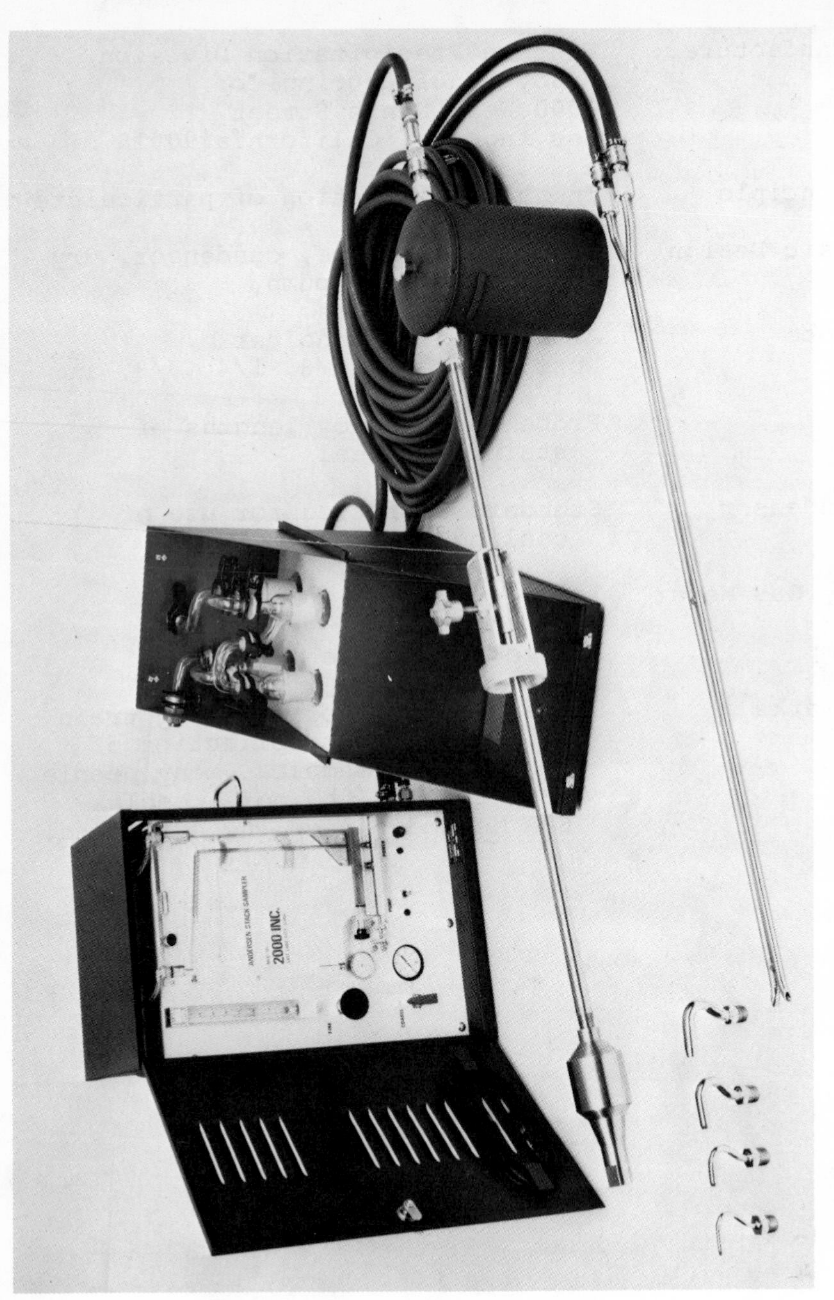

Figure 16.15. Anderson Stack Sampler (courtesy of 2000 Inc.).

Manufacturer	2000 Incorporated P.O. Box 20769 Atlanta, Georgia 30320
Principle	In-stack collection and sizing of particulate by inertial impaction
Basic Design	Stack impactor, condensor, control unit
Impactor	8-stage impactor with provision for final filter
Condenser	Replacable by a four impinger train
Control Unit	Rotameter for flow measurement Manometer for measurement of stack velocity head Vacuum gauge and dial thermometers
Remarks	The impaction sizing is based on a constant flow rate through the impactor. This conflicts with the requirement of flow adjustment to maintain isokinetic conditions in order to collect a representative sample.

GAS SAMPLING

The gas sampling commercial trains are designed to meet the requirements of Method 6, SO_2 sampling, and Method 7, NO_x sampling. Several glass manufacturers produce the glassware needed for these trains. It should be noted that the hardware for these trains is relatively simple, and consideration should be given to the construction of it. All information reported was obtained from manufacturers' bulletins.

Method 6 Trains

The Method 6 trains are presented in the same format used for the preceding Method 5 trains. No attempt has been made to compare prices, however.

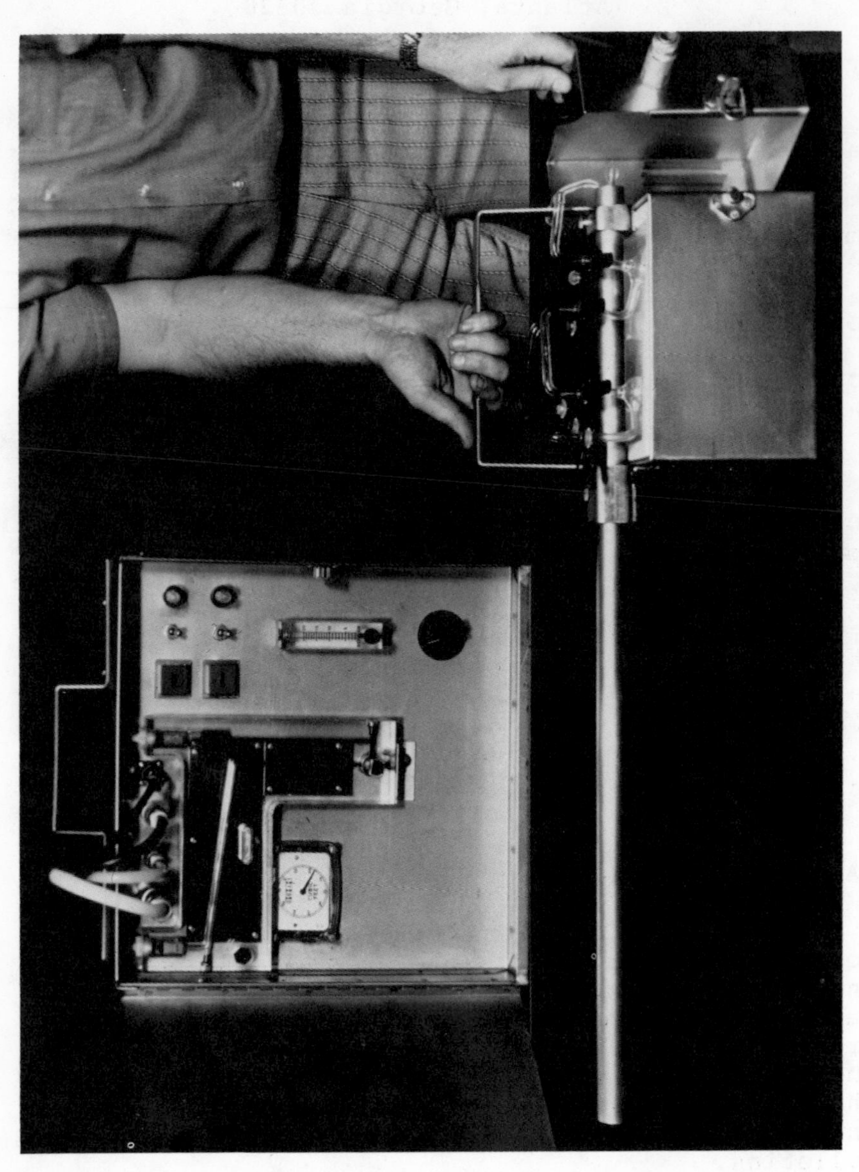

Figure 16.16. GII Gasampler Model GII-300 (courtesy of Glass Innovations).

Manufacturer	Glass Innovations, Inc. P.O. Box B Addison, New York 14801
Basic Design	Probe, sample unit and control unit
Probe	Liner: Pyrex Stainless steel jacket with pitot optional Standard lengths: 2 and 5 ft Probe heated standard
Sample Unit	Utilizes 4 midget impingers Umbilical cord: 10 ft
Control Unit	Vacuum pump: diaphragm Dry gas meter: 0.1 cf/rev Standard pitot manometer Heated control for probe
Special Features	Train can be used for Method 4 moisture determination
Specifications	Control unit: 110v, 7 amp, 45 lbs, 16H x 13½W x 10D in. Sample unit: 9 lbs with glassware, 11H x 12W x 6D in.
Remarks	This train can also be used for Method 4 moisture determination. All information provided by brochures and personal communications.[8]

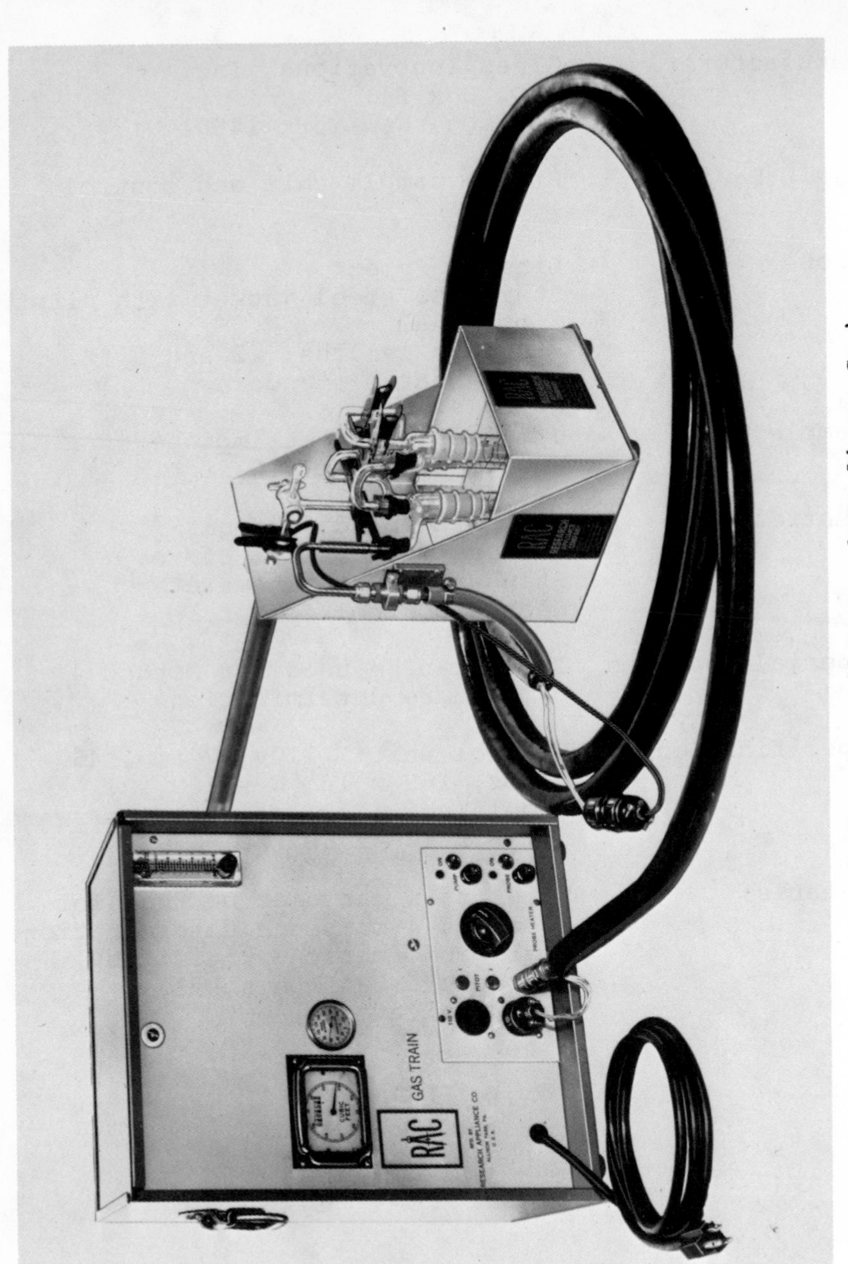

Figure *16.17*. RAC *Gas Train (courtesy of Research Appliance Co.).*

Manufacturer	Research Appliance Co. Route 8 Allison Park, Pennsylvania 15107
Basic Design	Probe, sample unit and control unit
Probe	Liner: Pyrex Stainless steel jacket with optional pitot Standard lengths: 1½, 3 and 4 ft Probe heated standard
Sample Unit	Utilizes 4 midget impingers Umbilical cord: 13 ft
Control Unit	Vacuum pump: diaphragm Dry gas meter: 0.1 cf/rev Optional pitot manometer Heater control for probe
Special Features	Train can be used for Method 4 moisture determination.
Specifications	Control unit: 115v, 7 amps, 36 lbs, 17H x 14W x 10½D in. Sample unit: 4 lbs with glassware, 12H x 7½W x 7½D in.
Remarks	The train can also be used for Method 4 moisture determination. All information provided by brochures and personal communications.[4]

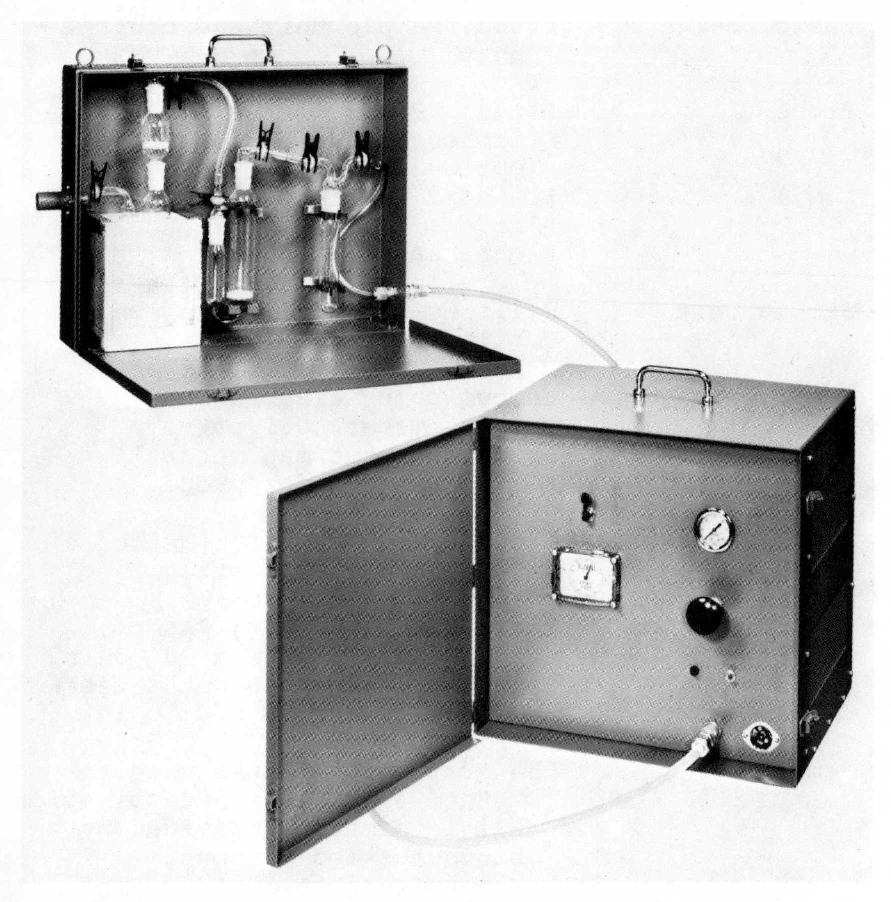

Figure 16.18. SO_2/SO_3 Sampling Train Model AP-2000 (courtesy of Scientific Glass Blowing Co., Inc.).

Manufacturer	Scientific Glass Blowing Company, Inc. P.O. Box 18306 Houston, Texas 77023
Basic Design	Probe, sample unit, and control unit
Probe	Liner: quartz or Vycor Stainless steel jacket with pitot optional Standard lengths: on order Probe heated: optional
Sample Unit	Utilizes lamp sulfur absorbers (ASTM D1266) Umbilical cord: 10 and 25 ft
Control Unit	Vacuum pump with fiber vanes Dry gas meter: 0.1 cf/rev Optional pitot manometer
Specifications	Control unit: 115v, 10 amps, 40 lbs, 17⅝H x 18W x 15⅛D in. Sample unit: 9¼ lbs with glassware, 18H x 22W x 5D in.
Remarks	This was the type of train used in the development of an SO_2 measurement technique. It should be noted that the impinger train for SO_2 is also available with the Stac-o-lator discussed previously. All information provided by brochures and personal communications.[5]

Method 7 Trains

These systems consist of grab sample containers which do not as such constitute a train. The glassware needed to conform to EPA Method 7 is available from several glass manufacturers. Since train construction is straight forward, consideration should be given to in-house construction of this sampling system.

Figure 16.19. CII-400 NO_x Sampler (courtesy of Glass Innovations Inc.).

Manufacturer	Glass Innovations, Inc. P.O. Box B Addison, New York 14801
Basic Design	Probe, grab sample flask, pump
Probe	Liner: glass with optional heater Stainless steel jacket
Flask	Foam incased Temperature measurement: dial thermometer Pressure measurement: glass mercury manometer
Special Features	Carrying case with provision for 8 flasks Holder for individual flask and probe Vacuum pump housed in case Rheostat for probe heater control located in case
Remarks	The large carrying case is a useful feature when transportation space is no problem.

REFERENCES

1. "Standards of Performance for New Stationary Sources," Federal Register, Vol 36, No. 247, December 23, 1971.
2. Martin, R. M. Construction Details of Isokinetic Source-Sampling Equipment, Publ. No. APTD-0581, Air Pollution Control Office, EPA, Research Triangle Park, North Caroline, 1971.
3. Baker, Wayne, Inside Salesman, RAC. Personal communication with C. David Turley on December 18, 1972.
4. Muller, C. B., Vice President/Market Manager, Scientific Glass Blowing Co. Personal communication with C. David Turley on December 18, 1972.
5. Luchessa, Charles E., Vice President, Misco Scientific. Personal communication with C. David Turley on Decmeber 18, 1972.
6. Bresie, D., President, LACE Engineering. Personal communication with C. David Turley on December 18, 1972.
7. Perry, Greg, Sales and Service Representative, Glass Innovations Inc. Personal communication with C. David Turley on December 18, 1972.
8. Lassiter, T. E., Service and Sales Engineer, Bendix, Environmental Science Division. Personal communication with C. David Turley on December 18, 1972.

CHAPTER 17

COMMERCIAL CONTINUOUS MONITORING EQUIPMENT

The previous chapter presented the various types of source testing equipment, and this chapter will examine the commercial types of source monitoring equipment. As prescribed by federal law (see Appendix A) all new Priority I sources must install such equipment; Table 17.1 is a summary of these requirements. The instruments and sampling systems installed and used must be capable of monitoring emission levels within ± 20% with a confidence level of 95%. The instruments must be calibrated in accordance with the methods prescribed by the manufacturers. During operation the instruments must be subjected to the manufacturers' recommended zero

Table 17.1

Continuous Monitoring Requirements for New or Modified Priority I Sources

Type of Source	*Monitoring Requirements*
Fossil-fuel fired steam generator	photoelectric smoke detector sulfur dioxide nitrogen oxides
Incinerators	none
Portland cement plants	none
Nitric acid plants	nitrogen oxides
Sulfuric acid plants	sulfur dioxide

adjust and calibration procedures at least once every 24 hours unless a shorter period is recommended by the manufacturer. The test methods presented in Chapters 10, 11, 12, 13, and 14 are the reference methods for the continuous monitors. The records of continuous monitoring measurements must be retained for a period of at least two years.

The commercial equipment described in this chapter operates on well-known measurement principles. In most cases the methods have received considerable use as industrial process analyzers or source monitors. This does not necessarily mean that such systems are not subject to operational problems. The methods presented in Chapter 18 are rather new to most readers and hence are covered in more detail. Some of these methods are also available commercially.

BASIC MONITORING METHODS

At the present time there are two basic approaches to source monitoring. The first is an automation of the source testing procedures already described. The second approach is *in-situ monitoring*. Measurement occurs at a point in the stack or it can be a spatially integrated measurement across some dimension of the stack. The *in-situ* method has the advantage of not being concerned with obtaining a *representative* sample. However, it is often difficult to calibrate *in-situ* methods. In many cases the extractive methods require extensive interface systems providing sample preservation and handling. Tables 17.2 and 17.3 show the present status of monitoring methods for gases and aerosols.[1] The following section provides a brief description of the various measurement principles. For more details the reader should use References 2 and 3.

Monitoring Gases

A number of principles have been applied for measuring pollutants. Schematic diagrams of these methods are shown in Figures 17.1 through 17.10. This section provides a brief description of each method.

Electrical Methods. Constituents soluble in aqueous solutions change the *electrical conductivity* of such solutions. This change in conductivity is proportional to the concentration of the pollutant. However, conductivity change is not due to a single

Table 17.2

Status of Instrument Development for Continuous Monitoring of Gases[1]

Sampling Method	*Measurement Technique*	*Status*	*Remark*
E	NDIR (gas filter)	C,R	CO,HC,SO_2,NO
E	NDIR (optical filter)	C*	CO,HC,SO_2,NO
E	UV spectroscopy	C	NO_2,SO_2
E	Coulometric titration	C	sulfur compounds
E	Conductimetric	C	SO_2,SO_3
E	Electrochemical	C	NO_x,SO_2
E	Colorimetric	C	SO_2,NO_2
E	2nd derivative spectroscopy	P*	NO_2,SO_2
E	Chemiluminescence	P*	NO,NO_2
E	Flame photometric	C*	sulfur compounds
I	UV correlation spectroscopy	C	NO_2,SO_2
I	IR and UV dispersive spectroscopy	C	NO,CO,SO_2

E denotes extractive method
I denotes *in-situ* method
C denotes commercial product
P denotes prototype development
R denotes research development
* denotes that monitoring method requires an interface system

Table 17.3

Status of Instrument Development for Continuous Monitoring of Aerosols[1]

Sampling Method	*Measurement Technique*	*Status*	*Remark*
E	Beta gauge	C*,R*	mass concentration
E	Peizoelectric	C*,R*	mass concentration
E	Optical	C	visible emissions
I	Optical	C,R	visible emissions

E denotes extractive method
I denotes *in-situ* method
C denotes commercial product
R denotes research development
* denotes that monitoring method requires an interface system

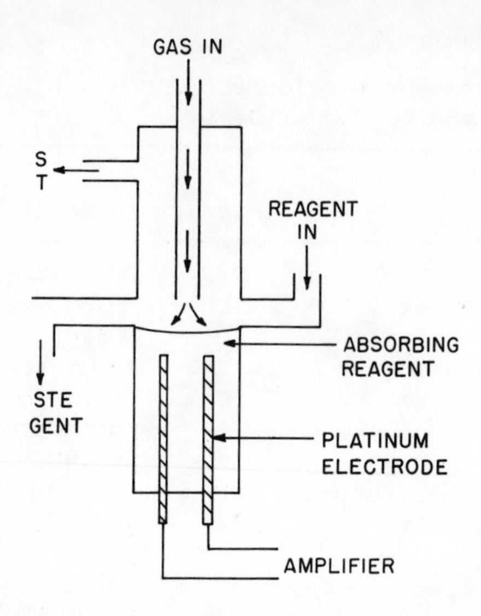

Figure 17.1. Conductivity monitor.

type of pollutant. In fact, it can be caused by any number of gases and aerosols. This method is often applied to the monitoring of gaseous constituents. Conductivity analyzers are highly sensitive, have a fairly fast response and due to their simple nature exhibit low maintenance. The major difficulty is interference. For example, the device will respond to both SO_2 and NO_x in a source stream and this would pose a problem.

Some gaseous constituents react with a collecting solution and change its pH. *Potentiometry* involves using this change of pH as a measure of pollutant concentration. As with conductivity, this method is not very specific; there are many gases which when dissolved in solution will cause a change in pH.

When electrical energy is added to a solution it causes chemical reactions to occur. When the number of electrons required is measured (in coulombs) for a specific reaction to go to completion, *coulometry* is being used. This method is generally subject to interference problems. For example H_2S, O_3, mercaptans, NO_x and NH_3 all interfere with the measurement of sulfur dioxide.

When some species enter into high energy processes such as combustion they take on a charge or become *ionized*. Ionization is a method whereby the number of ions formed are measured and related to the

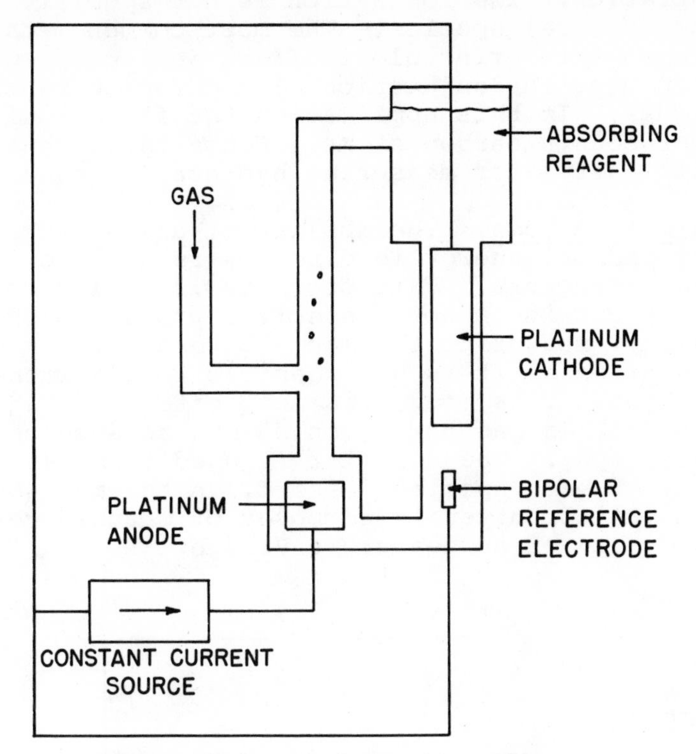

Figure 17.2. Coulometric monitor.

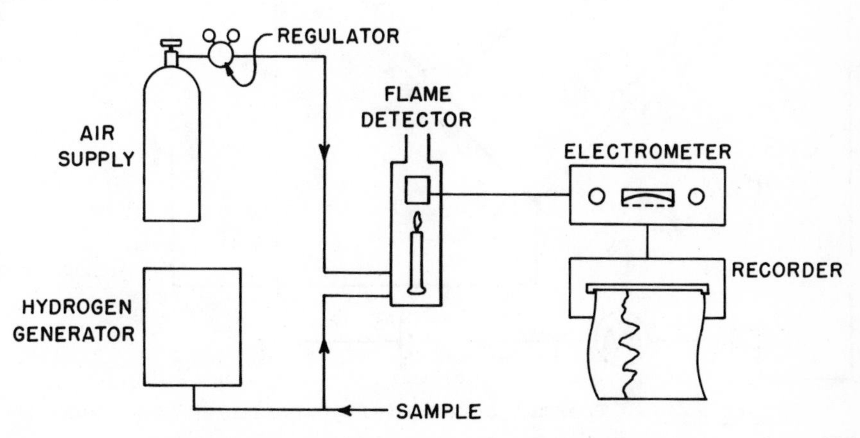

Figure 17.3. Flame ionization monitor.

concentration. The ionization is not specific to any one chemical species. The most common method based upon this principle is *flame ionization* which is often used in conjunction with chromatographic techniques. In this application the flame ionization detector counts carbon atoms. Hence the technique is quite useful for measuring hydrocarbon concentration.

Thermal Methods. The ability of a substance to conduct thermal energy is directly related to its chemical structure. This property is called conductivity but it is not a specific property. For example, by measuring the *thermal conductivity* of a mixture of gases it is not possible to determine the composition. This method is most often used in conjunction with a gas partitioning method such as chromatography. The detector is used to measure the quantity of each species present. A thermal conductivity detector often is composed of matched resistors and made part of a Wheatstone Bridge.

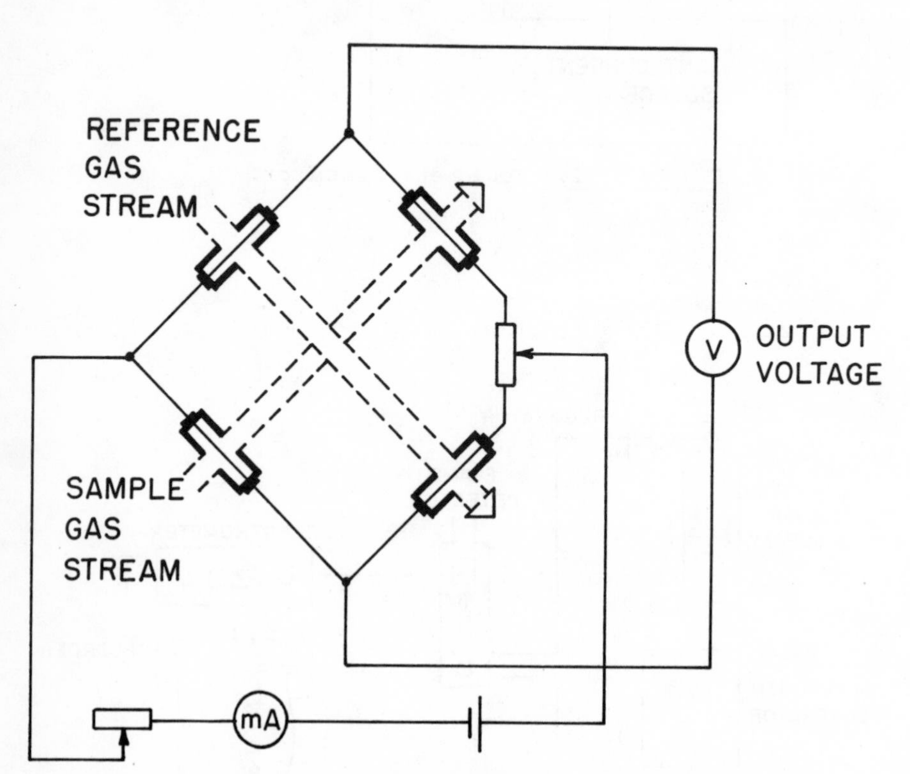

Figure 17.4. Thermal conductivity monitor (Wheatstone Bridge).

Sometimes a constituent is combustible, and by measuring the amount of heat evolved in the process the original concentration can be estimated. This combustion method is not specific because many types of compounds are reactive and would release thermal energy.

Electromagnetic Methods. Gases may absorb or scatter any electromagnetic energy incident upon them. This is called *attenuation*. The amount of absorption and scattering is dependent upon the physical and chemical characteristics of the pollutant and the wave length of the incident radiation. The method can be made quite specific by the proper selection of wave length and the use of optical filters.

There are several types of spectroscopic techniques. *Nondispersive* methods for gaseous constituents rely on the fact that gases and vapors absorb

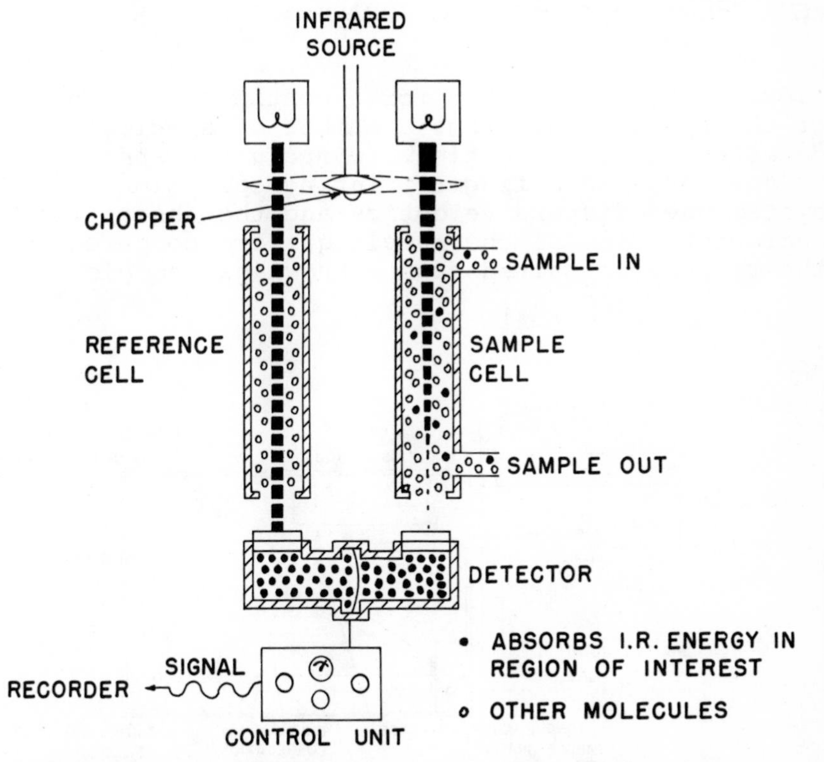

Figure 17.5. Nondispersive infrared analyzer (courtesy of Beckman Instruments, Fullerton, California).

electromagnetic energy only in specific band widths. These bands are a specific property of the molecular structure. For example, carbon dioxide and carbon monoxide absorb in the infrared region of the electromagnetic spectrum. *Dispersive* methods also rely upon molecular absorption but they contain optical systems capable of producing numerous incident wave lengths of electromagnetic energy. This

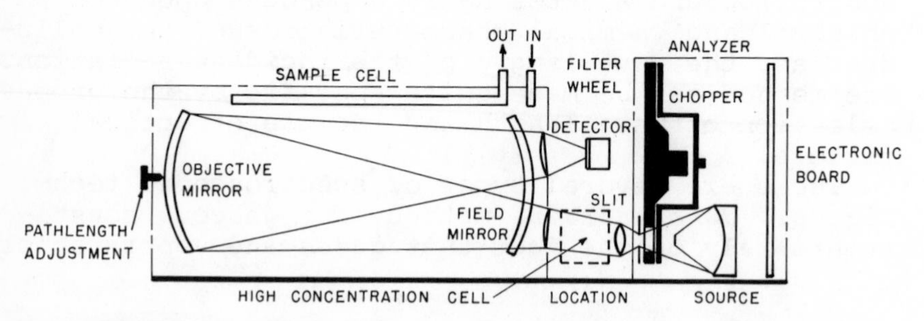

Figure 17.6. Dispersive infrared analyzer (courtesy of Wilks Scientific Corporation, South Norwalk, Connecticut).

flexibility allows then for a continuous scanning of the spectrum and thus a number of species can be detected at the same time. *Correlation spectroscopy* is analogous to a finger printing operation. The system uses dispersive optics and the absorption characteristics of the sample gas are compared to the response obtained from a known gas sample.

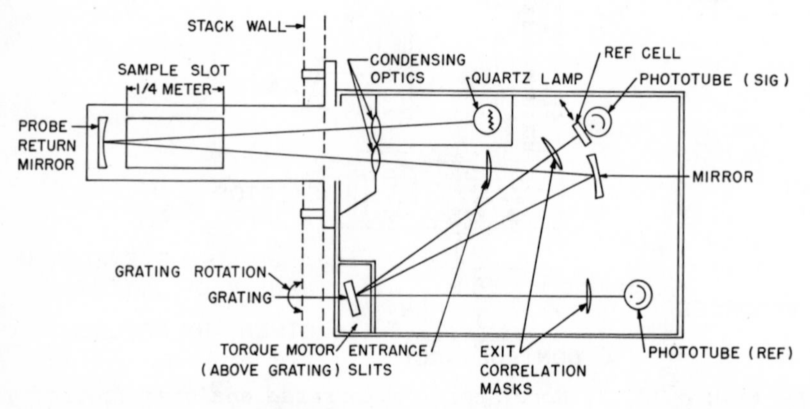

Figure 17.7. Correlation spectrometer.

Derivative spectroscopy involves examining the first and higher order derivatives of the transmission versus incident wave length of the light. It seems that the derivative signals tend to remove some of the interference caused by overlapping spectral bands of various gases and vapors.

Chemi-Electromagnetic Methods. These methods involve the occurrence of a chemical reaction and the subsequent measurement of electromagnetic radiation. *Colorimetry* is the collection of gaseous pollutants in a special absorbing solution. The specific desired reaction produces a product which has a specific color, with the intensity of the color being proportional to the amount of the original pollutant present. The color intensity is

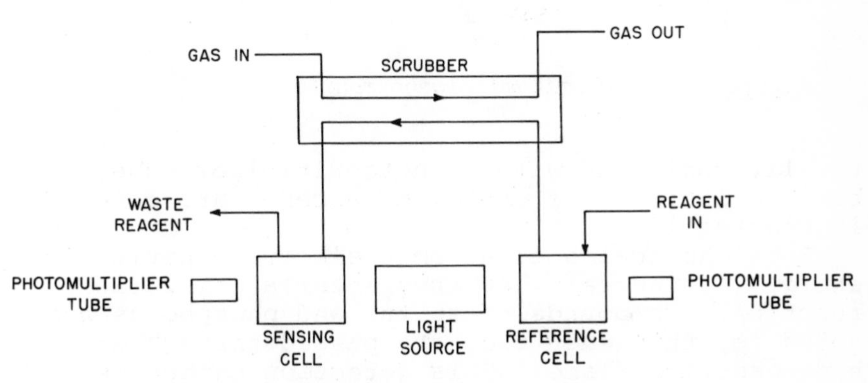

Figure 17.8. Colorimetric monitor.

measured by absorption of radiation of a specific wave length. This analysis method follows Beer's Law. The degree of light absorption by a colored solution is a function of the pollutant concentration and the wave length of the incident light. This technique has high sensitivity but it often suffers from interferences. The color intensity developed is also a function of temperature, pH, and the purity and age of the reagents.

Chemiluminescence is a property of certain chemical species whereby when reacted with a second species a characteristic light emission occurs. The wave length of this light emission is dependent upon chemical structure while the intensity is related to concentration. The chemiluminescent light emitted is

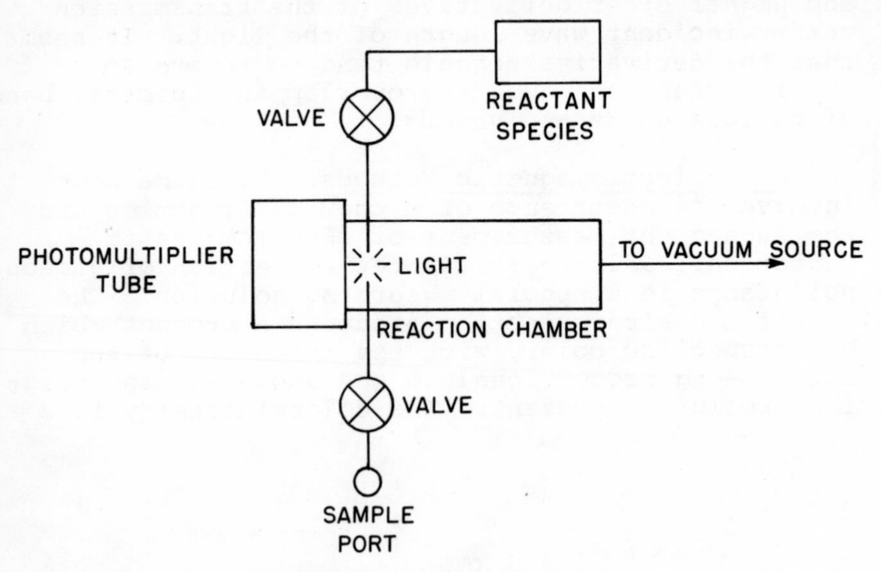

Figure 17.9. Chemiluminescent monitor.

usually monitored with a photomultiplier tube. The basic principles of chemiluminescence are covered in Chapter 18.

Some chemical species emit electromagnetic light energy (luminesce) when they enter a combustion reaction. Compounds of sulfur and phosphorus are noted for this when they are passed through a hydrogen-rich flame. This detection method is called *flame photometry*. The wave length of the

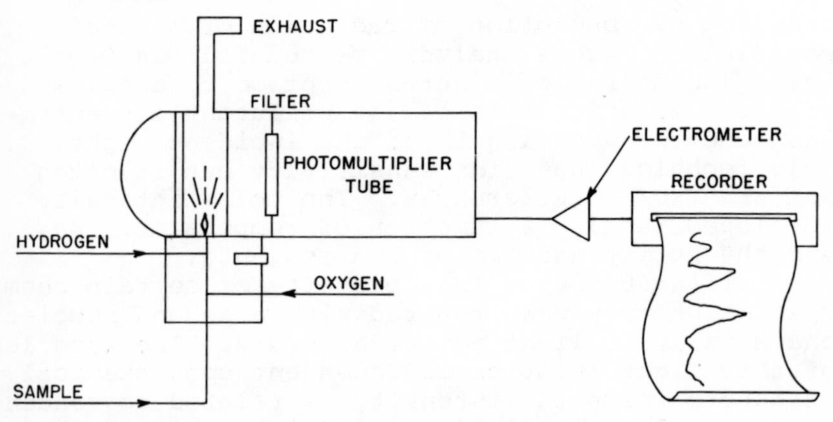

Figure 17.10. Flame photometric detector.

emitted light is characteristic of the chemical species and the intensity is proportional to the concentration present. This method is presented in greater detail in Chapter 18.

Magnetic Methods. The most important application of magnetic properties of species is mass spectrometry. In this method ionized species enter a magnetic field and are thereby deflected according to their mass and charge. This method is often used in connection with some partitioning method such as gas chromatography.

Monitoring Aerosols

There are many basic techniques which might be used to monitor aerosols. However, three show the greatest promise for source monitoring: optical, beta gauge and piezoelectric methods. These methods are briefly described in this section. Figures 17.11, 17.12, 17.13, and 17.14 show schematic diagrams of these methods.

Optical. Aerosols will absorb and scatter light, and these properties can be used to measure the quantity of aerosol present. Theoretically the light extinction follows the Beer-Lambert Law

$$I = I_o \text{ EXP } (-bL) \qquad (17\text{-}1)$$

where

- I_o is the initial light intensity
- I is the light transmitted
- b is the extinction coefficient of the aerosol
- L is the path length.

An increase in the aerosol concentration causes b to increase and thus produces a reduction of the light intensity. Hence the light reduction may be used to indicate the amount of aerosol present. It should be noted that the absorption is dependent upon the particle characteristics as well as size.

Optical monitoring methods can be used on *extractive* or *in-situ* methods. When the extractive method is used, generally the sample is filtered and the optical density of the sample measured. For *in-situ* measurements the light source and detector are located to minimize fouling of the optical surfaces. *In-situ* capacity measurements do not always agree with observations taken according to EPA Method 9 (see Chapter 14) because the gas stream may undergo

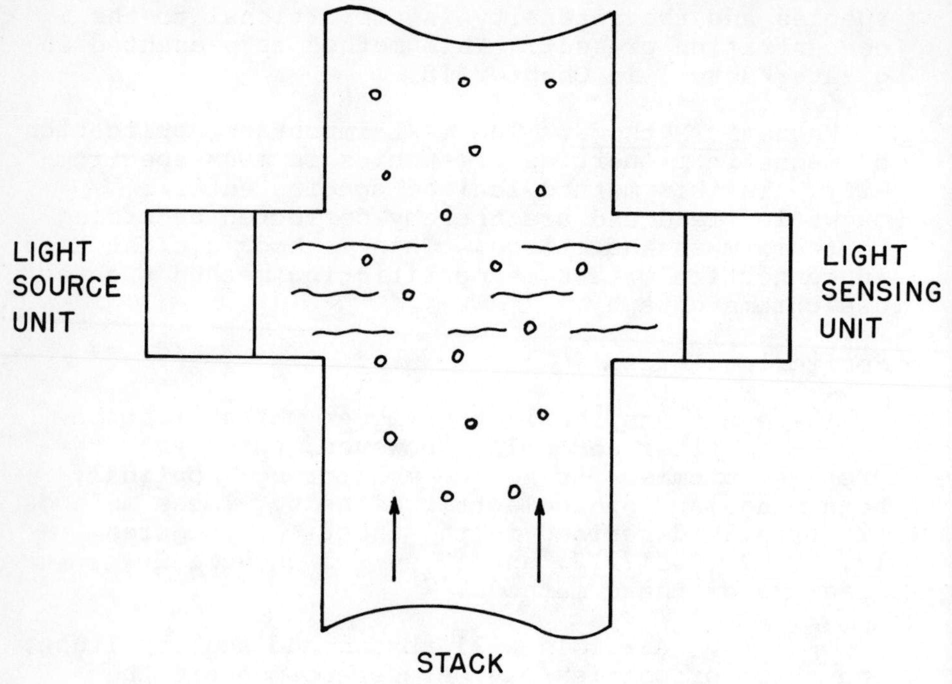

Figure 17.11. *Opacity meter (in-situ method).*

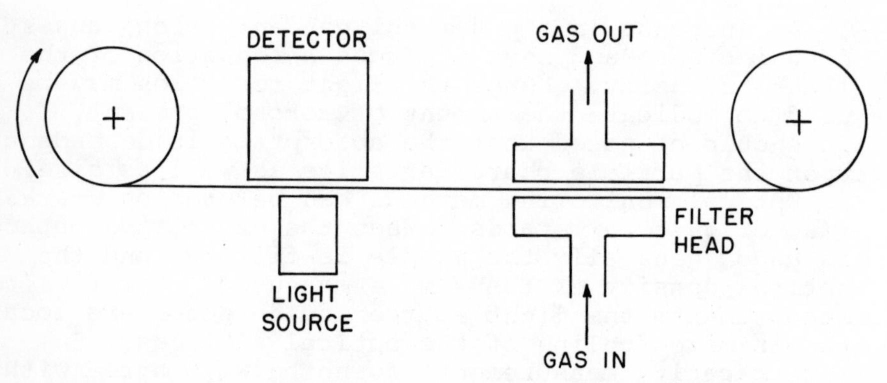

Figure 17.12. *Tape sampler (extractive method).*

physical and chemical changes after it is ejected into the outside air. Also opacity measurements do not always agree well with aerosol mass emission rates.

Beta Gauging. The mass of particulate matter collected on a filter paper can be determined by *beta absorption*. The amount of beta radiation absorbed is proportional to the mass. This method has been used successfully to measure the thickness of paper and metal sheet. Only recently have efforts been made to apply this technique to source monitoring. The theory of this method is presented in Chapter 18.

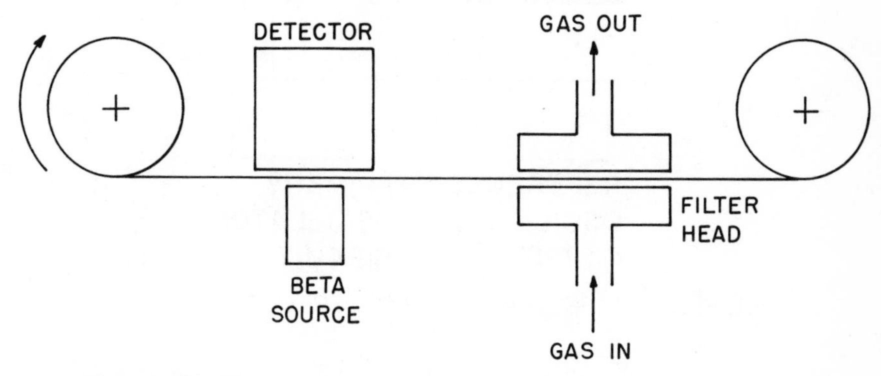

Figure 17.13. Beta gauge mass monitor.

Piezoelectric. A piezoelectric crystal will mechanically vibrate when placed in an oscillating field. This is the case when it is made part of an electrical driving circuit. If the electric field of the driving circuit oscillates at a frequency fairly close to the resonant frequency of the mechanical vibration of the crystal, the crystal forces the entire circuit to oscillate at that frequency. The natural frequency of the crystal depends on the mass of the crystal and any material which is attached rigidly to its surface. This property leads to the possibility of using the crystal to monitor increments of mass added to its surface. The natural frequency of oscillation of the crystal decreases in direct proportion to the amount of added mass. This method has been used to monitor ambient aerosols. However, only a few attempts have been made to apply this to sources. The theory of this method is presented in Chapter 18.

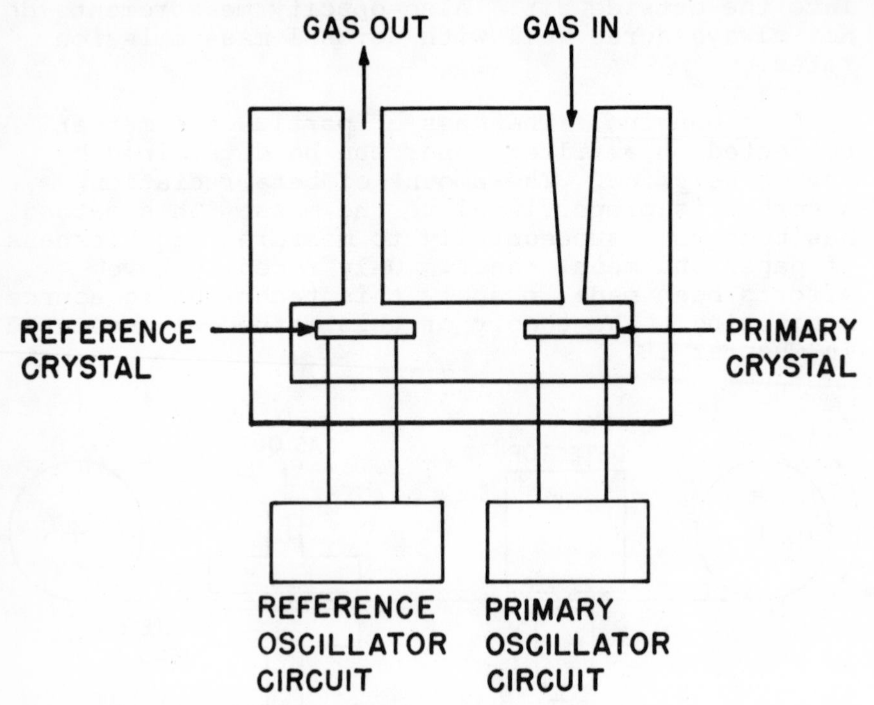

Figure 17.14. Piezoelectric mass monitor.

INSTRUMENT CHARACTERISTICS

What would you want in an ideal instrument? What are the parameters which characterize the performance of monitoring equipment? Before looking at source monitoring methods and evaluating their usefulness we must identify the desirable properties of the *ideal* instrument or method. With this as a standard we can assess the possible advantages and disadvantages of the methods. Table 17.4 provides a listing of items which the authors feel should be attributes of the ideal source monitoring method.

The design and operational characteristics of most importance are low cost, light weight, portability, low maintenance, easy adjustment and repair, durability in its environment, and simplicity of design and operation. The low cost and portability requirements are associated with the large numbers of instruments needed and the environmental location where the equipment must be located. Low maintenance and durability are essential as the instruments usually will be put in poorly accessible locations. The simplicity factor is basic to any design and helps accomplish the other requirements.

The performance characteristics most important in a system are sensitivity, specificity, stability and reliability, zero drift, span drift, reproducibility, precision, response time, calibration, linearity and accuracy.

Sensitivity

Sensitivity is the smallest change in the independent variable which the instrument can reliably measure. Sensitivity would then be the minimum detectable concentration of a pollutant. The sensitivity of a method is governed by the signal-to-noise ratio. Electrical noise is caused by the electrical components in the system.

Specificity

A response is highly specific if there are no interferences which also cause a similar response. If a method is not highly specific, interference effects must be taken into account by calibration. Sometimes this is an impossible accomplishment and the method is then rendered practically useless.

Stability and Reliability

A stable instrument is one which does not require frequent calibration; it retains its performance characteristics over rather long periods of time. A stable instrument would also be thought of as reliable in that sense. However, if key components fail often, instruments may still possess good stability characteristics but not be reliable.

Zero Drift

If the zero output signal changes with time, it is then said to drift. This is usually due to uncontrolled environmental conditions and may be adjusted electrically. If this is a problem, the instrument must be designed to automatically make zero drift corrections; otherwise an unaccountable error will be introduced into the measurement.

Table 17.4

Desirable Attributes for Ideal Source Monitoring Method

DESIGN CHARACTERISTICS

1. Simplicity of design and operation.
2. Light weight and portable if instrument must be removed for servicing.
3. Good accessibility for adjustments and repairs.
4. Easily installed with minimal manpower and alterations to existing facilities.
5. Low cost
6. Pollutant collection efficiency should be 100%. If it is less than 100% then it must be easily and accurately determined.

PERFORMANCE CHARACTERISTICS

1. Measurement must correlate with total mass emissions of pollutant emitted to the atmosphere.
2. Must have good sensitivity to pollutant.
3. Must cope with pollutant mass concentration profiles which vary by typically ± 50% spatially, by a factor of about 10 with time, and by a factor of about 10 during some special process conditions.
4. Reproducibility between two identical instruments, including both sampling and sensing error, of ± 20%.
5. Calibrated accuracy of ± 20%. Should be able to calibrate instrument while in use.
6. Should record such items as zero and span every 1 to 6 hours on data recordings.
7. Preferably have an instantaneous readout of the total mass of pollutant emitted into the atmosphere per unit time, or at least one average readout every 15 minutes.
8. Outputs compatible with standard strip chart or digital recorders and with computer data inputs.

ENVIRONMENTAL REQUIREMENTS

1. Operate with or take sample from flues with gas velocities of 30 to 120 feet per second.
2. Operate in a variety of stacks, each having widely different velocity profiles.
3. Operate with flue gas temperatures up to 1000°F. Instruments should not be adversely affected by temperature fluctuations.
4. Operate in corrosive flue gas.

5. Must not restrict the flue gas flow in any significant way.
6. Must withstand such environmental conditions as vibration and ambient weather conditions.
7. Must operate with positive and negative flue gas static pressures.

MAINTENANCE AND OPERATIONAL CONSIDERATIONS

1. Rugged enough to last several years without major repair.
2. Little or no maintenance except for regular weekly maintenance and for major maintenance only during regular plant maintenance shutdowns.
3. Little or no adjustment while in operation. Calibration should be easily performed while in use.
4. Easily accessible during and after installation for weekly maintenance and/or adjustments.
5. Easy for plant operators to understand, operate, and to make minor repairs.
6. Operate on 110V or 220V, 60 Hz electrical power and require little power, preferably no more than 1500 watts.

Span Drift

Span drift is the change with time of the instrument output signal when the system is exposed to a known (constant and specified) pollutant concentration.

Reproducibility

An instrument is reproducible if it gives the same output readings over time for identical input conditions.

Precision

The term precision when applied to an instrumental method is the amount of resolution which should be attached to a measurement. This is usually indicated by the number of significant digits contained in the measured value. Thus if the true value is 1.5261, an instrument which can detect 1.526 is more precise than one which can detect only 1.5.

Response Time

An instrument requires some specific time in responding to changes in the input. If a step input is applied to an instrument (*i.e.*, sudden change in pollutant concentration), the time required for the read out signal to reach some fixed percentage of the final value is called the response time. As shown in Figure 17.15 an instrument usually has a lag and rise time; the combination of these is called the response time. A 90% response time is often used. The effect of a slow response time is to smooth out the instantaneous fluctuation of detected concentrations. The fall time is that time required for the instrument to return to the 10% response point. The fall time is usually, but not necessarily, the same value as the rise time.

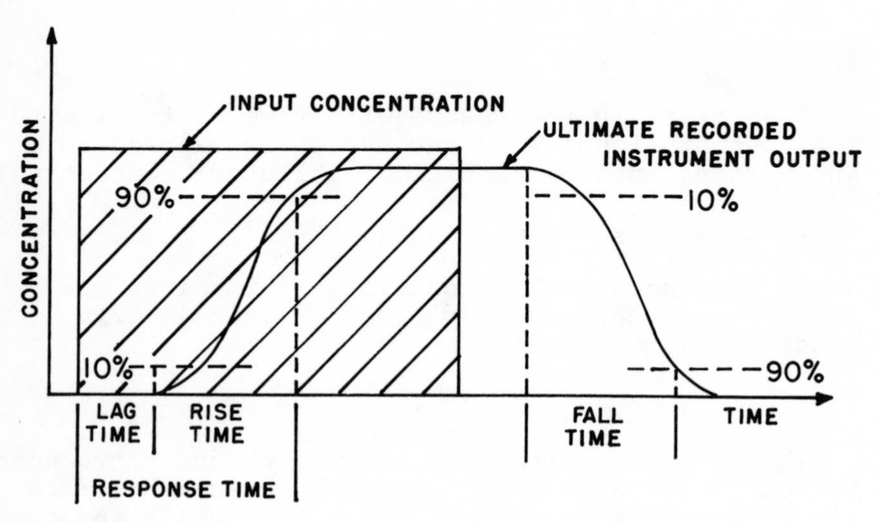

Figure 17.15. Instrument response characteristics.

Calibration

Calibration is the process of exposing the instrument to known concentrations of the pollutant and observing its response. Instruments which are not stable and have poor reproducibility must be subjected to frequent calibrations.

Linearity

A measure of the linearity of an instrument is the maximum deviation between an actual instrument reading and the reading predicted by drawing a straight line between the upper and lower calibration points.

Accuracy

Accuracy is the ability of an instrument to give the true value. An instrument which is not accurate can still be quite useful if the bias can be evaluated by calibration procedures.

COMMERCIAL EQUIPMENT

The following is a partial listing of equipment available for stationary source monitoring applications. All performance information reported in this chapter was provided by the various manufacturers and Reference 3. Several studies are being conducted to evaluate the performance of commercial particulate matter, SO_2 and NO_x extractive type monitors.[1]

Figure 17.16. Model 460 SO_2 (control station and recorder not shown)(courtesy of DuPont Instrument Products Division).

Manufacturer	E.I. du Pont de Nemours and Co. (Inc.) Instrument Products Division Wilmington, Delaware 19898
Class	Stationary
Principle of Operation	Nondispersive ultraviolet absorption spectroscopy: a beam of light is passed through sample, then split and monitored by two photomultiplier tubes. One tube monitors a reference wave length, the other monitors a detection wave length. The difference in intensities provides a measure of SO_2 in the sample.
Sensitivity	
Range	0-300 to 0-3000 ppm (up to 0-100% for Models 460 or 461)
Interferences	Slight or no interferences from NO, NO_2, H_2S, CO, CO_2, H_2O, O_2, or particulates
Multi-parameter Capability	Model 460 for SO_2 Model 461 for NO_x Model 460/1 for SO_2 and NO_x
Sampling	Method: continuous flow-through Volume: 250 cc measuring cell Maximum temperature input: 212°F
Performance	Accuracy: $\pm$ 2% of full scale Reproducibility: $\pm$ 1% of full scale Stability: Lag time: 1-2 sec Response time: about 15 sec including sample line Zero drift: $<1\%$ in 24 hrs Span drift: $<1\%$ in one week
Operation	Ambient temperature range: -20° to 125°F Temperature compensation: built-in Maximum relative humidity: 100% at 80°F

	Calibration: optical filter which mimics light absorption of pollutant; filter is calibrated against known sample and becomes the calibration standard. Operation: instrument is corrected for fluctuation of light, dirt, mists, since such effects are cancelled by split beam principle. Other pollutants do not interfere because absorption is specific to the pollutant being monitored. Maintenance: compressed air supply (30-80 psig) Output: linear 0-10 mV Options: automatic calibration, longer sample housing, 4-20 mA and 10-50 mA outputs available, integrally mounted recorder.
Specifications	Power: 115 V, 60 Hz, 2000 W Weight: 127 lb Dimensions: analyzer (11¼" x 32⅛" x 5⅝"); transformer (16" x 12" x 6½"); control station (6¾" x 9" x 9⅞")
Cost	$5000-$17,000
Remarks	Instrument consists of three separate components: analyzer, transformer and control station. Using these building blocks, manufacturer will design and fabricate a sampling system to be used with the analyzer for a specific application.

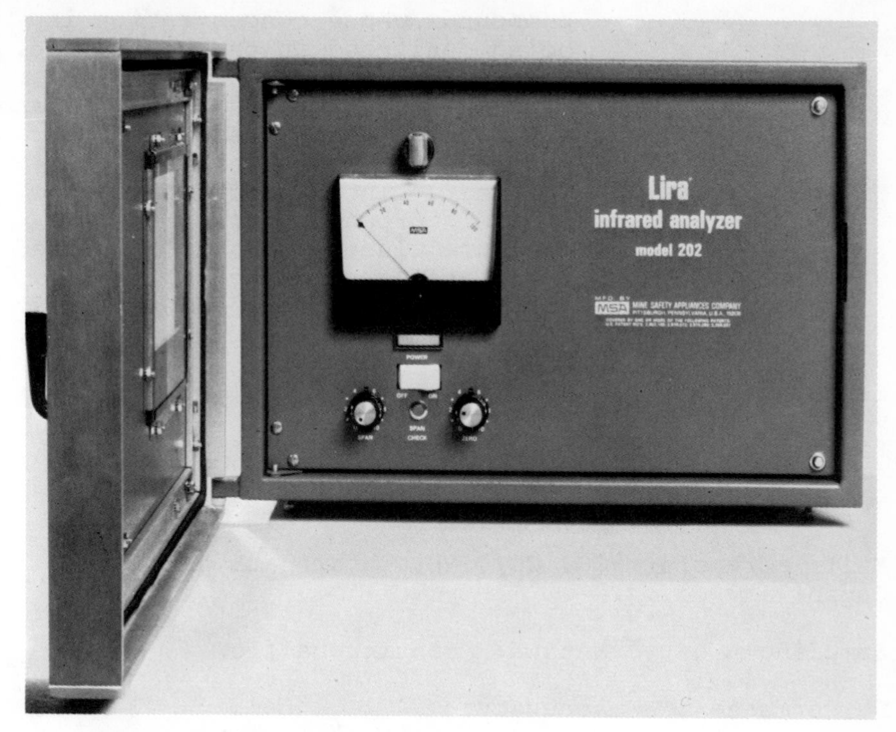

Figure 17.17. LIRA Model 202 Analyzer (courtesy of Mine Safety Appliance).

Manufacturer	Mine Safety Appliances Instrument Division 201 North Braddock Avenue Pittsburgh, Pennsylvania 15208
Class	Stationary
Principle of Operation	Nondispersive infrared absorption (NDIR): the absorption of infrared radiation, emitted by Nichrome filaments, is monitored in the SO_2 absorption spectral range. The two nondispersed infrared beams, which are periodically interrupted by the semicircular chopper, traverse the sample and the reference cell, respectively. The transmitted radiation is directed into the sensor compartment. This compartment contains the gaseous

compounds whose concentration is being monitored in the gas sample. Incorporated in the sensor is a microphone equipped with a membrane which responds to pressure fluctuations in the sensor gas compartment. The capacitance change of the microphone during the pressure fluctuations is converted to an analog electrical signal that is amplified and subsequently fed to a meter or recorded.

Sensitivity	
Range	
Interferences	
Multi-parameter Capability	CO, CO_2, NH_3, hydrocarbons
Sampling	Method: continuous flow
Performance	Accuracy: Reproducibility: ± 1% Stability: Lag time: Response time: 5 sec (in 0.4 sec optional) Zero drift: 1% in 24 hrs Span drift:
Operation	Ambient temperature range: Temperature compensation: Maximum relative humidity: Calibration: Operation: Maintenance: Output: meter and recorder output terminal 0-10, 0-25, 0-50, 0-100 mV
Specifications	Power: 110 V, 60 Hz, 200 W Weight: 76 lbs Dimensions: 12" x 19" x 12¾"
Cost	$3,570

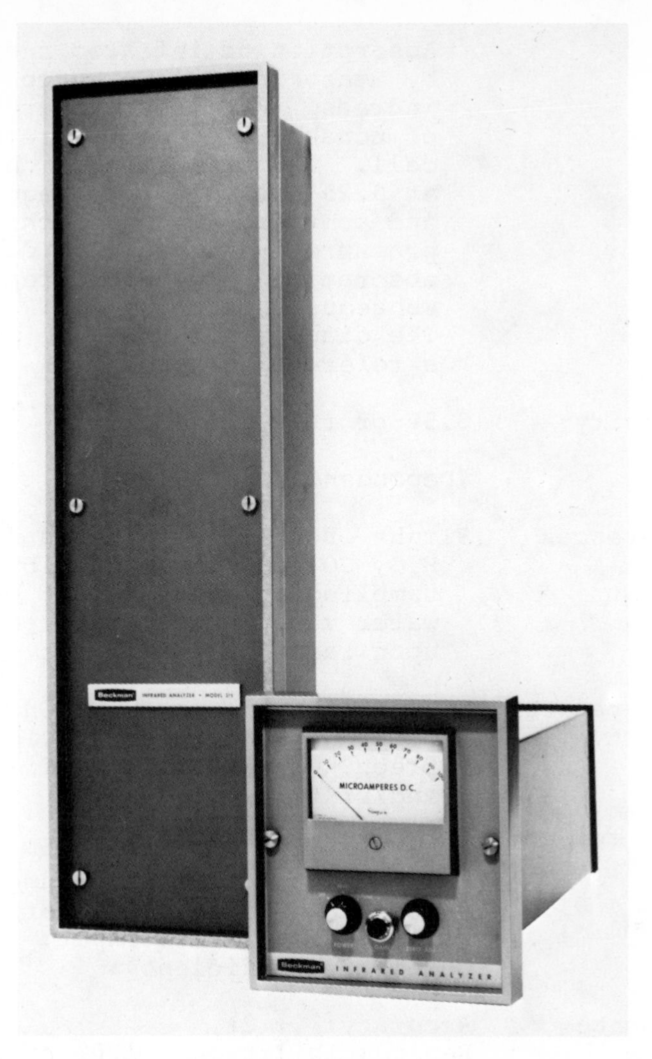

Figure 17.18. Beckman Model 315 NDIR Analyzer (photo courtesy of Beckman Instruments).

Manufacturer: Beckman Instruments, Inc.
2500 Harbor Boulevard
Fullerton, California 92634

Class: Portable

Principle of Operation: Nondispersive infrared absorption (NDIR): The detector consists of a measuring cell and reference cell separated by a diaphragm.

	Absorption of infrared radiation by measuring cell produces an increase in pressure over that of nonabsorbing gas in reference cell. The radiation is chopped at 9.25 Hz. The diaphragm expands and contracts with the changes in pressure in sample cell due to absorption, pressure effect and subsequent chopping of IR source. The diaphragm is used to modify a reference signal.
Sensitivity	0.5% of full scale
Range	(Dependent on application)
Interferences	Slight or no interferences from H_2O, CO_2 unless removed by sampling system. Effect of water vapor will vary depending upon range of instrument.
Multi-parameter Capability	CO_2, CO, acetylene, H_2O, CH_4, propane, isobutane, HCl, NH_3, freon, CS_2--by optical filters and/or filter cells
Sampling	Method: continuous flow Volume: typical cell volume <100 cc, flow 0.1 to 0.5 liter/min Maximum temperature input: Collection efficiency:
Performance	Accuracy: ± 2% Reproducibility: ± 1.0% of full scale Stability: Lag time: Response time: 0.5 sec for electronic response; sample flushing time dependent on cell volume Zero drift: ± 1% of full scale in 24 hrs Span drift: ± 1% of full scale in 24 hrs

Operation	Ambient temperature range: 30° to 120°F Temperature compensation: none needed--thermostat controlled Maximum relative humidity: (see Interferences) Calibration: preferably weekly, but at least monthly) Operation: may be purged with inert gas Maintenance: Output: choice of 0-10, 0-100 mV, 0-1, 0-5 V. Optional 0-5, 1-5, 4-20, 10-50 mA. Options: manual standardization equipment, sample line filters, sample handling systems, sealed sources, mounting kits
Specifications	Power: 115 V, 50/60 Hz, 300 W Weight: 85 lb Dimensions: analyzer (8⅛" x 29" x 9⅝"); control amplifier (⅞" x 8 5/32" x 20")
Cost	Model IR-315B approx. $2,800-$3,000 Sampling system approx. $1,000

(a) Probe

(b) Titration module

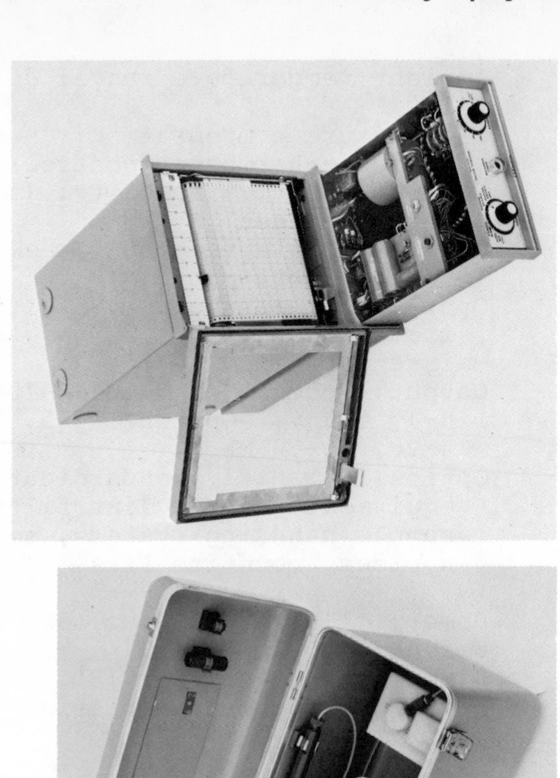

(c) Control module

Figure 17.19. Barton Model 400 Sulfur Titration (photo courtesy of Barton ITT Process Instruments Division).

Manufacturer	Barton ITT Process Instruments and Controls 580 Monterey Pass Road Monterey Park, California 91754
Class	Portable
Principle of Operation	Coulometric: Current flow is monitored between electrodes which are generating Br_2 from a HBr solution. Reaction with SO_2 changes the rate of Br_2 generation which is indicated by a change in current flow. This instrument is an adaptation of the Model 286 ambient sampler. The sample probe utilizes a fiberglass sock to remove particles from the gas train. This sock must be periodically purged by forcing high pressure air through the probe in the reverse direction.
Sensitivity	0.05 ppm
Range	0-1000 ppm (7 ranges)
Interferences	Appreciable interferences from H_2S, NO Slight or no interferences from O_3, CO_2, NH_3
Multi-parameter	H_2S, mercaptans, organic sulfides
Sampling	Method: continuous flow Volume: 200-600 ml/min
Performance	Accuracy: Reproducibility: Stability: Lag time: Response time: Zero drift: Span drift:
Operation	Ambient temperature range: Temperature compensation: Maximum relative humidity: Calibration:

	Operation: purge sample probe for a few seconds every ten minutes Output: strip chart recorder; 0-100 μA, 0-10 mV or 0-100 mV
Specifications	Power: 115 V, 60 Hz, <45 W Weight: 75 lbs Dimensions: titration module (10" x 9" x 22"); recording controller module (14½" x 9" x 13"); probe (11½" x 17" x 32")
Cost	$5,425

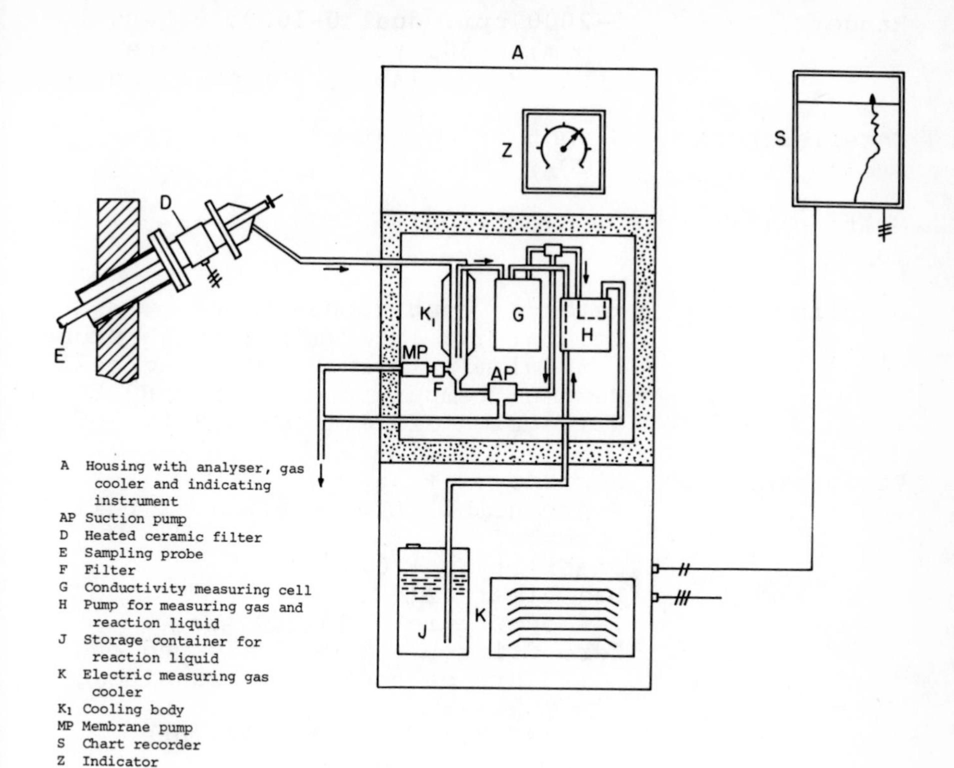

Figure 17.20. Schematic of Mikrogas-MSK Analyzer (courtesy of Calibrated Instruments).

Manufacturer	Calibrated Instruments Inc. 731 Saw Mill River Road Ardsley, New York 10502
Class	Stationary
Principle of Operation	Conductimetric: gas sample is filtered by heated ceramic filter to remove particulates. It is cooled to remove water vapor and for entry into measuring cell. The conductivity change of a hydrogen peroxide solution is measured as SO_2 is absorbed.
Sensitivity	1.4 ppm

Range	0-2000 ppm; dual 0-1000, 0-5000 ppm; 0-500 to 0-8000 ppm are other available, standard ranges
Interferences	Slight or no interferences from CO_2, CO
Multi-parameter Capability	CO_2 (optional)
Sampling	Method: continuous--pump Volume: ∿4 liter/min through probe; 80 ml/min through measuring cell Maximum temperature input: 800°F Collection efficiency: 99.8%
Performance	Accuracy: ± 3% Reproducibility: ± 0.01% of full scale Stability: ± 0.0% Lag time: 20-40 sec Response time: 90-100 sec Zero drift: ± 0.0% (measured over 523 hrs of continuous running), $<$ 0.5% Span drift: ± 0.8% in 24 hrs
Operation	Ambient temperature range: 32-125°F Temperature compensation: thermistor, $<$0.5% error within ambient Maximum relative humidity: 100% Calibration: performance check by electronic input provided Operation: probe may be up to 100 feet from analyzer Maintenance: replace solution every 8 days; blow out probe occasionally
Specifications	Power: 110 V, 60 Hz; 220 V, 50 Hz; analyzer 90 VA, gas cooler 250VA Weight: 215 lb Dimensions: 46" x 21" x 18"
Cost	About $6000

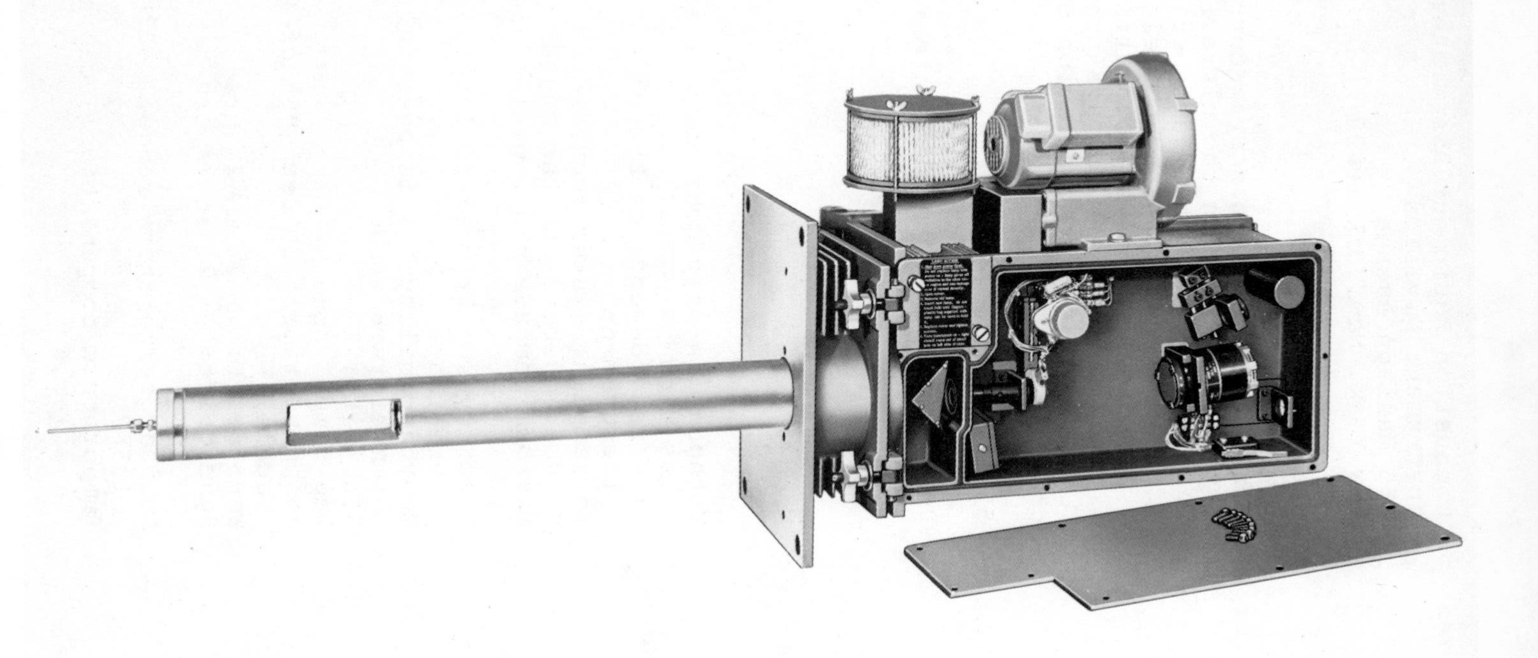

Figure 17.21. *CEA SO_2 Spectrometer (photo courtesy of Combustion Equipment Associates).*

Manufacturer	Combustion Equipment Associates Instrument Division 555 Madison Avenue New York, New York 10022
Class	Stationary
Principle of Operation	Correlation spectroscopy: correlation technique in the ultraviolet region in which the gas spectrum is modulated across a correlation mask in front of a photomultiplier tube, producing a voltage proportional to the concentration of SO_2.
Sensitivity	5% of full-scale reading
Range	0-200 to 0-5000 ppm with intermediate choices
Interferences	
Multi-parameter Capability	None
Sampling	Method: in-situ continuous stack gas flow Flow rate depends on breech measurement No sampling system required Volume: depends on the range Maximum temperature input: <750°F
Performance	Accuracy: ± 5% over full temperature range Reproducibility: ± 5% over full temperature range Stability: Lag time: none Response time: 20 sec (95% of full scale) Zero drift: ± 2% in 24 hrs Span drift: ± 2% in 24 hrs
Operation	Ambient temperature range: 40° to 135°F Temperature compensation: cabinet available

	Maximum relative humidity: unknown Calibration: dynamic calibration with internal gas standard. This device is thermistor-controlled and programmed to calibrate every 8 hours with manual override. Operation: the instrument probe is a slotted tube which is inserted through a port in the stack wall or breeching. An ultraviolet source in the instrument head shines through this tube and through the gas sample. It is reflected from a mirror at the end of the tube back through the gas sample and into the correlation spectrometer mounted in the head. An air curtain prevents flue gases from entering the spectrometer via the optical path. Maintenance: 30 days continuous operation possible Output: terminal outputs suitable for driving most chart recorders, 4-20 ma (other outputs available)
Specifications	Power: 108-122 V, 60 Hz Weight: 110 lbs Dimensions: 16" x 8" x 48"
Cost	About $8500

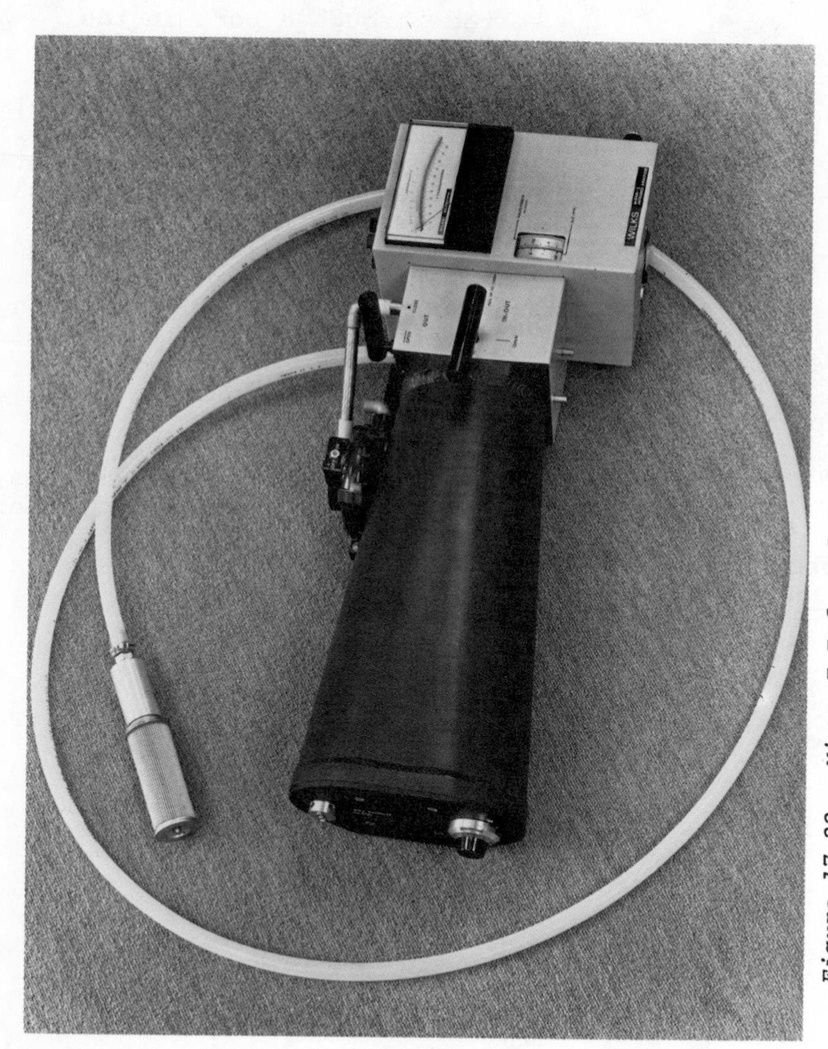

Figure 17.22. Miran I Infrared Analyzer (photo courtesy of Wilks Scientific Corporation).

Manufacturer	Wilks Scientific Corporation 140 Water Street, Box 449 South Norwalk, Connecticut 06856
Class	Portable
Principle of Operation	Dispersive infrared absorption spectroscopy (IR): variable IR range from 2.5 to 14.5 microns and variable path gas cell from 3/4 to 20 meters
Sensitivity	0.1 ppm
Range	0-3000 ppm
Interferences	Moderate interference from H_2O
Multi-parameter Capability	SO_2, NO_x, CO, CO_2, NH_3, hydrocarbons and other infrared absorbing gases
Sampling	Method: continuous flow with pump; intermittent sampling also Volume: 5 liter cell capacity Maximum temperature input: 475°F
Performance	Accuracy: Reproducibility: Stability: $\pm$ 1% in 24 hrs Lag time: approximately the amplifier time const. (0.15, 0.6 or 2.5 sec) Response time: approximately 30 sec at a flow rate of 3 cfm Zero drift: $< \pm$ 1% in 24 hrs Span drift: $< \mp$ 3% in 24 hrs
Operation	Ambient temperature range: 32° to 113°F Temperature compensation: none Maximum relative humidity: 95% Calibration: can be permanently calibrated using standard gas mixtures Operation: manual Maintenance: modular design with plug-in components; built-in voltage test points
Specifications	Power: 115/230 V, 50/60 Hz, 25 W Weight: 27 lbs Dimensions: 27" x 11" x 6.5"
Cost	About $4200

Figure 17.23. *LSI Model RM4 Smoke Density Instrument (photo courtesy of Lear Siegler Inc.).*

Manufacturer	Lear Siegler Inc. 32 Denver Technology Center Englewood, Colorado 80110
Class	Stationary
Principle of Operation	The light from the source in the optical sensor is divided into two paths. The measuring beam is transmitted across the smoke channel to the reflector which is mounted on the opposite wall. At the reflector side an image of the evenly illuminated aperture of the optical system is obtained. Since the reflector area is only a small fraction of the transmitted field of view, the amount of reflected light is insensitive to the usual alignment changes due to thermal shifts in the duct walls. The second or reference light beam is used for comparison purposes and compensates for light source fluctuations, temperature drifts, and aging of the lamp and electronic components. In addition, a rotating disc modulates the two light beams at two different frequencies so that both beams can be focused on a common photocell element and amplifier. In front of the photocell is an optical filter which selects

	monochromatic light to correspond with the spectral sensitivity of the element and thereby eliminates any erroneous contribution from longer wave length light.
Sensitivity	
Range	Opacity: 8 ranges available in 0 to 18.7 and 33.9; 0 to 33.9 and 64.5; 0 to 64.5 and 87.4; 0 to 87.4 and 98.4
Interferences	
Multi-parameter Capability	Responds to all aerosol materials
Sampling	Analysis is in-situ to the source
Performance	Accuracy: ± 3% Reproducibility: Stability: Lag time: none Response time: Zero drift: ± 3% over 3 months Span drift: ± 3% over 3 months Linearity: ± 2%
Operation	Ambient temperature range: -22° to 131°F Temperature compensation: Maximum relative humidity: Calibration: The calibration checks are accomplished by automatically inserting at periodic time intervals into the light beam of the optical sensor a self-contained check reflector for zero reading and subsequently a gray filter for a reading at a preselected calibration point, which depends on the measuring range. Since the scale is linear, two data points are sufficient for calibration of the full measuring range. The zero reflector corrects for loss of light transmission due to accumulation of dirt on the lens of the optical

	sensor. In most installations, the loss on the reflector side can be assumed to be equal to that on the sensor side. If needed, an automatic check reflector on the reflector side can be provided on request. Calibration is performed hourly and is 90 seconds in duration. Maintenance: should be services twice a year
Specitications	Power: 115/230 V, 60 Hz, 55W Blowers require additional 700 W each Weight: optical sensor 59 lbs; reflector 26 lbs Dimensions: optical sensor unit (32" x 20" x 36"); reflector unit (24" x 20" x 36")
Cost	About $4000

CALIBRATION METHODS

Calibration is essential to the successful use of continuous monitoring equipment. It determines the relationship between the observed and true values. Most any system is subject to drift and variation and cannot be expected to maintain its calibration over a long period of time. This is particularly true of systems in the rather harsh source monitoring environments; therefore the federal government requires calibration at least every 24 hours (see Appendix A). These calibrations are to be performed to the manufacturer's specifications. There are many systems which could be used, but we shall briefly discuss only a few of them here. For detailed coverage see the work by Nelson.[4]

Extractive gas monitoring equipment can be calibrated using standard gas mixtures or dilution methods. If the pollutant is stable, the compressed mixture provides a good supply of calibration gas. Sometimes the calibration gas must be generated continuously, and this is done by dilution techniques. Figure 17.24 shows such a device which performs series dilutions using clean air and a cylinder of pure gas. A positive displacement pump and a series of electronically operated valves provide a wide range of concentrations. Another

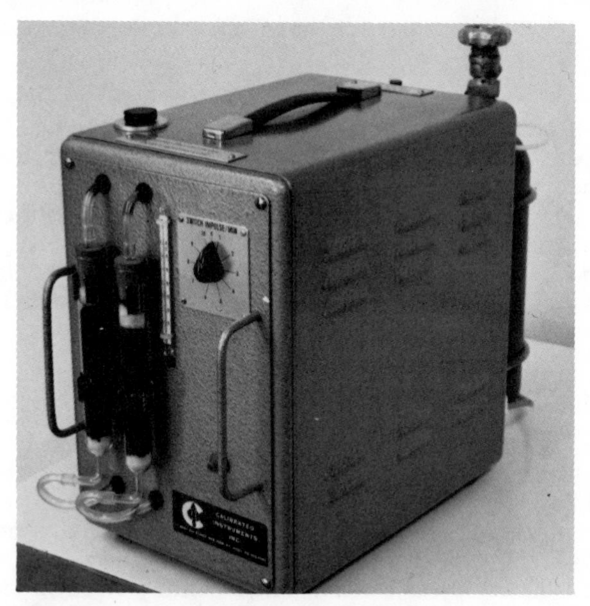

Figure 17.24. Stack gas calibrator (photo courtesy of Calibrated Instruments Inc.).

method to produce a known concentration of a gas is to use permeation tubes; this method works for those gases which are easily liquified. The pollutant is put into a Teflon tube and sealed. The tube is then kept at controlled temperature conditions. The gas permeates through the tube wall and into the gas stream. Figure 17.25 shows a system which uses the permeation tube method. The concentration is controlled by varying the gas flow rate and/or the bath temperature.

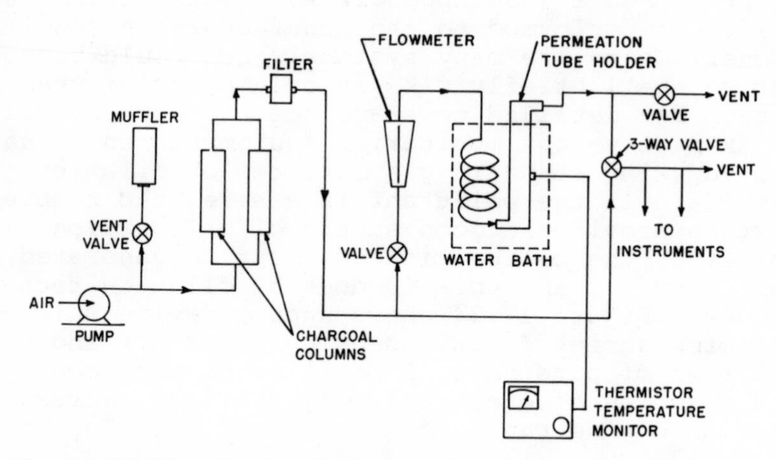

Figure 17.25. Permeation tube apparatus for calibration.

In-situ monitors are often difficult to calibrate for two reasons. First, the stack cannot be put out of service to perform the calibration. Second, it is difficult to produce known pollutant concentration conditions in the stack even if it is put out of service. Some instrument manufacturers have solved this problem by using special optical filters to provide calibration points.

SELECTING A SYSTEM

Before selecting a source monitoring system you must adequately identify your needs. Table 17.5 shows a listing of some of the information a manufacturer needs to modify his measurement system to your needs. A good portion of the data can be determined from plant drawings and process instrumentation. However, some of it may have to be obtained by the source testing procedures described

Table 17.5

Process Information Required for Selecting Source Monitor

Parameter	*Value* Ave	Max	Min	*Remarks*
Type of operation stack gas temperature				
Dew point of stack gas				
Pollutant concentration				
Gas velocity				
Total gas volume flow rate				
Stack dimensions (show sketch)				
Approximate stack gas composition				
Stack gas pressure (at desired monitoring point)				

in the previous chapters. Before contacting instrument companies, collect your process data and specifications.

There are many factors which must be considered in selecting a system. If you are instrumenting a new Priority I source the most basic requirement is: "Is it EPA approved?" This most important piece of information must be obtained from the manufacturers and/or the various control agencies. Beyond this consideration, the factors of cost, ease of use, serviceability, etc. can be evaluated. Table 17.6 shows a check list which is useful in evaluating the suitability of a monitoring method for a specific source. If you assign a rating scale to each consideration, an overall rating can be obtained. Depending upon your specific needs, a different weighting factor should be assigned to each consideration. The overall rating of a monitoring system would be

$$OR = \Sigma\, w_i\, r_i \tag{17-2}$$

Table 17.6

Check List for Rating Various Monitoring Systems

Manufacturer ____________________ Model ____________________

Factor	*Weighting Factor** w_i	*Rating*** r_i	*Product* $w_i\, r_i$
1. Sensitivity			
2. Range			
3. Zero and span drift			
4. Interferences			
5. Reliability for continuous duty			
6. Response time			
7. Cost			
8. Portability			
9. Calibration method			
10. Interfacing system needs			
11. Power requirements			
12. Space and weight requirements			
13. Success of present applications			
14. Serviceability			
15. Ruggedness in monitoring environment			
16. Special operator training required			
17. Leasing and service contracts available			
18. Safety of equipment and reagents			
		sum	

*You might use a scale of 1 through 4, using 4 for the most important factors.

**The following rating system can be used: excellent 4, good 3, fair 2, poor 1, and fail 0.

where

- OR is the overall rating
- r_i is the rating assigned to factor i
- w_i is the weighting factor assigned to factor i.

Using Equation 17-2 the method with the highest rating would be best for the specific need.

An extremely important aspect of the monitoring system is the interfacing requirement. Some monitoring methods required extensive sample pretreatment before the analysis. For example, the gas sample may have to be diluted or heated to prevent condensation.

Most of the instrument companies will provide a maintenance service and some will also lease the equipment rather than sell it. Both of these features may be particularly useful. If you do not have adequate plant personnel and facilities, the service contract is a good way to obtain responsible operation and maintenance. A service contract for continuous monitoring may also be cheaper than hiring and/or training special personnel to do the work. The state of source monitoring technology is fairly young. For this and economic reasons it may be desirable to lease equipment. When and if newer and less expensive equipment becomes available it can be installed.

SUMMARY

Above all else be sure that the system you select meets the EPA guidelines for performance. This means you should contact your regional EPA office before making any final decisions regarding source monitoring equipment. Since source monitoring technology is still in its infancy, you should strongly consider the possibilities of leasing monitoring systems and using service contracts.

REFERENCES

1. Nader, J. S. "Developments in Sampling and Analysis Instrumentation for Stationary Sources," Paper 72-39, 65th Annual Meeting of the Air Pollution Control Association, Miami Beach, Florida. 1972.
2. Air Sampling Instruments. American Conference of Governmental Industrial Hygienists, Cincinnati, Ohio. 1972.
3. Instrumentation for Environmental Monitoring: Air. Lawrence Berkeley Laboratory, University of California, Berkeley, California. 1972.

4. Nelson, G. A. Controlled Test Atmospheres. Ann Arbor Science Publishers, Ann Arbor, Michigan. 1971.
5. Rodes, C. E. *et al.* J. of Air Poll. Cont. Assoc., Vol 19, p 575. 1969.

CHAPTER 18

ADVANCED METHODS FOR SOURCE TESTING AND MONITORING

This chapter presents a number of promising methods which in the future will find many applications in source testing and monitoring. Some of the methods discussed are still in the development stages while others are in the early commercial stages. Each has some decided advantage over the wet chemical and gravimetric methods now used. The purpose of this chapter is to provide a more detailed coverage of these methods because they are fairly new to the air pollution monitoring field.

The present methods for aerosol sampling usually rely upon filtration for collection with subsequent gravimetric methods of analysis. The latter requires laboratory measurements and thus rules out any instantaneous or direct reading of aerosol concentrations. The methods of the future will hinge upon being instantaneous and some will provide information concerning size distribution as well as total mass emission rates. Sem *et al.*[1] have completed a literature review of possible methods for monitoring aerosols from power plants. However, their findings, assessments, and comments are generally applicable to a wide range of sources. In the following sections *beta attenuation* and *piezoelectric crystal methods* will be discussed in detail because they show the greatest promise at the present time.

The present source testing procedures rely exclusively on wet chemical methods for determining the concentrations and mass emission rates of gaseous constituents. However, there are some direct physical and chemical methods which can be used. Some of these have been applied to ambient monitoring and also show promise for application to sources. The methods covered here are *chemiluminescence*, *electrochemical transducers*, and *flame photometry*. The works of

Yarmac,[2] Stevens and O'Keefe,[3] Shaw,[4] and Reference 5 should be referred to for further details.

BETA ATTENUATION

Beta attenuation has been applied to measuring the thickness of materials. Nader and Allen[6] suggested its use for air pollution measurements. But most of the early work on this method was performed in Europe by Dresia *et al.*[7] Recently Herling[8] has discussed its use for monitoring particulate matter in auto exhaust.

Principle of Operation

When beta particles (electrons) pass through a medium, some are absorbed and some reflected, resulting in a net reduction in the beam intensity. The reduction in beam intensity, known as *beta radiation attenuation*, depends statistically on the electron density of the medium. Thus, correlation of attenuation with mass depends on a constant relationship between the number of electrons per molecule (atomic number) and the mass of the molecular nucleus (atomic weight). This ratio is between 0.45 and 0.50 for essentially all elements found in coal and oil combustion effluents except hydrogen which does not contribute enough to particle mass to cause any significant error. Therefore, beta radiation attenuation is a more direct measure of particle mass than any other known technique except gravimetric weighing, vibrational weighing, and centrifugal sensing.

Instruments using this technique consist of a beta radiation source (a radioisotope) and detector. Figure 17.13 shows a typical instrument arrangement. Carbon-14 is often used as a beta source. Particles from a known volume of effluent gas are collected on a filter tape and then placed between the radiation source and detector. The difference in the count rate of the detector before and after the particles are collected is a measure of the mass of the particles. Common detectors are Geiger-Muller counters, proportional counters, scintillation counters, or semiconductor counters.

Beta radiation attenuation is unable to sense airborne effluent particles without first collecting the particles and concentrating them because gas molecules also attenuate beta radiation. Since the mass of gas molecules in effluent streams is several orders-of-magnitude greater than the corresponding mass of particles, radiation attenuation caused by suspended particles cannot be accurately separated from attenuation caused by gas molecules.

The equation for beta attenuation for the instrument shown in Figure 17.13 is given

$$I = I_o \exp(-\mu \rho x) \tag{18-1}$$

where

- I_o is the radiation flux from the source
- I is the radiation flux reaching the detector
- μ is the mass absorption coefficient
- ρ is the density of the absorbing media
- x is the thickness of the absorbing media.

The mass absorption coefficient, μ, is actually dependent upon the specific beta source used and can be determined experimentally. Since we actually have attenuation by several different substances, Equation 18.1 can be expanded as follows:

$$I = I_o \exp[-\mu(\rho x)_{air} - \mu(\rho x)_{filt} + \mu(\rho x)_{misc}] \exp[-\mu(\rho x)_{part}] \tag{18-2}$$

where

- $(\rho x)_{air}$ is the mass per unit area (area density) of air absorbing beta radiation
- $(\rho x)_{filt}$ is the area density of the filter media absorbing beta radiation
- $(\rho x)_{misc}$ is the area density of the miscellaneous items such as the detector window absorbing beta radiation
- $(\rho x)_{part}$ is the area density of particulate matter absorbing beta radiation.

If we apply Equation 18-2 to a clean filter $[(\rho x)_{part}=0]$ we obtain

$$I_1 = I_o \exp[-\mu(\rho x)_{air} + \mu(\rho x)_{filt} + \mu(\rho x)_{misc}]$$

Then we can write Equation 18-1

$$I = I_1 \exp[-\mu(\rho x)_{part}] \tag{18-3}$$

Rearranging we obtain

$$(\rho x)_{part} = \frac{1}{\mu} \log \left(\frac{I_1}{I} \right) \tag{18-4}$$

Now incorporating Equation 18-4 with other source sampling parameters, the particulate matter concentration can be determined

$$C = \frac{A\ (\rho x)_{part}}{Q_s\ t} \qquad (18\text{-}5)$$

where

- C is the concentration
- A is the collecting area of the filter
- Q is the sampling flow rate
- t is the total sampling time.

Application

Only limited source sampling data has been taken with instruments using this technique. Schnitzler, *et al.*[9] recently reported evaluating several instruments, including two beta attenuation devices, in a modern coal-fired plant. The beta instruments showed an excellent correlation with the gravimetric mass measurements. Table 18.1 lists the advantages and disadvantages of this method.

The basic problems associated with application of the beta attenuation method are condensation, selection of proper filter media, and temperature limitations. However, these are engineering problems which can be solved. The problem with the filter paper arises because papers which have the desired strengths and filtration efficiency also are efficient beta attenuators. This tends to reduce the sensitivity of the instrument.

The sensitivity of the method requires that at least 0.05 mg of particulate matter be deposited per square centimeter of area. In practice this means that the minimum sampling time is about 15 minutes.

Equipment

Figure 18.1 shows a stack monitor which was developed based upon the work of Bulba.[10-12] Several other U.S. and European firms also have beta gauging instrumentation available.

Table 18.1

The Advantages and Disadvantages of Beta Gauging Method

Advantages

1. Directly senses a parameter closely related to particle mass.
2. Commercially available with some stack experience.
3. Needs little calibration.
4. Appears to have capability for fair reliability during long-term use.
5. Can probably operate at high temperatures.
6. Can be used with several particle collectors giving the designer greater flexibility in developing reliable instruments.
7. Problems appear to be basically engineering problems which can be solved.

Disadvantages

1. Requires sampling probe and is subject to probe loss errors.
2. Requires conditioning of the sampling stream.
3. Requires particle deposition.
4. Requires an automatic advancing mechanism for the sample deposit.
5. Readout not instantaneous or continuous: 1 data point every 1 to 15 minutes.
6. Filter tape in present models needs replacement every 1 to 4 weeks.
7. Moderately expensive: ≃ \$8,000-\$15,000.
8. Little reliable data has been reported although most reports are positive.

Figure 18.1. AISI Stack Monitor (courtesy of Research Appliance Co.).

Manufacturer	Research Appliance Co. Route 8 and Craighead Road Allison Park, Pennsylvania 15101
Class	Stationary
Principle of Operation	A sample of gas is collected isokinetically from the stack. It is then diluted to prevent condensation and finally filtered. The

	mass of particulate matter collected on the filter is then determined by the beta radiation absorption technique. A Carbon-14 isotope of 1 Millicurie activity is used.
Sensitivity	
Range	
Interferences	
Multi-parameter Capability	Responds to filterable materials only
Sampling	Method: isokinetic sample withdrawal with air dilution used Volume: Maximum temperature input: 1000°F
Performance	Accuracy: for system $\pm$ 0.002 grains per standard cubic foot Reproducibility: Stability: Lag time: Response time: Zero drift: Span drift:
Operation	Ambient temperature range: Temperature compensation: Maximum relative humidity: Calibration: Maintenance:
Specifications	Power: 115 V, 60 Hz Weight: stack console 285 lbs; computer console 250 lbs Dimension: stack console 51" x 24" x 25"; computer console 40" x 24" x 25" Air supply: ejector air - 12 cmf; dilution air - 3 cfm
Cost	\$7000-\$15,000

Summary

Beta radiation attenuation is presently the best extractive technique for monitoring the mass of particulate emissions from stacks. Beta attenuation senses a particle parameter closely related to mass, and then the feasibility of using this for stack monitoring is proven. Only the piezoelectric microbalance and the capacitance impact techniques offer as much promise for such measurements at this time. Considerable testing must be done to fully evaluate the beta radiation attenuation technique, but basic feasibility has been proven. Some commercial units are now available.

PIEZOELECTRIC MICROBALANCE

The piezoelectric microbalance has been used to effectively measure the thickness of vacuum-deposited metal films.[13] Recently the method has been applied to monitoring of ambient aerosols.[14] Brenchley and Carpenter[15] have reviewed the method and discussed its application to source monitoring.

Principle of Operation

When certain types of crystals are put under a mechanical stress they respond by having electrical charges appear on certain faces of the crystal. Conversely a mechanical stress results when the crystal is subjected to external electrical fields at the crystal surface. This phenomenon is called *piezoelectricity*. It has been understood for many years and one of the best known applications is the quartz pressure transducer.[16]

The piezoelectric effect for mass measurements is useful when the crystal is made a part of an electrical oscillator circuit. The crystal will vibrate mechanically when placed in the oscillating electrical field. If the electric field oscillates at a frequency close to the mechanical resonant vibration frequency of the crystal, the crystal forces the circuit to oscillate at precisely the resonant frequency. It is this important effect which allows the piezoelectric crystal to be used as a mass monitor. When a small quantity of foreign material is deposited on the surface of a crystal, the natural resonant frequency of the crystal decreases. The frequency of vibration is expressed

$$f = \frac{N}{L} \tag{18-6}$$

where

- f is the frequency of vibration
- N is the constant for fundamental mode
- L is characteristic dimension for vibration.

Thus the characteristic dimension for vibration is the crystal thickness and it is this parameter which is changed when a layer of particulate matter is deposited on the surface. Since frequency can be measured so precisely, the resulting decrease in resonant frequency of the crystal allows a corresponding precise mass measurement.

The fact that an oscillating quartz crystal may be used to detect a change in mass has been known almost as long as quartz crystal oscillators have been in use. For example, in early radio days it was common practice to lower the transmitter frequency by marking the surface of the controlling quartz plate with a pencil, thus adding an adhering mass of graphite.[13]

Thin Film Theory

The application of the piezoelectric microbalance stems from the consideration of the deposited foreign material as a thin film. The theory was originally presented by Sauerbrey[17] in conjunction with metal film evaporation and later confirmed by Olin and Sem[14] for particle deposition on a crystal. Figure 18.2 shows a side view of a crystal with particulate matter attached.

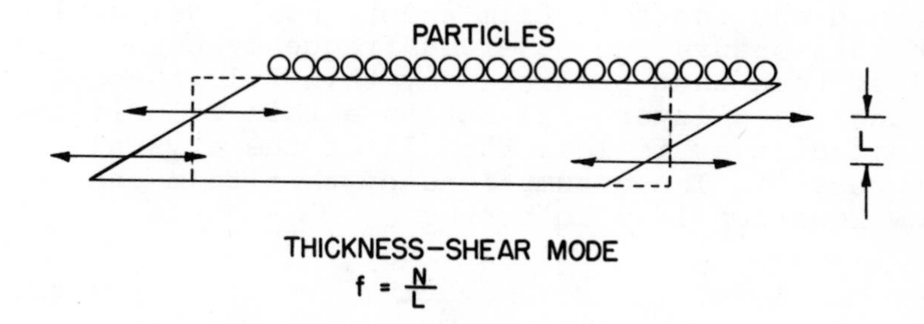

Figure 18.2. Piezoelectric crystal.

The thickness of a crystal may be expressed as

$$L = \frac{M_q}{A \rho_q} \tag{18-7}$$

where

M_q is the mass of the electrically-driven portion of the crystal

A is the area of the electrically-driven portion of the crystal and

ρ_q is the mass density of the crystal

Differentiating Equation 18-6, we obtain

$$\frac{\Delta f}{f} = - \frac{\Delta L}{L} \tag{18-8}$$

where Δf is the change in resonant frequency caused by a change in the thickness ΔL of the crystal. Differentiating Equation 18-7 and substituting into Equation 18-8 we obtain

$$\frac{\Delta f}{f} = \frac{\Delta M_q}{\rho_q A\ L} = - \frac{\Delta M_q}{M_q} \tag{18-9}$$

At this point an assumption is made that the frequency shift caused by the change in mass, ΔM_q, will also be caused by an identical foreign mass, ΔM, deposited on the surface. This then also implies that the foreign particles strictly adhere to the crystal surface. This necessarily means that the layer must be very thin.

Under the thin film condition the foreign material has a negligible contribution to the elastic properties of the crystal. For thick layers a composite resonator exists and the collected material is also strained and the thin film theory no longer applies. When this occurs the observed frequency change will be less than that predicted by thin film theory. In general, a layer will act as a thin film if the deposited layer is less than 1% of the crystal thickness.[14] The assumptions of thin film theory allow Equation 18-9 to become

$$\frac{\Delta f}{f_o} = - \frac{\Delta M}{M_{q_o}} \tag{18-10}$$

where subscript "o" refers to the initial condition before any foreign mass is added. This equation indicates that the resonant frequency decreases linearly with the addition of foreign mass. In practice it has been found that the deviation from linearity is less than 1% as long as Δf is less than 1/2% of f_o. Substituting Equation 18-6 into Equation 18-10 and rearranging we obtain

$$\frac{\Delta f}{\Delta M} = - \frac{C_f}{A} \qquad (18\text{-}11)$$

where C_f is a constant for a specific type crystal and is defined

$$C_f = \frac{f_o^{\,2}}{\rho_q N} \qquad (18\text{-}12)$$

Equations 18-11 and 18-12 indicate that crystals which have higher resonant frequencies, f_o, will have correspondingly higher mass sensitivities. However, as will be discussed later, there is an upper limit to this because at higher frequencies particle reentrainment becomes more of a problem.

Material Requirements

The piezoelectric crystals must possess several important characteristics in order to be successfully used as part of a microbalance. These are mechanical and chemical in nature. First, the crystal must have a low internal friction, and second, the material must be essentially inert to its environment. This latter factor is critical because a surface chemical reation of the crystal could cause its weight to increase or decrease. It would be impossible to separate this weight change from that collected on the surface.

With the foregoing consideration in mind, quartz is a material which best meets these requirements. Also, it has a very high frequency stability, *i.e.*, 1 part in 10^9. Thus at present all piezoelectric microbalances use some sort of quartz crystal. The properties of the crystal vary somewhat depending on how it is cut. The AT-cut crystal finds the greatest use in aerosol monitoring because it is a high frequency cut possessing a low temperature coefficient. This latter property is desirable if frequency change is to occur only as a function of collected mass. When the material properties associated with the AT-cut quartz crystal are applied to Equation 18-12, we obtain

$$C_f = 2.27\ f_o^{\ 2} \qquad (18\text{-}13)$$

ρ_q is 2.654 g cm^{-3},
N is 0.166 MH_z - cm, and thus,
C_f is in $\frac{H_z - cm^2}{\mu g}$ and
f_o is in MH_z

Components

The basic components of a piezoelectric microbalance aerosol detector are the collection method, the crystal and the detection electronics. The aerosol must be collected or deposited efficiently on the crystal surface. The crystals should have the properties previously discussed and the electronic detection system is composed of the oscillation circuits, a mixer, and a read-out apparatus for the frequency. The approach is to optimize the unit by designing for the most efficient removal mechanism, selecting a crystal with a sufficiently high mass sensitivity and then operating the device so that the frequency change caused by all other factors is negligible compared to that caused by the collected mass.

Crystal Characteristics

The quartz crystals take the shape of flat plates or circular wafers. They are quite small and have an electrode on each side. The electrodes are gold, silver, nickel or aluminum and are deposited by vacuum evaporation techniques. Only that portion of the crystal which is covered by the electrode is sensitive to collected mass. This is true because the amplitude of vibration dampens very rapidly outside the electrode area. The crystal becomes a part of the electrical circuit by attaching leads or clips to each side. While in operation the results of the mass measurements are not affected by normal mechanical shock or vibration. In addition, the crystal will function equally well in any mounted position.

Particle Collection

The aerosol may be deposited upon the crystal surface by the following mechanisms: inertial impaction, electrostatic precipitation, thermal precipitation, centrifugal separation and gravity. Of these, the first two mechanisms are most commonly used. Figure 18.3 shows the side view of devices used to collect the aerosol. In one case the sample stream enters through a jet and the particles impinge on the crystal surface. In the other device the crystal is actually the collecting electrode of an electrostatic precipitator. The needle valve is the discharge electrode and is located in the incoming gas stream. The particles thus become charged and are attracted to the crystal surface. The important point here is that either the aerosol should be completely collected or the exact collection efficiency of the device must be known. In most cases it is necessary to know the collection efficiency as a function of particle size.

Particle Adhesion

Once the particle has been effectively removed from the gas stream it remains on the surface due to adhesion forces. Inertial forces tend to reentrain the particle because of crystal vibration. These forces are represented as follows:[11]

$$F_a \propto D_p \qquad (18\text{-}14)$$

$$F_i \propto D_p^3 \; f_o^2 \qquad (18\text{-}15)$$

where

- F_a is the force causing the particle to adhere to the crystal surface
- F_i is the force tending to reentrain the particle
- D_p is the particle diameter
- f_o is the resonant frequency of the crystal.

Equations 18-14 and 18-15 indicate that the larger particles will tend to be reentrained, but this can be alleviated somewhat by using crystals with lower resonant frequencies, f_o. However, as indicated in Equations 18-11 and 18-12 this will result in a lower mass sensitivity of the system.

The particle collection and containment process can be enhanced by conditioning the aerosol and/or altering the crystal surface. Certain gas phase constituents such as water vapor, ammonia and sulfur

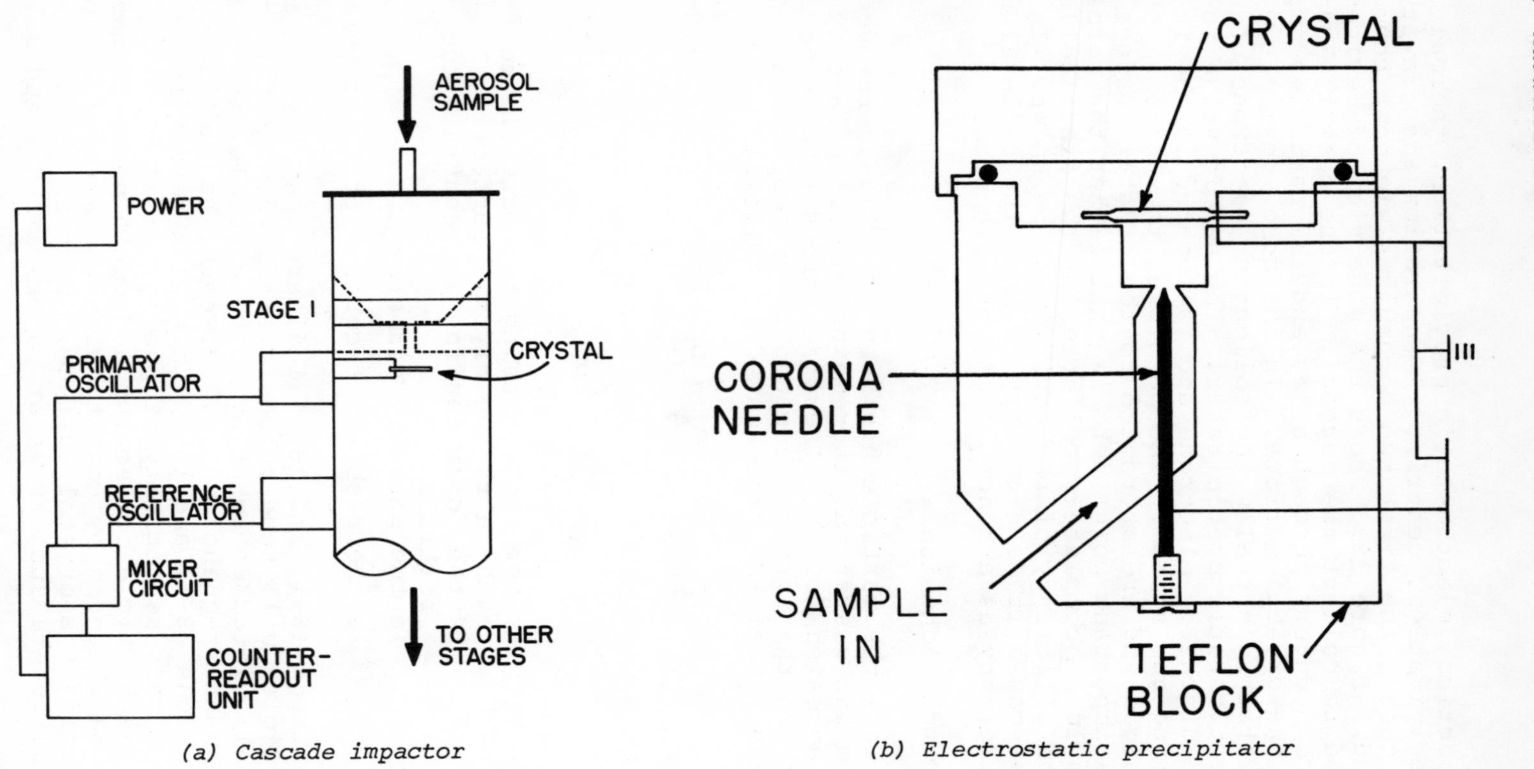

(a) *Cascade impactor*

(b) *Electrostatic precipitator*

Figure 18.3. *Devices for collecting aerosols on piezoelectric crystals.*[15]

dioxide, can be used to enhance the particle collection and adhesion characteristics. Also a thin coating of adhesive material such as grease can be applied to the crystal. These methods may help to decrease particle reentrainment but they raise other uncertainties which must be considered. Care must be taken not to overload the crystal with the coating and cause it to be in a nonlinear response regime. Also, the gas conditioning agents which are added are undoubtedly adsorbed onto the aerosol and will therefore be measured along with the collected aerosol. The coating put onto the crystal may also off-gas or adsorb gases which may then cause error. These problems can only be handled by using careful calibration procedures.

The piezoelectric crystals must have their surfaces cleaned periodically. This requirement stems from the thin film theory and the possibility of deviating from linearity. The crystals otherwise simply become overloaded. The cleaning is accomplished by using a piece of tissue paper possibly in conjunction with a washing solution. It is not necessary to clean it up so that the original resonant frequency is obtained. Mass measurement is based upon frequency change and thus it is acceptable if some residual mass is still present.

Electronics System

The piezoelectric microbalance may consist of one or two crystals. If one crystal is used, a high frequency counter capability is a necessity. The resonant frequencies of the clean crystals are usually $5MH_Z$ or $10\ MH_Z$ and require high frequency counters. This problem can be avoided if two crystals are used. One crystal is used as a reference, while the second collects the aerosol material. The outputs of these two oscillator circuits are fed into a mixer which basically subtracts the two signals. The frequency difference is then read as the output. Hopefully the reference crystal accounts for any changes in frequency due to environmental factors other than added mass. Details of the various electronic circuits cannot be presented here; however, they have been quite adequately described in references 1 and 18.

Operating Conditions

The operating conditions of a microbalance must be known and held constant. As mentioned previously the crystals are sensitive to temperature. This variable can be controlled by proper selection of crystal cut and maintaining conditions to keep the temperature coefficient low. For example, the AT-cut quartz crystal exhibits a frequency deviation of less than 0.001% in the 20°C to 60"C range. This change is therefore negligible compared to that caused by the addition of the micrograms of material.

Some care must also be taken with regard to pressure. The gas pressure may affect crystal response due to stressing and gas adsorption-desorption processes. Olin and Sem[19] performed tests and indicated a 12 Hz change for a static pressure change of 25 cm of Hg. This should not be a serious source of error but it should be considered. Frequency readings should be made at the same pressure if possible.

Aerosol Concentration

The concentration, C, of an aerosol in a gas stream can be calculated from the following expression:

$$C = \frac{\Delta f}{\Delta t} \frac{1}{S\,Q\,E_C\,E_W} \tag{18-16}$$

where

- C is the concentration in $\mu g\ m^{-3}$
- Δf is the change in frequency, Hz
- Δt is the sampling time, sec.
- Q is the sample flow rate, $m^3\ sec^{-1}$
- S is the theoretical mass sensitivity of the crystal, Hz μg^{-1} (from Equation 18-11)
- E_C is the efficiency of particle collection by the collector
- E_W is the efficiency of the piezoelectric microbalance in weighing the deposited particles.

In this expression Δf, Δt, and Q are quite easily measured. The theoretical mass sensitivity is calculated based upon the known characteristics of the crystal. The efficiencies, E_C and E_W, are not known usually, without performing calibration tests. The collection efficiency for a given device may be estimated from theory based upon design and operating conditions. The electrostatic precipitator technique will usually have a higher collection efficiency than inertial impactor techniques. The

E_W term will most likely be close to a value of 1.0 if the thin film theory conditions are met. It should be pointed out there that both E_C and E_W are dependent upon the particle size.

Particle Sizing

The piezoelectric microbalance has some capability for determining aerosol size distributions. There are two approaches available. One is the method of Carpenter[18] in which several crystals were used in conjunction with a multiple stage cascade impactor that separated the aerosol into various size ranges. In the second approach the sampled gas must be diluted so that no two particles are collected at the same instant. Thus a change of frequency over a short period of time can be attributed to a single particle, and if it is assumed to be spherical with a constant and known density its size can be calculated. All of this can be automatic if the digital output of the frequency counter is converted to analog and then differentiated. Additional electronics must be used to count the number of frequencies associated with each size category. For a material with a density of 2 g cm^{-3}, particles as small as 0.5 microns can be detected.

Calibration

If the design and operation of the microbalance meets the theoretical assumptions, no calibration is necessary. The response is linear and predictable. If the layer becomes thick and/or the collected material is dissipative, the method can still be used, but a laboratory calibration is necessary. This means that the method can actually be extended to the detection of liquids and gases, but calibration is then a necessity.

Equipment

There have been a number of piezoelectric mass monitors constructed, some of which are available commercially. Most of them have been designed for a specific application.

Thermo-Systems, Inc., of St. Paul, Minnesota, has developed a line of instruments designed for monitoring aerosol concentrations under ambient conditions and from emission sources. These instruments are single stage collectors using the electrostatic precipitator design (see Figure 18.3). The precipitator operates at 5000 vdc. The 5 MHz quartz crystals are housed in a Teflon chamber and have a mass sensitivity of 180 Hz μg^{-1}. A *monitoring* and a *reference* crystal are used. The sample flow rate through the instrument is 1 LPM. Applying this data and assuming E_C and E_W to be unity to Equation 18-16, we obtain the manufacturer's expression for aerosol concentration

$$C = 333 \frac{\Delta f}{\Delta t} \quad (18\text{-}17)$$

The particle mass monitors are designed to collect particles in the 0.01 to 20 micron range. The maximum amount of material which can be collected before the crystal must be cleaned is 40 micrograms. This restraint in conjunction with the 1 LMP sample flow rate means that the crystals must be cleaned every several hours under high concentration conditions. Of course the very low concentrations would allow the crystals to be used a couple days before cleaning.

Celesco Industries of Costa Mesa, California, has developed a single state impactor which uses 10 MHz crystals and has a mass sensitivity of 1000 Hz μg^{-1}.[20] A two crystal circuit is used with the second one being a reference crystal. The collection efficiency of the monitoring crystal is increased by placing a thin 10^{-4} cm layer of adhesive on the surface. The instrument is designed to collect particles in the 0.1 to 100 micron size range and can be used for aerosol concentrations of 10 $\mu g\ m^{-3}$. The sample flow rate used is 0.15 LPM.

Carpenter[18] fabricated a four stage cascade impactor after the design of Mitchell and Pilcher[21] and equipped each stage with 10 MHz crystals. A dual crystal oscillator circuit was employed with one reference crystal and a monitoring crystal located at each stage. A single mixer and digital read-out unit was used. A switch was used to monitor each stage separately. The impactor operated at a flow rate of 0.5 LMP and for an aerosol with density of 1 $\mu g\ cm^{-3}$ had stage constants of 18.9, 12.6, 6.3, and 2.5 microns respectively for the four stages. The unit was calibrated using a uranine dye aerosol which was analyzed by a simple fluorescence analysis using a spectrophotometer. Figure 18.5 shows the results of the calibration tests for the jet which had a stage constant of 6.3 microns. Note that the

Figure 18.4. Model PM 37 Piezoelectric Mass Monitor (photo courtesy of Celesco Industries).

Manufacturer	Celesco Industries 3333 Harbor Boulevard Costa Mesa, California 92626
Class	Portable
Principle of Operation	Aerosol is collected on the surface of the piezoelectric crystal by inertial impaction. This added mass causes a change in frequency. A thin layer of adhesive material is used to enhance the particle collection efficiency.
Sensitivity	
Range	10 micrograms per cubic meter to 100 milligrams per cubic meter
Interferences	
Multi-parameter Capability	Responds to aerosols collected
Sampling	Continuous

Performance	Accuracy: ± 10% Reproducibility: Stability: Lag time: Response time: 0.1 second Zero drift: Span drift:
Operation	Ambient temperature range: 0 to 120°F Relative humidity: 0-99%
Specifications	Power: 115 V, 60 Hz, 20 W Weight: 5 lbs
Cost	About $3000

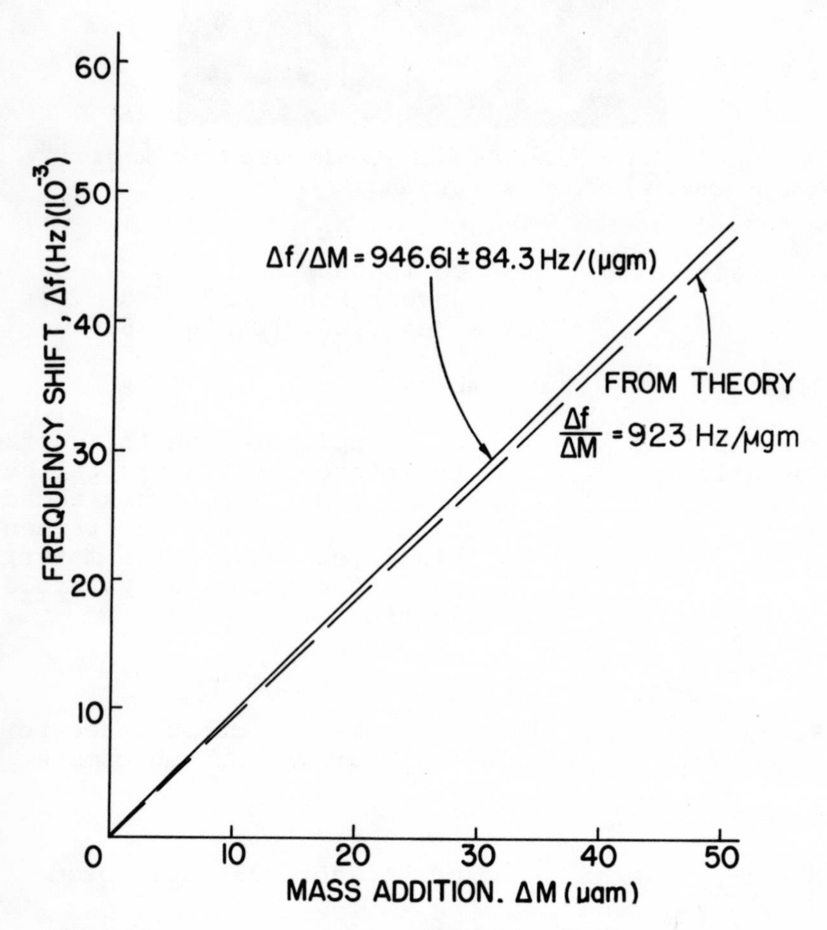

Figure 18.5. Calibration curve for Piezoelectric Mass Monitor.[15]

calibration line agrees quite well with the calculated theoretical mass sensitivity of 923 Hz μg^{-3}. This was not the case for the other stages, however, as particle deposition became a problem. At the top two stages, particles were collected over an area greater than that covered by the electrodes, thus causing the observed mass sensitivity to be lower than theoretical. On the fourth stage the aerosol was not being uniformly distributed over the crystal surface. Although still in a linear response regime, the observed mass sensitivity was greater than theoretical. This situation is explained by reexamining Equation 18-11 and realizing that "A" was effectively being decreased due to the smaller jets at the lower stages. In stage 3 the jet size and the electrode area were essentially the same. All this does not pose a problem as long as the unit is properly calibrated.

Applications

The piezoelectric microbalance has found applications for aerosol monitoring in ambient air and various sources. These applications have met with varying degrees of success. Table 18.2 shows the advantages and disadvantages of piezoelectric method.

Both the Thermo-Systems and Celesco units have been successfully used to monitor aerosols in the ambient air.[22,23] Thermo-System[24] reports a good correlation of the piezoelectric device with a filtration method. Chuan[22] reports that ambient aerosol concentrations were successfully measured with an airborne instrument; in a test in Los Angeles, concentrations ranged from 197 $\mu g\ m^{-3}$ in the wake of a jet aircraft taking off to 2 $\mu g\ m^{-3}$ at an altitude of 5000 feet.

The microbalance instruments have been applied to source testing of automobiles and powerplants with some success. The major problems in this type of application are the high aerosol concentrations and the presence of interfering substances such as water vapor. Herling[25] has reported tests on automobile exhaust and indicated a $\pm$ 30% deviation from filtration measurements. Herling attributes this variation to the presence of organics and water vapor. Sem[26] has outlined a method for using the piezoelectric crystal to monitor particulate matter from power plant stacks.

Table 18.2

The Advantages and Disadvantages of the Piezoelectric Method

Advantages

1. Very high mass sensitivity
2. Linear response
3. Light weight and portable
4. Fast response
5. Response unaffected by shock and vibration
6. Moderate cost
7. Can provide total mass or size distribution data

Disadvantages

1. Crystals are not cleaned automatically and hence requires attention
2. Particle reentrainment may be a problem
3. Volatile components are lost
4. Condensation may be a problem
5. Temperature limitations
6. Difficulty getting even particle distribution on crystals
7. Response may be affected by gaseous constituents also.

This system is shown in Figure 18.6. Notice that the isokinetic sampling process must be done twice. First, the sampling rate from the stack must be sufficient to provide an adequate sample. Second, the sample must be split because the particle mass monitor can accommodate only a small flow rate. The sample conditioning, which usually consists of heating or dilution, is needed to prevent condensation on the piezoelectric crystal. A detailed review of considerations for source monitoring has been presented by Sem *et al.*[1]

Summary

The piezoelectric microbalance shows promise for many types of applications for aerosol monitoring. It is particularly effective for ambient and work conditions where the temperature and water vapor problem is negligible. The application for

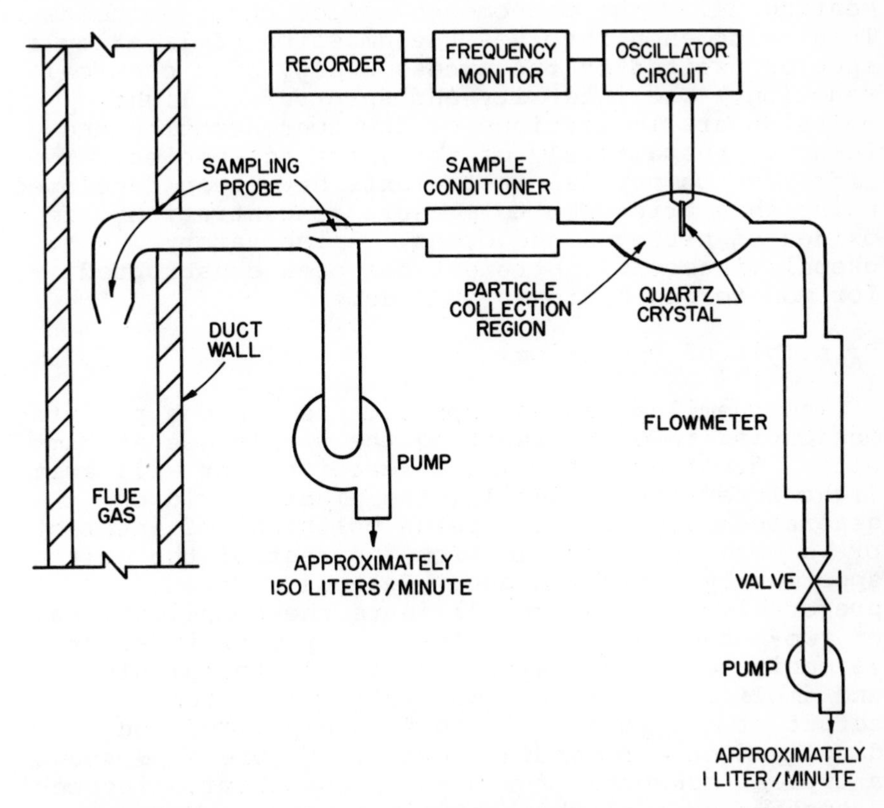

Figure 18.6. Source sampling scheme using Piezoelectric Mass Monitor.[1]

monitoring source concentrations of aerosols seems presently open to question. The eventual success of such an application hinges on the characteristics of the gas stream and the aerosol. The final evaluation of this method for a particular application stems from a consideration of the advantages and disadvantages. Basically the application of the piezoelectric microbalance for aerosol monitoring is in its infancy and many of its present problems will probably be worked out in the future.

CHEMILUMINESCENCE

In search of more direct methods for monitoring of gaseous pollutants researchers have been experimenting using the phenomenon called chemiluminescence. This phenomenon involves the emission of light by a species excited by the excess energy of a chemical reaction. The intensity and spectrum of light emission are indications of the concentration and identity respectively of the emitting species. Two different commercial instruments have been developed using this principle to measure concentrations of oxides of nitrogen and ozone. A laboratory chemiluminescent instrument has been constructed for monitoring hydrogen sulfide.[2]

Principle of Operation

The chemiluminescent approach to gaseous pollutant monitoring involves reacting the sample gas stream with a substance, forcing a reaction that will emit light intensity. Ideally, the light should be generated only by the gaseous pollutant of interest or be readily separable from the rest of the emitted spectrum by optical means. Gaseous reactants preferably are used to eliminate the complications of two-phase reactions. The spectrum of interest is monitored, translated into an electrical signal and amplified by a photomultiplier (PM) tube. The output of the PM tube in turn is amplified and displayed on a recording device. Figure 17.9 shows a typical schematic for a gas phase chemiluminescent detection system; this setup was used by Yarmac[2] for detecting hydrogen sulfide. The factors which affect the performance of such a system are reactor geometry, reaction pressure and temperature, mixing of reactants, presence of third bodies and contaminants, and of course the concentration of the pollutant species. Reactor geometry influences mixing and the ability of light emitted to be detected by the PM tube, and reactor pressure and temperature control the chemical kinetics and equilibrium conditions. Adequate mixing must occur or else the reaction will not take place in the desired manner. Third bodies may be either gaseous or solid and can either enhance or inhibit the chemiluminescent reaction; their presence and effect must be known. Contaminants in the system may also cause chemiluminescence and thus interfere; their effect may be removed by use of optical filters.

The chemiluminescent technique employed for pollutant measurement is demonstrated by the O_3-NO reaction

$$NO + 1/3\ O_3 = NO_2 + hv \qquad (18\text{-}18)$$

where

hv is the emitted radiation.

The intensity of the emitted radiation is expressed

$$I = K\ [NO]\ [O_3] \qquad (18\text{-}19)$$

where

K is a constant
NO is the concentration of nitric acid
O_3 is the concentration of ozone.

In practice a large excess but constant concentration of O_3 is always used. Under these conditions any change in intensity is due solely to changes in concentration of the pollutant species NO.

Ozone

Ozone reacts with ethylane at atmospheric pressure according to Equation 18-18. This method is reported by Warren and Babcock[27] and is now the standard method prescribed by EPA for ambient monitoring.[28] McMillan Electronics Corporation of Houston, Texas, and REM, Incorporated of Santa Monica, California, are two companies that have commercial ambient air monitors available based upon this method. The minimum detectable limit of the instruments is 1 ppb and no interferences are known. The response time is less than three seconds and the nonlinearity is less than 1%.

Other authors have reported the use of the chemiluminescent reaction of O_3 and Rhodamine B dye. Beris and Vassiliou[29] used a liquid solution of galic acid and Rhodamine B. Hodgeson *et al.*[30] used the solid surface of Rhodamine B supported on ac activated silica for chemiluminescent monitoring of O_3.

Oxides of Nitrogen

Nitric oxide reacts with ozone and emits radiation in a continuous spectrum between wave lengths of 0.5 and 2.5 microns. The early work with this chemiluminescent reaction was reported by Clyne *et al.*[31] and Clough *et al.*[32] Research instruments were built and reported by Fontajin *et al.*[33] and Niki *et al.*[34]

Several companies manufacture chemiluminescent instruments: REM Inc. of Santa Monica, California; Scott Research Laboratories of Plumsteadville, Pennsylvania; Aerochem Inc. of Princeton, New Jersey; and McMillan Instruments of Houston, Texas. The instruments generate ozone and use the O_3-NO reaction. Catalytic converters are available to convert NO_2 to NO for the NO_x measurement. The instruments can operate in a concentration range of 1 ppb to 1% NO. These instruments have been used mostly for ambient monitoring and source testing of emissions from internal combustion engines.

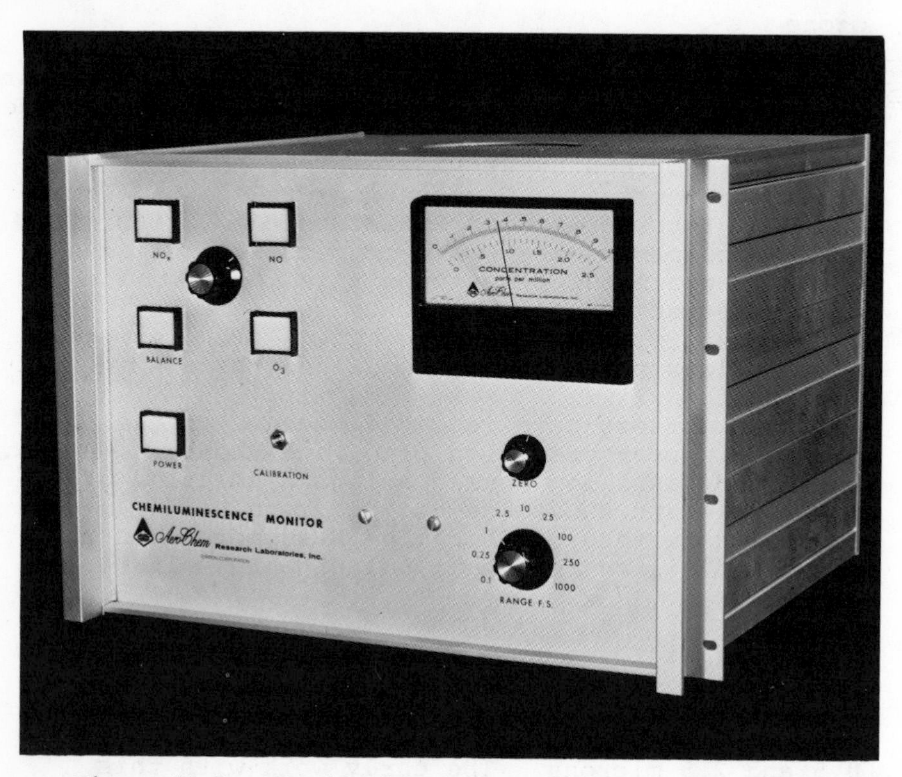

Figure 18.7. Chemiluminescent Monitor (courtesy of AeroChem Research Laboratories).

Manufacturer	AeroChem Research Laboratories P.O. Box 12 Princeton, New Jersey 08540
Class	Portable
Principle of Operation	Sample is drawn continuously into chamber where it is reacted with ozone. The chemiluminescent energy emitted is detected by a photomultiplier tube. The amount of energy emitted is proportional to the concentration present.
Sensitivity	
Range	0.002 to 1000 ppm
Interferences	None
Multi-parameter Capability	NO, NO_x and O_3
Sampling	
Performance	Accuracy: Reproducibility: Stability: Lag time: Response time: 1 second Zero drift: Span drift:
Operation	Ambient temperature range: Maximum relative humidity: Calibration:
Specifications	Power: 115 V, 60 Hz, 415 W Weight: 73 lbs Size: 12¼" x 19" x 17"
Cost	

Some of the first instruments operated at pressures of 1 to 5 mm Hg (torr) and thus required excellent vacuum pumps. These pumps have been the source of mechanical problems and hence recently Hodgeson *et al.*[35] and Martin *et al.*[36] reported using atmospheric pressure conditions. The present trend in commercial instruments is to operate at close to atmospheric pressure. A very small reactor volume was used and to increase sensitivity the photomultiplier tube was cooled. The instrument has a lower sensitivity of 0.005 ppm but its upper response is not linear above 15 ppm; this latter situation is caused by the low concentration of ozone presently being generated by the instrument, and therefore as shown in Equation 18-19 the intensity (response) becomes a function of both the NO_x and ozone concentration.

Hodgeson *et al.*[35] also reported using the O_3-NO reaction to measure ammonia. This is done by passing the sample through a converter containing gold wool at 600 C, the NH_3 is quantitatively converted to NO. It is then reacted with O_3 to allow indirect measurement of the NH_3.

Carbon Monoxide

Snyder and Wooten[37] reported work trying to monitor carbon monoxide via a chemiluminescent reaction with atomic oxygen. The oxygen atoms were produced by a microwave discharge through molecular oxygen. A very faint blue glow was observed. The method was sensitive only to CO concentration greater than 100 ppm. The authors feel that the sensitivity can be greatly improved if further work is done to optimize the system design and operating conditions.

Sulfur Oxides

The initial work regarding chemiluminescence of sulfur oxides was reported by Gaydon[38] and Halstead and Thrush.[39,40] Snyder and Wooten[37] reported using the SO_2-O reaction to detect SO_2 with a lower sensitivity of 100 ppb. The instrument was plagued with an extremely long fall time (> 2 minutes).

Hydrogen Sulfide

Kummer *et al.*[41] studied chemiluminescent reactions involving O_3 with olefins and sulfur compounds.

Yarmac[2] reported being able to detect H_2S in argon in the concentrations as low as 0.1 ppm; the maximum concentration used was 500 ppm. Although the H_2S-O_3 reaction is a gas phase one, Yarmac reported some difficulty due to wall effects. Apparently the stainless steel reactor walls were still aging and hence the sensitivity of the instrument improved over the time it was in use.

Applications

Chemiluminescent devices can definitely be applied to ambient or source monitoring. The commercially available units are first generation instruments which have performed quite well for ambient[42] and source testing[34] applications. These instruments have not as yet received any widespread use on a multitude of stationary sources, but it is felt they will in the future. Table 18.3 shows the advantages and disadvantages of chemiluminescent methods.

Table 18.3

Advantages and Disadvantages of Chemiluminescent Methods

Advantages

Where applicable, the chemiluminescent instruments offer the following advantages:

1. Direct continuous read out
2. Excellent sensitivity
3. Linear over a large concentration range
4. Minimal interferences
5. Good realiability: fewer moving parts and no need for liquid absorbing agents.

Disadvantages

The presently developed chemiluminescent instruments have the following disadvantages:

1. May require rather bulky pump capable of maintaining high vacuums.
2. May require a compressed gas supply.
3. Most instruments do not have a multi-parameter capability.

Summary

The chemiluminescent methods show great promise for source testing and monitoring. They are definitely superior to the present wet chemical methods. Most of the present problems which exist for these instruments can be solved easily in the development stages.

ELECTROCHEMICAL TRANSDUCERS

Analyzers using electrochemical transducers measure the current induced by the electrochemical oxidation of the pollutant at a sensing electrode. Electrochemical transducer (ECT) analyzers, introduced commercially around 1970, avoid the wet chemistry of the traditional conductimetric, colorimetric, and coulometric analyzers by using a sealed module, the electrochemical transducer, inside which all chemical reactions occur. Sensors are available for measurement of SO_2, CO, H_2S, NO, and NO_2. They are intended for both ambient and source monitoring applications. However, descriptive literature and field test data are very limited. The best reference is that of Chand and Marrote.[43]

Principle of Operation

Figure 18.8 shows a simplified schematic of an electrochemical transducer. The pollutant diffuses through the semipermeable membrane into the transducer. The rate of diffusion is proportional to the concentration. At the sensing electrode, the pollutant undergoes an electrochemical oxidation, or reduction which results in a current directly proportional to the partial pressure of the gas being monitored. For example, SO is oxidized as follows:

$$SO_2 + 2\ H_2O \rightarrow SO_4^{-2} + 4H^+ + 2e^- \qquad (18\text{-}20)$$

The production of electrons at the sensing electrode causes this electrode to be at a lower potential relative to the counterelectrode. Thus, an electron current can flow from the sensing electrode through the amplifier to the counterelectrode. The current is proportional to the sample SO_2 concentration.

The rate at which the gas species reaches the sensing electrode is controlled by diffusion through

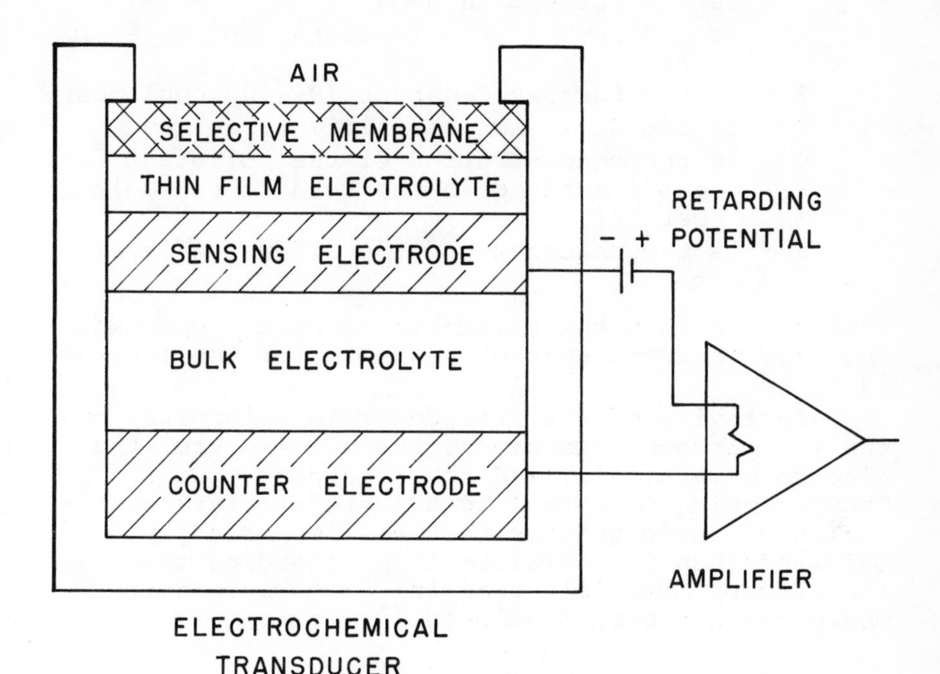

Figure 18.8. Electrochemical transducer.

the membrane, and the thin film of electrolyte according to Fick's law of diffusion, the rate at which the pollutant species will reach the sensing electrode, is directly proportional to the concentration gradient. This gradient results from the removal of the gas at the sensing electrode by electrooxidation or electroreduction, so that the concentration at the electrode surface is much lower than that in the bulk of the thin layer electrolyte. The magnitude of this gradient is directly proportional to the partial pressure and thus concentration of the pollutant in the gas mixture. The diffusion current is expressed by the Equation 18-21.

$$i = \frac{n\ F\ A\ D\ C}{S} \qquad (18\text{-}21)$$

where

- i is the current in amps
- n is the number of electrons per mole of gas
- F is the Faraday Constant (96,500 coulombs)
- D is the diffusion coefficient, cm^2/sec
- C is the concentration of the pollutant gas dissolved in the thin layer (mole per cc)
- S is the thickness of the diffusion layer, cm.

Thus we see that the thickness is quite important and also that the current response for the instrument is linear.

Selectivity of the transducer is determined by the semipermeable membrane, the electrolyte, the electrode materials, and the retarding potential. The retarding potential is adjusted to retard oxidation of those species that are less readily oxidized than SO_2. Information regarding the identify of materials used in commercial instruments has not been disclosed.[44]

ECT Equipment

Commercial ECT analyzers are available for ambient air and stationary source monitoring. Some units are capable of both applications by using different plug-in-transducers. Table 18.4 shows the advantages and disadvantages of the electrochemical transducer method. Figure 18.11 shows a system which uses ECT analyzer for monitoring SO_x and NO_x emissions from a power plant.

Table 18.4

The Advantages and Disadvantages of the Electrochemical Transducer Method

Advantages

1. Simple to operate
2. Low maintenance
3. Fast response
4. Excellent portability
5. Light weight
6. No reagents needed
7. Has multi-pollutant capability by use of different transducers.

Disadvantages

1. Must replace or rejuvenate the transducer periodically.
2. Deterioration of the transducer is gradual and hence frequent calibrations are necessary.
3. Usually transducers must be replaced or rejuvenated every six months. New transducers cost several hundred dollars and rejuvenation exceeds fifty dollars.
4. The gas sample must be pushed through the transducer; this puts severe requirements on the pump not to contaminate the sample. If the transducer is operated under a vacuum, the membrane is pulled away from the thin film electrolyte and the device becomes essentially inoperable.
5. Recovery time may be slow due to pollutant stored in the transducer membrane and thin film electrolyte.
6. Some species in the gas stream may adhere or clog the membrane and thus render it ineffective.

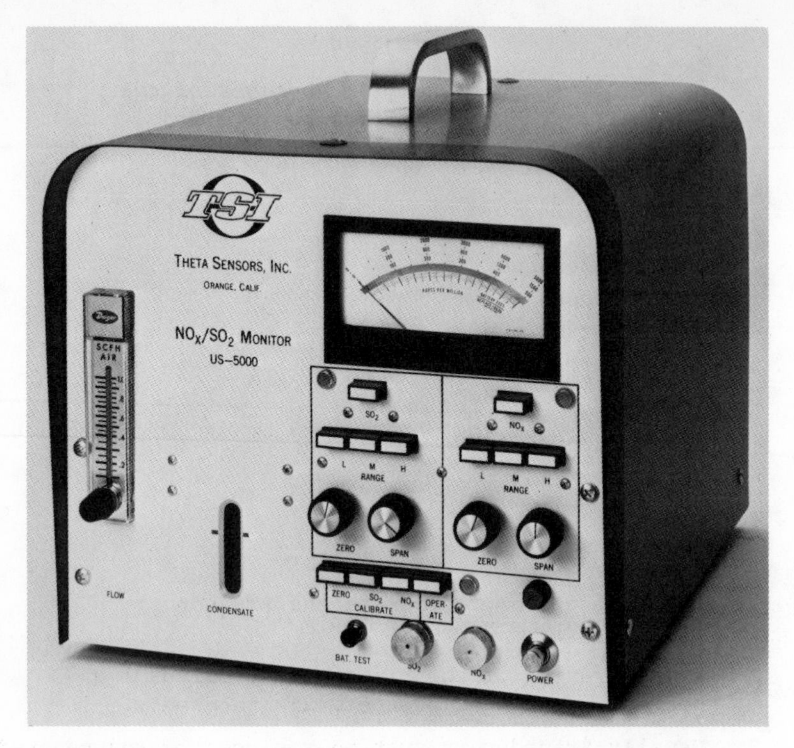

Figure 18.9. NO_x/SO_2 *Model US-5000 Monitor (photo courtesy of Theta Sensors, Inc.).*

Manufacturer	Theta Sensors, Inc. 1015 North Main Street Orange, California 92667
Class	Portable
Principle of Operation	Electrochemical Transducers: Principle of operation is based on the combined use of semi-permeable membranes and liquid state controlled oxidation-reduction.
Sensitivity	
Range	0-500, 0-1500, 0-5000 ppm
Interferences	

Multi-parameter Capability	NO_x
Sampling	Method: continuous-pump Volume: 4.72 liters/min Maximum temperature input: 300°F
Performance	Accuracy: 2% of full scale Reproducibility: 1% of full scale Stability: Lag time: Response time: 20 sec Zero drift: 1/2% in 24 hr, 1% in one week Span drift: 1% in 24 hr, 2% in one week
Operation	Ambient temperature range: 35° to 110°F Temperature compensation: ± 1% over any 20°F range Maximum relative humidity: Calibration: solenoid valves for introducing calibration gases Operation: transducers can be changed and renewed. Maintenance: transducers can be changed and subsequent calibration performed in 15 minutes. Total is less than 2.5 hours per month. Output: meter and 0-10 mV recorder conversion
Specifications	Power: 115 V, 60 Hz Weight: 32 lbs Dimensions: 10½" x 12" x 14¼"
Cost	US-5000, $3,500 Spare Theta Sensor Transducer $250 Charge for renewing transducer $75

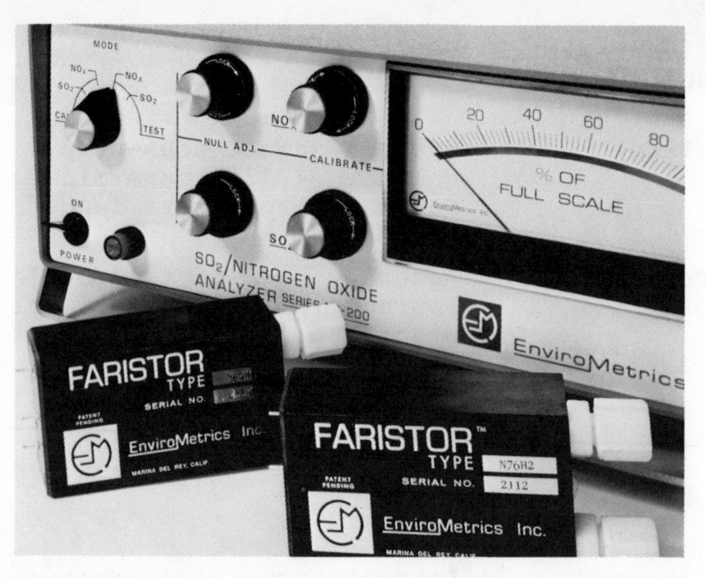

Figure 18.10. Multigas Analyzer Series NS-200 (photo courtesy of EnviroMetrics Inc.).

Manufacturer	EnviroMetrics Inc. 13311 Beach Avenue Marina Del Rey, California 90291
Class	Portable
Principle of Operation	Electrochemical transducer: fuel cell cartridge sensors in which electrocatalytic oxidation or reduction of pollutant species generates a current under diffusion limited conditions giving output linear to concentration.
Sensitivity	0.01 ppm
Range	Faristor cartridges 0.01-100 ppm; 0.02-200 ppm; 0.8-5000 ppm; 2-10,000 ppm; 5-20,000 ppm
Interferences	Moderate interferences from NO_2, H_2S*; slight or no interferences from all other gases *Series NS-280 and NS-290 minimize these interferences.
Multi-parameter Capability	NO_x, NO_2, H_2S, CO, RCHO

Sampling	Method: continuous-pump (Sampler Control Series SC-400 contains pump, filter, heat exchanger) Volume: 0.5-1.5 liter/min Maximum temperature input: 180°F Collection efficiency:
Performance	Accuracy: ± 2% Reproducibility: ± 1% Stability: see drift Lag time: zero Response time: 5-10 sec Zero drift: less than 1% in 24 hrs** Span drift: less than 1% in 24 hrs** **Data for special SO_2 Faristor Ambient Type S64E3. Drift varies with Faristor type.
Operation	Ambient temperature range: 30°R to 150°F Temperature compensation: 30°F to 150°F Maximum relative humidity: 0-85% with no sample conditioning; higher humidity with conditioning Calibration: calibrates against known gas Operation: calibrate to any desired full scale range within Faristor capability (see Range). Range switch gives second range at fixed ratio. Maintenance: Faristor replacement Output: analog for strip chart recorder or meter
Specifications	Power: 115 V, 60 Hz, 5 W Weight: 8 lbs Dimensions: 12" x 4.25" x 8.25"
Cost	About $2500

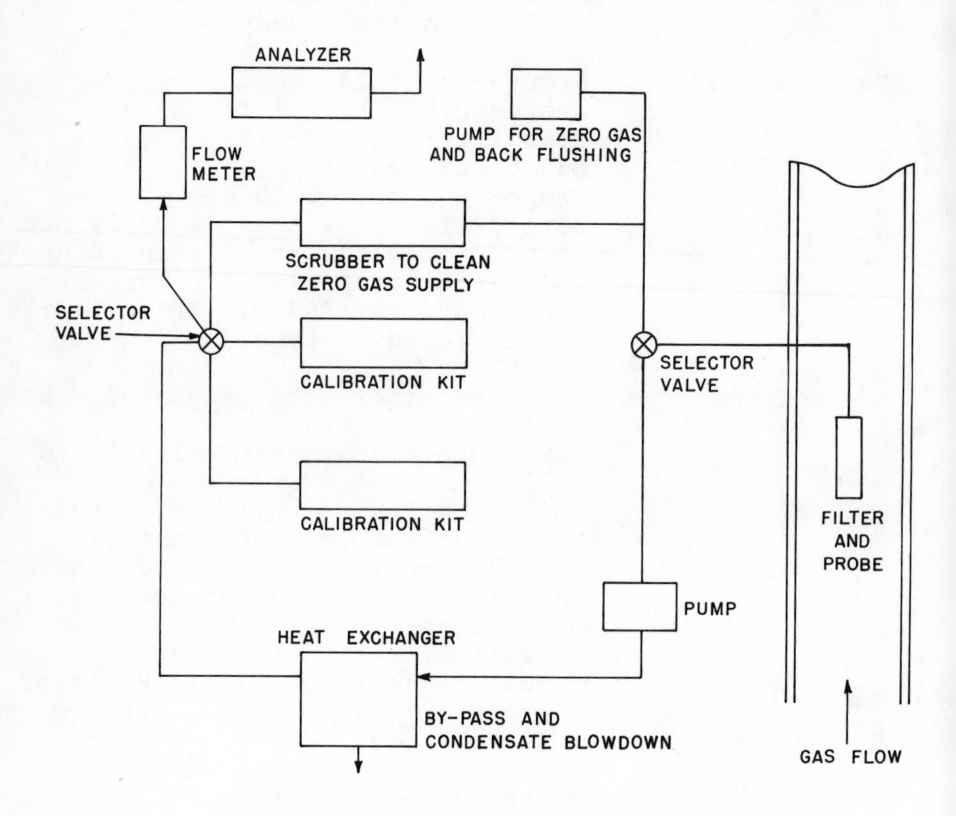

(a) Schematic of monitoring system

Figure 18.11. Monitoring station for power plant (courtesy of EnviroMetrics Inc.).

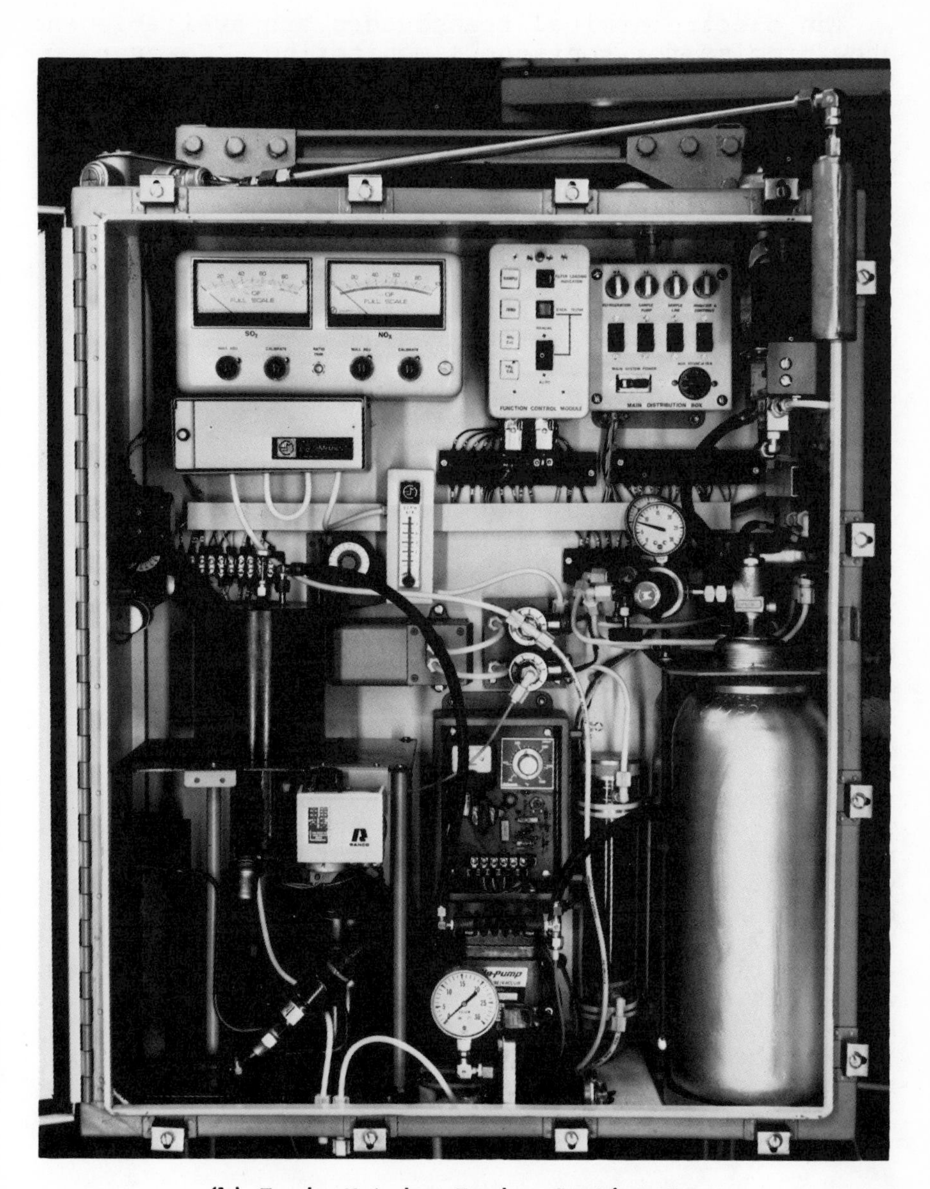

(b) EnviroMetrics Enviro-Station System

Summary

The electrochemical transducers are available and show promise for continuous monitoring. However, their true value needs to be substantiated by many field tests and applications.

FLAME PHOTOMETRY

Sulfur and phosphorus compounds, when introduced into a hydrogen-rich flame, produce strong luminescent emissions. As with chemiluminescence, the wave length is characteristic of the species while the intensity of the radiation is dependent upon pollutant concentration. There are presently several commercial units available.

Principle of Operation

Crider[45] and Brody and Chaney[46] describe flame photometric detectors (FPD) designed to produce a semispecific response to volatile phosphorus and sulfur compounds. Crider applied flame photometry to air monitoring for SO_2 and H_2SO_4 while Brody and Chaney used a flame photometric detector and gas chromotography to analyze for organic sulfur compounds. These compounds become chemically excited when introduced into a hydrogen-rich flame. The molecules release some of this energy by emitting radiation. The sulfur compounds, which are of interest as pollutants, emit in the 300 to 423 nanometer range. A specificity ratio for sulfur to nonsulfur compounds of about 20,000 to 1 is obtained by using a narrow band filter at 394 nanometers. Figure 17.10 shows a schematic of a typical flame photometric (FPD) sulfur monitor. The main problem with this method is that it cannot distinguish between SO_2, H_2S, CS_2, and CH_3SH. However, recently Stevens *et al.*[3,47] have reported the feasibility of using chromatographic technique in conjunction with the flame photometric detector to separate the responses due to SO_2 and H_2S etc. in the sampled gas.

Flame Photometric Equipment

There are numerous commercial units available. Commercial equipment is manufactured by Meloy Laboratories of Springfield, Virginia; Tracor Inc. of Austin, Texas; Bendix Corporation of Ronceverte,

West Virginia; and Analytical Instruments Development of West Chester, Pennsylvania. However, nearly all models are intended for ambient monitoring. Most of these would require a dilution system to lower the sulfur concentration to within detector range.

Figure 18.12. SO_2 Flue Gas Monitor Model FSA 190 (photo courtesy of Meloy Laboratories, Inc.).

Manufacturer	Meloy Laboratories, Inc. 6631 Iron Place Springfield, Virginia 22151
Class	Stationary
Principle of Operation	Flame photometric detector (FPD): utilizes the photometric detection of the 394/nanometers-centered band emitted by sulfur-containing compounds in a hydrogen rich air flame. Photomultiplier tube is optically shielded from the primary flame and the unit uses a narrow band-pass interference filter.
Sensitivity	25 ppm

Range	25 to 2,500 ppm up to 100 to 10,000 ppm (preset at factory)
Interferences	Slight or no interferences from common ambient constituents such as CO, CO_2, hydrocarbons, ozone, particulates, or oxides of nitrogen when used in common stack monitoring applications.
Multi-parameter Capability	Phosphorous compounds (by changing to optional 526 mμ filter).
Sampling	Method: continuous heated trace line for up to 4 stacks or ports to be sampled in automatic sequence Volume: 4 liters/min Maximum temperature input: 700°F Collection efficiency: 100%
Performance	Accuracy: ± 2% Reproducibility: ± 2% Stability: Lag time: 5 sec Response time: 12 sec Zero drift: ± 1% Span drift: ± 1%
Operation	Ambient temperature range: 0° to +120°F Temperature compensation: vortex cooling system Maximum relative humidity: 100% (trace line maintained at 212°F) Calibration: internal calibration signals and sampling sequency includes zero air on each cycle. Calibrated span gas can be programmed into automatic sequence. Operation: stack samples are drawn through a filter and heated sample lines to the dilution system where they are mixed with precise quantities of air to bring the sample concentration into the operating range of the flame photometric detector. Programable switching for multiple

	in-process or multiple stack sampling. An external supply source of compressed electrolytic or equivalent grade hydrogen is required. Output: meter and 0-100 mV output terminal to data recorder or readout device (other outputs to 1 V avilable)
Specifications	Power: 120 V, 60 Hz, 30 A Weight: 130 lbs Dimensions: 27" x 28" x 12"
Cost	$8000-$12,000

Table 18.5

The Advantages and Disadvantages of Flame Photometric Methods

Advantages

1. High sensitivity
2. Fast response
3. Excellent selectivity for sulfur compounds
4. Low maintenance; few moving parts
5. Have good potential for unattended operation.

Disadvantages

1. The need for a compressed hydrogen supply
2. Inability to discriminate among sulfur compounds without use of some separating method
3. Electrical output is not linear; the photomultiplier output is logarithmically proportional to the sulfur concentration
4. No field experience for source testing applications.

Applications

The present applications of flame photometric detection have been mostly for ambient monitoring. Few results have been published for source monitoring. Recently Stevens *et al.*[42] that the FPD analyzer had a correlation of 0.874 with the West-Gaeke procedure. The chromatographic-FPD had a poorer correlation but this was due to equipment malfunctions and not the method itself. Table 18.5 lists the merits of FPD.

Summary

The FPD is highly suitable for ambient monitoring and may also be similarly suited for source monitoring. However, at present there definitely is an inadequate amount of field experience for an adequate assessment of the source monitoring applications.

REFERENCES

1. Sem, G. J. *et al.* State of the Art: 1971, Instrumentation for Measurement of Particulate Emissions from Combustion Sources. Volume I: Particulate Mass - Summary Report. Clearinghouse Report PB 202 655, NTIS, Springfield, Virginia.
2. Yarmac, R. "The Design Construction, and Evaluation of a Chemiluminescent Hydrogen Sulfide Detector," M.S. Thesis, Purdue University, June, 1972.
3. Stevens, R. K. and O'Keefe, A. E. "Modern Aspects of Air Pollution Monitoring," Anal. Chem. 42: 143A-148A (1970).
4. Shaw, M. "Plug-In Sensors for Air Pollution Monitoring," 12th Conference on Methods in Air Pollution and Industrial Hygiene Studies, Los Angeles, California, April, 1971.
5. Instrumentation for Environmental Monitoring, LBL-1 Vol. 1: Air, Lawrence Berkeley Laboratory, University of California, Berkeley. 1972.
6. Nader, J. S. and Allen, D. R. "A Mass Loading and Radioactivity Analyzer for Atmospheric Particulates," American Industrial Hygiene Association Journal 21:300 (1960).
7. Dresia, H., Fischotter, P., and Felden, G. "Kontinuierliches Messen des Staubegehaltes in Luft und Abgasen mit Betastrahlen," VDI-Z, Vol 106, No. 24, pp. 1191-1195, August, 1964.
8. Herling, R. "Beta Gauge and Filter Collection System for Determination of Automobile Particulate Emissions," Paper 71-1032, Joint Conference on Sensing of Environmental Pollutants, Palo Alto, California. 1971.

9. Schnitzler, H., Maier, O., and Jander, K. "Messtand fur die Prufund und Kalibrierung von Registrierenden Staub - und Gasmessgeraten in einem Steinkohlengefeuerten Kraftwerk," Schr Reihe Ver. Wass. - Boden Lufthyg. Berlin-Dahlem, Vol 33, Stuttgart, 1970.
10. Bulba, E. and Silverman, L. "A Mass Recording Stack Monitoring System for Particulates," 58th Annual Meeting of the Air Pollution Control Association, Paper 65-141, Toronto, Canada, June, 1965.
11. McShane, W. P. and Bulba, E. "Automatic Stack Monitoring of a Basic Oxygen Furance," Paper 67-120, 60th Annual Meeting of the Air Pollution Control Association, Cleveland, Ohio, June, 1967.
12. Bulba, E. and Wallace, D. "Automatic Monitoring of Dust Concentration in Cleaned Blast Furnace Gas," 27th Iron-making Conference of the Metallurgical Society, Atlantic City, New Jersey, April, 1968.
13. Wolsky, S. P. and Danuk, E. J. Z. (editor). Ultra Micro Weight Determination in Controlled Environments, Interscience Publishers, New York, New York. 1969.
14. Olin, J. G. and Sem, G. J. "Piezoelectric Microbalance for Monitoring the Mass Con-entration of Suspended Particles," Atmospheric Environment 5:653 (1971).
15. Brenchley, D. L. and Carpenter, T. E. "Monitoring Aerosol Concentrations with Piezoelectric Crystals," Proceedings of the 12th AEC Air Cleaning Conference, Oak Ridge, Tenn. October, 1972 (in press).
16. Beckwith, T. G. and Buck, N. L. Mechanical Measurements. Addison-Wesley Publishing Co., Reading, Massachusetts. 1961.
17. Sauerbrey, G. Z. "Verwedung von Schwinquarzen Sur Wagung dunner Schichten und zur Mickro-wagung," Ziets. Phys. 155, 206 (1959).
18. Carpenter, T. E. "The Design, Construction and Calibration of a Piezoelectric-Cascade Impactor for Monitoring Aerosols," M. S. Thesis, School of Civil Engineering, Purdue University, 1972.
19. Olin, J. G. and Sem, G. J. "Piezoelectric Aerosol Mass Concentration Monitoring," Advances in Instrumentation for Air Pollution Control, Symposium, Cincinnati, Ohio, May, 1969.
20. Chuan, Raymond L. "An Instrument for the Direct Measurement of Particulate Mass," Aerosol Science, Vol. 1, 111 (1970)
21. Mitchell, R. I. and Pilcher, J. M. "Improved Cascade Impactor for Measuring Aerosol Particle Size," Ind. and Eng. Chem., 51, 1039 (1959).
22. Chuan, R. L. "Measurement of Particulate Pollutants in the Atmosphere," AIAA Paper No. 71-1100, Joint Conference on Sensing of Environmental Pollutants, Palo Alto, California, 1971.

23. Olin, J. G. "Airborne Particle Monitoring Applications of the Particle Mass Monitor System," AIAA Paper No. 71-1100, Joint Conference on Sensing of Environmental Pollutants, Palo Alto, California, 1971.
24. "Air Quality Monitoring Experiments with Particle Mass Monitor System," Technical Note No. 6, Thermo-Systems Inc. St. Paul, Minnesota.
25. Herling, R. "A Comparison of Automotive Particle Mass Emission Measurement Techniques," Central States Meeting of the Combustion Institute, Ann Arbor, Michigan.
26. Sem, G. J., Porgos, J. A., and Olin, J. G. "Monitoring Particulate Emissions," Chem. Engr. Prog. Vol 67, No. 10 (1971).
27. Warren, G. J. and Babcock, G. "Portable Ethylene Chemiluminescence Ozone Monitor," Review of Scientific Instruments 41:280 (1970).
28. "National Primary and Secondary Ambient Air Quality Standards," Federal Register, Vol 36, No. 84, April 30, 1971.
29. Beris, D. and Vassiliou, E. "A Chemiluminescent Method for Determining Ozone," Analyst, 91: 499-505 (1966).
30. Hodgeson, J. A., Krost, K. J., O'Keefe, A. E., and Stevens, R. K. "Chemiluminescent Measurement of Atmospheric Ozone," Journal of Analytical Chemistry 42: 1795 (1970).
31. Clyne, M. A., Thrush, B. A., Wayne, R. P. "Kinetics of the Chemiluminescent Reaction Between Nitric Oxide and Ozone," Transactions of the Faraday Society, 60:359 (1964).
32. Clough, P. N. and Thrush, B. A. "Mechanism of the Chemiluminescent Reaction Between Nitric Oxide and Ozone," Transactions of the Faraday Society, 63:915 (1967).
33. Fontajen, A., Sabadell, A. J. and Ronco, R. J. "Homogeneous Chemiluminescent Measurement of Nitric Oxide with Ozone," Journal of Analytical Chemistry 42:575-579 (1970).
34. Niki, H., Warnick, A., and Lord, R. R. "An Ozone-NO Chemiluminescent Method for NO Analysis in Piston and Turbine Engines," Paper No. 710072, Society of Automotive Engineers, January 1971.
35. Hodgeson, J. A., Rehme, K. A., Martin, B. E., and Stevens, R, K. "Measurements for Atmospheric Oxides of Nitrogen and Ammonia by Chemiluminescence," Paper 72-12, 65th Annual Meeting of the Air Pollution Control Association, Miami Beach, Florida, June, 1972.
36. Martin, B. E., Hodgeson, J. A., and Stevens, R. K. "Chemiluminescence Detection of Nitric Oxide at Atmospheric Pressure," Preprint. 164th National ACS Meeting, New York, New York, September, 1972.

37. Snyder, A. D. and Wooten, G. W. "Feasibility Study for the Development of a Multifunctional Emission Detector for NO, CO, and SO_2," Contract No. CPA 22-69-8, National Air Pollution Control Administration, Cincinnati, Ohio. 1969.
38. Gaydon, A. G. "Spectrum of the Afterglow of Sulfur Dioxide," Proceedings of the Royal Society of London A146:901 (1934).
39. Halstead, C. J. and Thrush, B. A. "Chemiluminescent Reactions of Sulfur Monoxide with Oxygen Atoms and Ozone," Nature 204:992 (1964).
40. Halstead, C. J. and Thrush, B. A. "Chemiluminescent Reactions of Sulfur Monoxide," Photochemistry and Photobiology 4:1007 (1965).
41. Kummer, W. A., Pitts, J. N. and Steer, R. P. "Chemiluminescent Reactions of Ozone with Olefins and Sulfides," Environmental Science and Technology 5:1045 (1971).
42. Stevens, R. K., Clark, T. A., Decker, C. E. and Ballard, L. F. "Field Performance Characteristics of Advanced Monitors for Oxides of Nitrogen, Ozone, Sulfur Dioxide, Carbon Monoxide, Methone, and Nonmethone Hydrocarbons," Paper 72-13, 65th Annual Meeting of the Air Pollution Control Association, Miami Beach, Florida, June, 1972.
43. Chand, R. and Marcote, R. J. "Evaluation of Portable Electrochemical Monitors and Associated Stack Sampling for Stationary Source Monitoring," 68th National Meeting of the American Institute of Chemical Engineers, Houston, Texas, February 28-March 4, 1971.
44. Parts, L. P., Sherman, P. L., and Snyder, A. D. "Instrumentation for the Determination of Nitrogen Oxides Content of Stationary Source Emissions," APTD-0847, Environmental Protection Agency, Research Triangle Park, North Carolina, 27711.
45. Crider, W. L. Anal. Chem. 37:1770 (1965).
46. Brody, S. S., and Chaney, J. E. J. Gas Chromatography 4:42 (1966).
47. Stevens, R. K., O'Keefe, A. E., Mulik, J. D., and Krost, K. J. "Gas Chromatography of Reactive Sulfur Compounds in Air at the ppb Level," 157th National ACS Meeting, Minneapolis, Minnesota, 1969.

CHAPTER 19

AND LET THERE BE LIGHT

This chapter primarily contains the authors' ideas, thoughts, and opinions regarding the present and future of industrial source sampling. We are now at the end of the beginning of industrial source sampling. Federal emission standards and testing procedures have been promulgated by the EPA. Presently these performance standards for new or modified installations apply only to five types of air pollution sources. However, they are expected to be extended to many other new sources in the near future. Currently many states are applying the federal testing procedures to existing air pollution sources and the trend is expected to continue. The authors therefore feel that source testing, as a method of quantifying air pollution emissions for compliance, is here to stay. It is hoped, however, that the present methods will continue to be improved and thus better protect both the public and industry. The remainder of this chapter consists of a critique of the existing source sampling methods and a section containing the authors' suggestions to individuals faced with the problem of having source tests conducted.

PRESENT EPA TEST METHODS

The current federal EPA source testing methods are an important step in the right direction. Since the federal performance standards are law, all new affected sources are required to prove compliance by performing source tests. Necessarily then, the source testing procedures were standardized. The methods promulgated are now also being specified for use on many old and new sources by many state and local air pollution authorities. In this regard caution must be exercised to be sure that the methods are applicable. The agencies should give explicit directions concerning required process operating conditions and delegation of responsibilities as to site preparation, actual sampling and the control agency's right of entry and testing. These items will be particularly important in the advocation of permit regulations.

The authors feel that the new federal source testing methods are basically sound, but there is still much room for improvement. The test procedures were designed to be applied to new (after August 17, 1971) sources. Considering the number of already existing sources in operation, some mention should be made on how to best apply the EPA source testing methods to other sources. The federal test procedures are at times not explicit enough. All assumptions made when developing the methods are not clearly stated. For instance no mention is made of the theory and assumptions used to develop Method 1, Sample and Velocity Traverses for Stationary Sources. All assumptions used in developing the methods should be clearly stated. This provides the user with some assessment of the advantages and limitations of the procedure.

Although the federal emission standards are written in a fashion normalized to process input or output (*e.g.*, lb $SO_2/10^6$ BTU), procedures are not very clearly specified for determining process variables. In some cases thousands of dollars are wasted getting an accurate estimate of a pollutant mass emission rate, that has to be divided by a questionable process throughput quantity in order to ascertain compliance.

The reliability of some of the mass emission data may also be in doubt. Results from EPA Method 5, Determination of Particulate Emissions from Stationary Sources, have shown that a significant amount of the total particulate collected can be collected in the probe prior to the filter. However, there is no apparent well-defined procedure for cleaning the probe. The only requirement is that the source test team have some sort of a probe brush that is at least as long as the probe.

The reliability of EPA Method 7, Determination of Nitrogen Oxide Emissions from Stationary Sources, may also be of some question. Is it reasonable to assume that the concentration determined by four 2-liter grab samples taken over a two-hour period will always be representative (of the concentration during that period) when the actual flow rate may be hundreds of thousands of cubic feet per minute and varying by thousands of cubic feet per minute?

Sections of the federal performance standards other than the test methods also need to be upgraded. New facilities affected by the performance standards are usually required to install continuous air pollution monitoring instrumentation. The standards only give general specifications for such equipment. Continuous monitoring equipment need not be approved by the EPA. On the other hand the standards state that prescribed source testing methods may be replaced by equivalent methods as approved by the Administrator of the EPA. However, no mention is made of how equivalency is actually determined.

Overlooking the imperfections of the source testing methods, there are other obvious problems encountered during source sampling. Due to a simple lack of knowledge or professional integrity, source tests are often conducted with little regard for the purpose of the test and thus the accuracy required. An example might be the determining of the efficiency of a control device as indicated in Figure 19.1.[1] The efficiency of the device may be

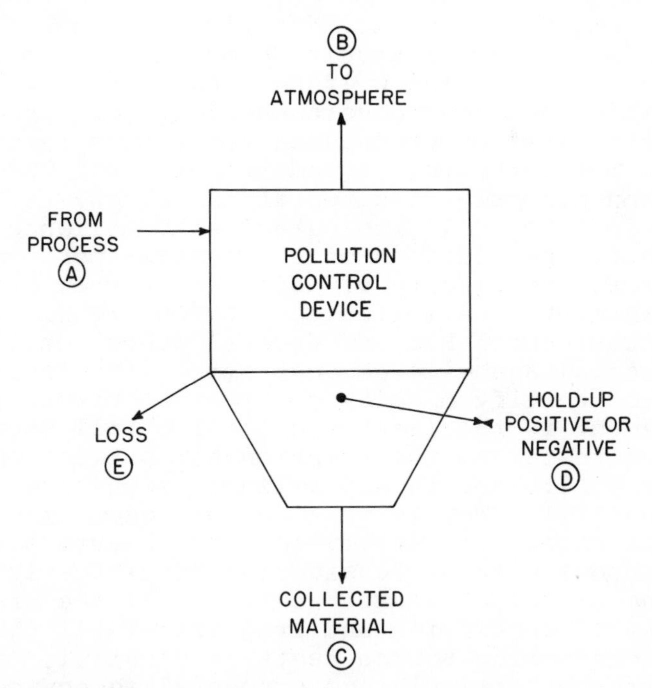

Figure 19.1. Calculating the efficiency of an air pollution control device.[1]

calculated in several ways:

$$\text{Efficiency} = 100 \left(\frac{C}{A}\right) \quad (19\text{-}1)$$

$$\text{Efficiency} = 100 \left(\frac{C}{C+B}\right) \quad (19\text{-}2)$$

$$\text{Efficiency} = 100 \left(\frac{A-B}{A}\right) \quad (19\text{-}3)$$

or

$$\text{Efficiency} = 100 \left(\frac{A-B}{C+B}\right) \quad (19\text{-}4)$$

The above calculations all assume that A=B+C and that both D and E equal zero. If these two assumptions are correct, it makes no difference which of the quantities A, B, or C are used for the calculation. However, the trained source tester should be aware that such ideal situations are not generally found in the field. Thus any calculations using only the measured values of A and B, B and C, A and C, or A, B, and C, will probably be in error. Furthermore, the efficiency calculated by using A and B would differ from that calculated using B and C, etc.

Governmental agencies have air pollutant emission standards and source testing procedures to protect the public health and welfare. Does it not seem reasonable then that governmental agencies should have certification procedures for source testers to insure the testing procedures are well understood and properly executed at all times? As discussed in Chapter 14, to establish compliance with the opacity regulations of the federal performance standards, the opacity of affected sources must be determined by a certified opacity observer. Therefore, the federal EPA and several other local governmental agencies conduct smoke schools to train and certify opacity observers. Visual opacity measurement is the least technical of all source emission tests and the relationship between opacity and the public health and welfare is obscure (of high opacity). Why is it then that governmental agencies choose to certify opacity observers for EPA Method 9 without considering certification and training of actual source testers? If the practicality of certifying and training of all the people conducting source tests is doubtful, then at least the particular governmental agency concerned with compliance should send a trained and certified observer to witness all source tests.

Finally, when the government, state, local, or federal, does not act fairly in the interest of the public or industry, it is the duty of professional organizations such as APCA, ASME, ASTM, AIChE, and ASCE to act. These organizations possess a great wealth of technical talent from all phases of life. Such organizations must shoulder a greater responsibility in the development and certification of source testing equipment, procedures and personnel.

SO YOU HAVE A SOURCE THAT NEEDS TESTING?

If as a plant manager you think you might have an air pollution source needing testing, make an effort to define precisely the purpose of the test. Typical purposes for source testing are:

1. to determine compliance with emission regulations
2. to obtain process data
3. to determine control efficiency, possibly to check the equipment guarantee
4. to quantify process and/or control equipment degradation
5. combinations of the above.

Once the exact purpose for source testing is determined, the type, number, and frequency of tests are constrained. Given enough time, resources, and possibly outside help, the specific type, number, and frequency of source tests needed to accomplish the intended goal can be evaluated.

There are effectively three means by which a source test can be done. Source tests can be accomplished using in-plant personnel, an outside consultant, or both. The decision as to who should perform the required source tests should be straight forward and should take into account personnel availability, cost, reliability, and knowledge gained. For example of an outside consultant does all the work involved:

1. The plant will have to provide no significant personnel other than that needed to prepare the sample location.
2. The cost will be about $175-$225 per man-day plus $30 per man-day expenses plus transportation for the consultant and equipment plus the cost of safety equipment and insurance if applicable.

3. The reliability of results will be good if the consultant is chosen properly.
4. The knowledge gained by the plant staff will probably be limited to test results only.

On the other hand if the source testing is done solely by in-plant personnel:

1. The plant will have to provide at least 2-3 full-time personnel during the testing and analysis of samples and data. Generally the nature of the work requires at least one engineer and one chemist, both preferably young and in good condition for hard work.
2. The cost will include personnel salaries and cost of training, the capital investment for sampling equipment and analytical laboratory equipment (rental is sometimes possible), and the operating cost during sampling (maintenance and calibration of equipment, chemical supplies, and possibly insurance).
3. The reliability of results will generally only be good after extensive training or experience. In addition the results might be viewed as biased since all work was done by plant personnel.
4. The plant staff will have a good working knowledge of the source test and results.

Generally speaking, it is the authors' experience that the ideal solution to the problem of who should conduct the source test is the combination of outside consultant and plant personnel. In most cases, assigning a plant engineer to oversee the source sampling project works well. The consultant gets a third pair of hands to help when necessary and the plant engineer obtains first-hand knowledge. In fact it is the authors' experience that most source sampling consultants realize that it is to their benefit to have a direct plant contact to handle problems that arise during sampling.

If the decision to employ a source testing consultant is made, make sure that a consultant of proven capability is chosen. It is good practice to ask for references as to the consultant's capability. Past clients and the local air pollution control agency should be contacted with respect to reliability. When possible it is also advisable to visit the consultant's facilities. Using Chapter 4 and the various chapters in this book on specific source testing procedures, check to see if the consultant has the required equipment necessary to conduct the source tests required. Do not neglect calibration and analytical laboratory equipment. Finally, settle the contract before any work is done. When difficulties are encountered there is often no other method of recourse for client or consultant.

YOUR DAY IN COURT

The end result of your source testing efforts could well be a day in court. Legal action may be taken against a company and source testing results may decide if compliance has been met. What do you do to be prepared to make such a testimony? First, use the proper test procedures required by law. Second, adequately document all of your work. This means using good data sheets and writing an excellent report. Remember, your written report and oral testimony provide facts regarding the case. There is no such thing as too much data.

There are many things to consider when being an expert witness. If you are awkward and uncertain, there is only one thing certain--the attorneys will make you look like an amateur and a fool. Why spend thousands of dollars on a source test and then let the fruits of your effort slip away on the witness stand? The following are a few thoughts and suggestions which may make your day in court less traumatic and more fruitful.

1. Your impression on others
   - A. Dress in a conservative fashion, do not wear excessive amount of clothing accessories.
   - B. Take the oath in a dignified manner; keep your poise.
   - C. Be yourself. Act natural and feel as though you are telling a story to some friends.
   - D. Remain calm. Keep your temper under all conditions.
2. Your speech
   - A. Make your words heard clearly. Speak out in a confident fashion.
   - B. Select your words carefully. Be prepared to explain the meaning of technical terms.
   - C. Be respectful. Show respect for courtroom situation. Be confident but not dramatic.

3. Your responses
   A. Be sure that you understand questions before answering. Ask for question to be repeated if necessary.
   B. Beware of double questions in one sentence from an attorney. You may wish to restate the questions and then answer them separately. Allow adequate time to think over questions before answering them. Don't allow attorneys to put words in your mouth.
   C. Beware of hypothetical questions. Do not be anxious to answer such questions because such theoretical discussions can easily lead beyond your facts, observations, and knowledge.
   D. Give direct answers. Answer in a straightforward fashion. Do not seem prejudiced. You are relating facts and are not the judge or jury.
   E. Keep within your limits of knowledge. Do not be afraid to say "I do not know" or "I do not remember." Such statements are not equated to ignorance on your part.
   F. Treat both attorneys alike. Be courteous and do not lose your temper even though you are being criticized in cross-questioning. Do not defend yourself against personal attacks; remember this is just a tool to be used by both sides in trying to win the case. Respond slowly and never volunteer information.
   G. Give accurate testimony. If necessary refer to notes which you have prepared beforehand. You should study your case thoroughly before court. Make a list of possible questions you may be asked and try to determine the best answers.
   H. Opinion Evidence. Give your opinion only when it is asked for. Before giving an opinion be sure that you have adequately identified your expertise--experience and qualifications. Always state the facts upon which your opinion is based.

REFERENCES

1. Boubel, R. W. "Air Pollution Measurements--Numbers, Errors and Decisions," Paper 72-AP-3, 1972 Annual Meeting of the Air Pollution Control Association, Pacific Northwest International Section, Eugene, Oregon, November, 1972.

APPENDIXES

APPENDIX A

FEDERAL EPA STANDARDS OF PERFORMANCE FOR NEW STATIONARY SOURCES*

TITLE 40--PROTECTION OF ENVIRONMENT
Chapter I--Environmental Protection Agency
Subchapter C--Air Programs
Part 60--Standards of Performance for New Stationary Sources

On August 17, 1971 (36 F.R. 15704) pursuant to section 111 of the Clean Air Act as amended, the Administrator proposed standards of performance for steam generators, portland cement plants, incinerators, nitric acid plants, and sulfuric acid plants. The proposed standards, applicable to sources the construction or modification of which was initiated after August 17, 1971, included emission limits for one or more of four pollutants (particulate matter, sulfur dioxide, nitrogen oxides, and sulfuric acid mist) for each source category. The proposal included requirements for performance testing, stack gas monitoring, record keeping and reporting, and procedures by which EPA will provide preconstruction review and determine the applicability of the standards to specific sources.

Interested parties were afforded an opportunity to participate in the rule making by submitting comments. A total of more than 200 interested parties, including Federal, State, and local agencies, citizens groups, and commercial and industrial organizations submitted comments. Following a review of the proposed regulations and consideration of the comments, the regulations, including the appendix, have been revised and are being promulgated today. The principal revisions are described below:

1. Particulate matter performance testing procedures have been revised to eliminate the requirement for impingers in the sampling train. Compliance will be based only on material collected in the dry filter and the probe preceding the filter. Emission limits have been adjusted as appropriate to reflect the change in test methods. The adjusted standards

*Reproduced from the Federal Register, 36, No. 247, 24876, December 23, 1971.

require the same degree of particulate control as the originally proposed standards.

2. Provisions have been added whereby alternative test methods can be used to determine compliance. Any person who proposes the use of an alternative method will be obliged to provide evidence that the alternative method is equivalent to the reference method.

3. The definition of modification, as it pertains to increases in production rate and changes of fuels, has been clarified. Increases in production rates up to design capacity will not be considered a modification nor will fuel switches if the equipment was originally designed to accommodate such fuels. These provisions will eliminate inequities where equipment had been put into partial operation prior to the proposal of the standards.

4. The definition of a new source was clarified to include construction which is completed within an organization as well as the more common situations where the facility is designed and constructed by a contractor.

5. The provisions regarding requests for EPA plan review and determination of construction or modification have been modified to emphasize that the submittal of such requests and attendant information is purely voluntary. Submittal of such a request will not bind the operator to supply further information; however, lack of sufficient information may prevent the Administrator from rendering an opinion. Further provisions have been added to the effect that information submitted voluntarily for such plan review or determination of applicability will be considered confidential, if the owner or operator requests such confidentiality.

6. Requirements for notifying the Administrator prior to commencing construction have been deleted. As proposed, the provision would have required notification prior to the signing of a contract for construction of a new source. Owners and operators still will be required to notify the Administrator 30 days prior to initial operation and to confirm the action within 15 days after startup.

7. Revisions were incorporated to permit compliance testing to be deferred up to 60 days after achieving the maximum production rate but no longer than 180 days after initial startup. The proposed regulation could have required testing within 60 days after startup but defined startup as the beginning of routine operation. Owners or operators will be required to notify the Administrator at least 10 days prior to compliance testing so that an EPA observer can be on hand. Procedures have been modified so that the equipment will have to be operated at maximum expected production rate, rather than rated capacity, during compliance tests.

8. The criteria for evaluating performance testing results have been simplified to eliminate the requirement that all

values be within 35 percent of the average. Compliance will be based on the average of three repetitions conducted in the specified manner.

9. Provisions were added to require owners or operators of affected facilities to maintain records of compliance tests, monitoring equipment, pertinent analyses, feed rates, production rates, etc. for 2 years and to make such information available on request to the Administrator. Owners or operators will be required to summarize the recorded data daily and to convert recorded data into the applicable units of the standard.

10. Modifications were made to the visible emission standards for steam generators, cement plants, nitric acid plants, and sulfuric acid plants. The Ringelmann standards have been deleted; all limits will be based on opacity. In every case, the equivalent opacity will be at least as stringent as the proposed Ringelmann number. In addition, requirements have been altered for three of the source categories so that allowable emissions will be less than 10 percent opacity rather than 5 percent or less opacity. There were many comments that observers could not accurately evaluate emissions of 5 percent opacity. In addition, drafting errors in the proposed visible emission limits for cement kilns and steam generators were corrected. Steam generators will be limited to visible emissions not greater than 20 percent opacity and cement kilns to not greater than 10 percent opacity.

11. Specifications for monitoring devices were clarified, and directives for calibration were included. The instruments are to be calibrated at least once a day, or more often if specified by the manufacturer. Additional guidance on the selection and use of such instruments will be provided at a later date.

12. The requirement for sulfur dioxide monitoring at steam generators was deleted for those sources which will achieve the standard by burning low-sulfur fuel, provided that fuel analysis is conducted and recorded daily. American Society for Testing and Materials sampling techniques are specified for coal and fuel oil.

13. Provisions were added to the steam generator standards to cover those instances where mixed fuels are burned. Allowable emissions will be determined by prorating the heat input of each fuel, however, in the case of sulfur dioxide, the provisions allow operators the option of burning low-sulfur fuels (probably natural gas) as a means of compliance.

14. Steam generators fired with lignite have been exempted from the nitrogen oxides limit. The revision was made in view of the lack of information on some types of lignite burning. When more information is developed, nitrogen oxides standards may be extended to lignite fired steam generators.

15. A provision was added to make it explicit that the sulfuric acid plant standards will not apply to scavenger

acid plants. As stated in the background document, APTD 0711, which was issued at the time the proposed standards were published, the standards were not meant to apply to such operations, e.g., where sulfuric acid plants are used primarily to control sulfur dioxide or other sulfur compounds which would otherwise be vented into the atmosphere.

16. The regulation has been revised to provide that all materials submitted pursuant to these regulations will be directed to EPA's Office of General Enforcement.

17. Several other technical changes have also been made. States and interested parties are urged to make a careful reading of these regulations.

As required by section 111 of the Act, the standards of performance promulgated herein "reflect the degree of emission reduction which (taking into account the cost of achieving such reduction) the Administrator determines has been adequately demonstrated." The standards of performance are based on stationary source testing conducted by the Environmental Protection Agency and/or contractors and on data derived from various other sources, including the available technical literature. In the comments on the proposed standards, many questions were raised as to costs and demonstrated capability of control systems to meet the standards. These comments have been evaluated and investigated, and it is the Administrator's judgment that emission control systems capable of meeting the standards have been adequately demonstrated and that the standards promulgated herein are achievable at reasonable costs.

The regulations establishing standards of performance for steam generators, incinerators, cement plants, nitric acid plants, and sulfuric acid plants are hereby promulgated effective on publication and apply to sources, the construction or modification of which was commenced after August 17, 1971.

Dated: December 16, 1971.
William D. Ruckelshaus, Administrator,
Environmental Protection Agency.

A new Part 60 is added to Chapter I, Title 40, Code of Federal Regulations, as follows:

Subpart A--General Provisions

Sec.
60.1 Applicability.
60.2 Definitions.
60.3 Abbreviations.
60.4 Address.
60.5 Determination of construction or modification.
60.6 Review of plans.
60.7 Notification and recordkeeping.
60.8 Performance tests.

60.9 Availability of information.
60.10 State authority.

Subpart D--Standards of Performance for Fossil Fuel-Fired Steam Generators

60.40 Applicability and designation of affected facility.
60.41 Definitions.
60.42 Standard for particulate matter.
60.43 Standard for sulfur dioxide.
60.44 Standard for nitrogen oxides.
60.45 Emission and fuel monitoring.
60.46 Test methods and procedures.

Subpart E--Standards of Performance for Incinerators

60.50 Applicability and designation of affected facility.
60.51 Definitions.
60.52 Standard for particulate matter.
60.53 Monitoring of operations.
60.54 Test methods and procedures.

Subpart F--Standards of Performance for Portland Cement Plants

60.60 Applicability and designation of affected facility.
60.61 Definitions.
60.62 Standard for particulate matter.
60.63 Monitoring of operations.
60.64 Test methods and procedures.

Subpart G--Standards of Performance for Nitric Acid Plants

60.70 Applicability and designation of affected facility.
60.71 Definitions.
60.72 Standard for nitrogen oxides.
60.73 Emission monitoring.
60.74 Test methods and procedures.

Subpart H--Standards of Performance for Sulfuric Acid Plants

60.80 Applicability and designation of affected facility.
60.81 Definitions.
60.82 Standard for sulfur dioxide.
60.83 Standard for acid mist.
60.84 Emission monitoring.
60.85 Test methods and procedures.

Appendix--Test Methods

Method 1--Sample and velocity traverses for stationary sources.
Method 2--Determination of stack gas velocity and volumetric flow rate (Type S pitot tube).
Method 3--Gas analysis for carbon dioxide, excess air, and dry molecular weight.
Method 4--Determination of moisture in stack gases.
Method 5--Determination of particulate emissions from stationary sources.
Method 6--Determination of sulfur dioxide emissions from stationary sources.
Method 7--Determination of nitrogen oxide emissions from stationary sources.
Method 8--Determination of sulfuric acid mist and sulfur dioxide emissions from stationary sources.
Method 9--Visual determination of the opacity of emissions from stationary sources.

Authority: The provisions of this Part 60 issued under sections 111, 114, Clean Air Act; Public Law 91-604, 84 Stat. 1713.

SUBPART A--GENERAL PROVISIONS

§60.1 Applicability

The provisions of this part apply to the owner or operator of any stationary source, which contains an affected facility the construction or modification of which is commenced after the date of publication in this part of any proposed standard applicable to such facility.

§60.2 Definitions

As used in this part, all terms not defined herein shall have the meaning given them in the Act:

(a) "Act" means the Clean Air Act (42 U.S.C. 1857 et seq., as amended by Public Law 91-604, 84 Stat. 1676).

(b) "Administrator" means the Administrator of the Environmental Protection Agency or his authorized representative.

(c) "Standard" means a standard of performance proposed or promulgated under this part.

(d) "Stationary source" means any building, structure, facility, or installation which emits or may emit any air pollutant.

(e) "Affected facility" means, with reference to a stationary source, any apparatus to which a standard is applicable.

(f) "Owner or operator" means any person who owns, leases, operates, controls, or supervises an affected facility or a stationary source of which an affected facility is a part.

(g) "Construction" means fabrication, erection, or installation of an affected facility.

(h) "Modification" means any physical change in, or change in the method of operation of, an affected facility which increases the amount of any air pollutant (to which a standard applies) emitted by such facility or which results in the emission of any air pollutant (to which a standard applies) not previously emitted, except that:

(1) Routine maintenance, repair, and replacement shall not be considered physical changes, and

(2) The following shall not be considered a change in the method of operation:

(i) An increase in the production rate, if such increase does not exceed the operation design capacity of the affected facility;

(ii) An increase in hours of operation;

(iii) Use of an alternative fuel or raw material if, prior to the date any standard under this part becomes applicable to such facility, as provided by §60.1, the affected facility is designed to accommodate such alternative use.

(i) "Commenced" means that an owner or operator has undertaken a continuous program of construction or modification or that an owner or operation has entered into a binding agreement or contractual obligation to undertake and complete, within a reasonable time, a continuous program of construction or modification.

(j) "Opacity" means the degree to which emissions reduce the transmission of light and obscure the view of an object in the background.

(k) "Nitrogen oxides" means all oxides of nitrogen except nitrous oxide, as measured by test methods set forth in this part.

(l) "Standard of normal conditions" means 70° Fahrenheit (21.1° centigrade) and 29.92 in. Hg (760 mm. Hg).

(m) "Proportional sampling" means sampling at a rate that produces a constant ratio of sampling rate to stack gas flow rate.

(n) "Isokinetic sampling" means sampling in which the linear velocity of the gas entering the sampling nozzle is equal to that of the undisturbed gas stream at the sample point.

(o) "Startup" means the setting in operation of an affected facility for any purpose.

§60.3 Abbreviations

The abbreviations used in this part have the following meanings in both capital and lower case:

B.t.u.--British thermal unit.
cal.--calorie(s).
c.f.m.--cubic feet per minute.
CO_2--carbon dioxide.
g.--gram(s).
gr.--grain(s).
mg.--milligram(s).
mm.--millimeter(s).
l.--liter(s).
nm.--nanometer(s),--10^{-3} meter.
μg.--microgram(s), 10^{-6} gram.
Hg.--mercury.
in.--inch(es).
K--1,000.
lb.--pound(s).
ml.--milliliter(s).
No.--number.
%--percent.
NO--nitric oxide.
NO_2--nitrogen dioxide.
NO_x--nitrogen oxides.
$NM.^3$--normal cubic meter.
s.c.f.--standard cubic feet.
SO_2--sulfur dioxide.
H_2SO_4--sulfuric acid.
SO_3--sulfur trioxide.
$ft.^3$--cubic feet.
$ft.^2$--square feet.
min.--minute(s).
hr.--hour(s).

§60.4 Address

All applications, requests, submissions, and reports under this part shall be submitted in triplicate and addressed to the Environmental Protection Agency, Office of General Enforcement, Waterside Mall S.W., Washington, DC 20460.

§60.5 Determination of construction or modification.

When requested to do so by an owner or operator, the Administrator will make a determination of whether actions taken or intended to be taken by such owner or operator constitute construction or modification or the commencement thereof within the meaning of this part.

§60.6 Review of plans.

(a) When requested to do so by an owner or operator, the Administrator will review plans for construction or modification for the purpose of providing technical advice to the owner or operator.

(b) (1) A Separate request shall be submitted for each affected facility.

(2) Each request shall (i) identify the location of such affected facility, and (ii) be accompanied by technical information describing the proposed nature, size, design, and method of operation of such facility, including information on any equipment to be used for measurement or control of emissions.

(c) Neither a request for plans review nor advice furnished by the Administrator in response to such request shall (1) relieve an owner or operator of legal responsibility for compliance with any provision of this part or of any applicable State or local requirement, or (2) prevent the Administrator from implementing or enforcing any provision of this part or taking any other action authorized by the Act.

§60.7 Notification and record keeping.

(a) Any owner or operator subject to the provisions of this part shall furnish the Administrator written notification as follows:

(1) A notification of the anticipated date of initial startup of an affected facility not more than 60 days or less than 30 days prior to such date.

(2) A notification of the actual date of initial startup of an affected facility within 15 days after such date.

(b) Any owner or operator subject to the provisions of this part shall maintain for a period of 2 years a record of the occurrence and duration of any startup, shutdown, or malfunction in operation of any affected facility.

§60.8 Performance tests.

(a) Within 60 days after achieving the maximum production rate at which the affected facility will be operated, but not later than 180 days after initial startup of such facility and at such other times as may be required by the Administrator under section 114 of the Act, the owner or operator of such facility shall conduct performance test(s) and furnish the Administrator a written report of the results of such performance test(s).

(b) Performance tests shall be conducted and results reported in accordance with the test method set forth in this part or equivalent methods approved by the Administrator; or where the Administrator determines that emissions from the affected facility are not susceptible of being measured by such methods, the Administrator shall prescribe alternative test procedures for determining compliance with the requirements of this part.

(c) The owner or operator shall permit the Administrator to conduct performance tests at any reasonable time, shall cause the affected facility to be operated for purposes of such tests under such conditions as the Administrator shall specify based on representative performance of the affected facility, and shall make available to the Administrator such records as may be necessary to determine such performance.

(d) The owner or operator of an affected facility shall provide the Administrator 10 days prior notice of the performance test to afford the Administrator the opportunity to have an observer present.

(e) The owner or operator of an affected facility shall provide, or cause to be provided, performance testing facilities as follows:

(1) Sampling ports adequate for test methods applicable to such facility.

(2) Safe sampling platform(s).

(3) Safe access to sampling platform(s).

(4) Utilities for sampling and testing equipment.

(f) Each performance test shall consist of three repetitions of the applicable test method. For the purpose of determining compliance with an applicable standard of performance, the average of results of all repetitions shall apply.

§60.9 Availability of information.

(a) Emission data provided to, or otherwise obtained by, the Administrator in accordance with the provisions of this part shall be available to the public.

(b) Except as provided in paragraph (a) of this section, any records, reports, or information provided to, or otherwise obtained by, the Administrator in accordance with the provisions of this part shall be available to the public, except that (1) upon a showing satisfactory to the Administrator by any person that such records, reports, or information, or particular part thereof (other than emission data), if made public, would divulge methods or processes entitled to protection as trade secrets of such person, the Administrator shall consider such records, reports, or information, or particular part thereof, confidential in accordance with the purposes of section 1905 of title 18 of the United States Code, except that such records, reports, or information, or particular part thereof, may be disclosed to other officers, employees, or authorized representatives of the United States concerned with carrying out the provisions of the Act or when relevant in any proceeding under the Act; and (2) information received by the Administrator solely for the purposes of §§60.5 and 60.6 shall not be disclosed if it is identified by the owner or operator as being a trade secret or commercial or financial information which such owner or operator considers confidential.

§60.10 State authority.

The provisions of this part shall not be construed in any manner to preclude any State or political subdivision thereof from:

(a) Adopting and enforcing any emission standard or limitation applicable to an affected facility, provided that such emission standard or limitation is not less stringent than the standard applicable to such facility.

(b) Requiring the owner or operator of an affected facility to obtain permits, licenses, or approvals prior to initiating construction, modification, or operation of such facility.

SUBPART D--STANDARDS OF PERFORMANCE FOR FOSSIL-FUEL FIRED STEAM GENERATORS

§60.40 Applicability and designation of affected facility.

The provisions of this subpart are applicable to each fossil fuel-fired steam generating unit of more than 250 million B.t.u. per hour heat input, which is the affected facility.

§60.41 Definitions.

As used in this subpart, all terms not defined herein shall have the meaning given them in the Act, and in Subpart A of this part.

(a) "Fossil fuel-fired steam generating unit" means a furnace or boiler used in the process of burning fossil fuel for the primary purpose of producing steam by heat transfer.

(b) "Fossil fuel" means natural gas, petroleum, coal and any form of solid, liquid, or gaseous fuel derived from such materials.

(c) "Particulate matter" means any finely divided liquid or solid material, other than uncombined water, as measured by Method 5.

§60.42 Standard for particulate matter.

On and after the date on which the performance test required to be conducted by §60.8 is initiated no owner or operator subject to the provisions of this part shall discharge or cause the discharge into the atmosphere of particulate matter which is:

(a) In excess of 0.10 lb. per million B.t.u. heat input (0.18 g. per million cal.) maximum 2-hour average.

(b) Greater than 20 percent opacity, except that 40 percent opacity shall be permissible for not more than 2 minutes in any hour.

(c) Where the presence of uncombined water is the only reason for failure to meet the requirements of paragraph (b) of this section such failure shall not be a violation of this section.

§60.43 Standard for sulfur dioxide.

On and after the date on which the performance test required to be conducted by §60.8 is initiated no owner or operator subject to the provisions of this part shall discharge or cause the discharge into the atmosphere of sulfur dioxide in excess of:

(a) 0.80 lb. per million B.t.u. heat input (1.4 g. per million cal.), maximum 2-hour average, when liquid fossil fuel is burned.

(b) 1.2 lbs. per million B.t.u. heat input (2.2 g. per million cal.), maximum 2-hour average, when solid fossil fuel is burned.

(c) Where different fossil fuels are burned simultaneously in any combination, the applicable standard shall be determined by proration. Compliance shall be determined using the following formula:

$$\frac{y(0.80) + z(1.2)}{x + y + z}$$

where:

x is the percent of total heat input derived from gaseous fossil fuel and,

y is the percent of total heat input derived from liquid fossil fuel and,

z is the percent of total heat input derived from solid fossil fuel.

§60.44 Standard for nitrogen oxides.

On and after the date on which the performance test required to be conducted by §60.8 is initiated no owner or operator subject to the provisions of this part shall discharge or cause the discharge into the atmosphere of nitrogen oxides in excess of:

(a) 0.20 lb. per million B.t.u. heat input (0.36 g. per million cal.), maximum 2-hour average, expressed as NO_2, when gaseous fossil fuel is burned.

(b) 0.30 lb. per million B.t.u. heat input (0.54 g. per million cal.), maximum 2-hour average, expressed as NO_2, when liquid fossil fuel is burned.

(c) 0.70 lb. per million B.t.u. heat input (1.26 g. per million cal.), maximum 2-hour average, expressed as NO_2, when solid fossil fuel (except lignite) is burned.

(d) When different fossil fuels are burned simultaneously in any combination the applicable standard shall be determined by proration. Compliance shall be determined by using the following formula:

$$\frac{x(0.20) + y(0.30) + z(0.70)}{x + y + z}$$

where:

x is the percent of total heat input derived from gaseous fossil fuel and,

y is the percent of total heat input derived from liquid fossil fuel and,

z is the percent of total heat input derived from solid fossil fuel.

§60.45 Emission and fuel monitoring.

(a) There shall be installed, calibrated, maintained, and operated, in any fossil fuel-fired steam generating unit subject to the provisions of this part, emission monitoring instruments as follows:

(1) A photoelectric or other type smoke detector and recorder, except where gaseous fuel is the only fuel burned.

(2) An instrument for continuously monitoring and recording sulfur dioxide emissions, except where gaseous fuel is the only fuel burned, or where compliance is achieved through low sulfur fuels and representative sulfur analysis of fuels are conducted daily in accordance with paragraph (c) or (d) of this section.

(3) An instrument for continuously monitoring and recording emissions of nitrogen oxides.

(b) Instruments and sampling systems installed and used pursuant to this section shall be capable of monitoring emission levels within ±20 percent with a confidence level of 95 percent and shall be calibrated in accordance with the method(s) prescribed by the manufacturer(s) of such instruments; instruments shall be subjected to manufacturers recommended zero adjustment and calibration procedures at least once per 24-hour operating period unless the manufacturer(s) specifies or recommends calibration at shorter intervals, in which case such specifications or recommendations shall be followed. The applicable method specified in the appendix of this part shall be the reference method.

(c) The sulfur content of solid fuels, as burned, shall be determined in accordance with the following methods of the American Society for Testing and Materials.

(1) Mechanical sampling by Method D 2234065.

(2) Sample preparation by Method D 2013-65.

(3) Sample analysis by Method D 271-68.

(d) The sulfur content of liquid fuels, as burned, shall be determined in accordance with the American Society for Testing and Materials Methods D 1551-68, or D 129-64, or D 1552-64.

(e) The rate of fuel burned for each fuel shall be measured daily or at shorter intervals and recorded. The heating value and ash content of fuels shall be ascertained at least once per week and recorded. Where the steam generating unit is used to generate electricity, the average electrical output and the minimum and maximum hourly generation rate shall be measured and recorded daily.

(f) The owner or operator of any fossil fuel-fired steam generating unit subject to the provisions of this part shall maintain a file of all measurements required by this part. Appropriate measurements shall be reduced to the units of the applicable standard daily, and summarized monthly. The record of any such measurement(s) and summary shall be retained for at least 2 years following the date of such measurements and summaries.

§60.46 Test methods and procedures.

(a) The provisions of this section are applicable to performance tests for determining emissions of particulate matter, sulfur dioxide, and nitrogen oxides from fossil fuel-fired steam generating units.

(b) All performance tests shall be conducted while the affected facility is operating at or above the maximum steam production rate at which such facility will be operated and while fuels or combinations of fuels representative of normal operation are being burned and under such other relevant conditions as the Administrator shall specify based on representative performance of the affected facility.

(c) Test methods set forth in the appendix to this part or equivalent methods approved by the Administrator shall be used as follows:

(1) For each repetition, the average concentration of particulate matter shall be determined by using Method 5. Traversing during sampling by Method 5 shall be according to Method 1. The minimum sampling time shall be 2 hours, and minimum sampling volume shall be 60 $ft.^3$ corrected to standard conditions on a dry basis.

(2) For each repetition, the SO_2 concentration shall be determined by using Method 6. The sampling site shall be the same as for determining volumetric flow rate. The sampling point in the duct shall be at the centroid of the cross section if the cross sectional area is less than 50 $ft.^2$ or at a point no closer to the walls than 3 feet if the cross sectional area is 50 $ft.^2$ or more. The sample shall be extracted at a rate proportional to the gas velocity at the sampling point. The minimum sampling time shall be 20 min. and minimum sampling volume shall be 0.75 $ft.^3$ corrected to standard conditions. Two samples shall constitute one repetition and shall be taken at 1-hour intervals.

(3) For each repetition the NO_x concentration shall be determined by using Method 7. The sampling site and point shall be the same as for SO_2. The sampling time shall be 2 hours, and four samples shall be taken at 30-minute intervals.

(4) The volumetric flow rate of the total effluent shall be determined by using Method 2 and traversing according to Method 1. Gas analysis shall be performed by Method 3, and moisture content shall be determined by the condenser technique of Method 5.

(d) Heat input, expressed in B.t.u. per hour, shall be determined during each 2-hour testing period by suitable fuel flow meters and shall be confirmed by a material balance over the steam generation system.

(e) For each repetition, emissions, expressed in lb./10^6 B.t.u. shall be determined by dividing the emission rate in lb./hr. by the heat input. The emission rate shall be determined by the equation, lb./hr.=Q_sXc where, Q_s = volumetric flow rate of the total effluent in ft.^{3}/hr. at standard conditions, dry basis, as determined in accordance with paragraph (c)(4) of this section.

(1) For particulate matter, c = particulate concentration in lb./ft.^{3}, at determined in accordance with paragraph (c)(1) of this section, corrected to standard conditions, dry basis.

(2) For SO_2, c = SO_2 concentration in lb./ft.^{3}, as determined in accordance with paragraph (c)(2) of this section, corrected to standard conditions, dry basis.

(3) For NO_x, c = NO_x concentration in lb./ft.^{3}, as determined in accordance with paragraph (c)(3) of this section, corrected to standard conditions, dry basis.

SUBPART E--STANDARDS OF PERFORMANCE FOR INCINERATORS

§60.50 Applicability and designation of affected facility.

The provisions of this subpart are applicable to each incinerator of more than 50 tons per day charging rate, which is the affected facility.

§60.51 Definitions.

As used in this subpart, all terms not defined herein shall have the meaning given them in the Act and in Subpart A of this part.

(a) "Incinerator" means any furnace used in the process of burning solid waste for the primary purpose of reducing the volume of the waste by removing combustible matter.

(b) "Solid waste" means refuse, more than 50 percent of which is municipal type waste consisting of a mixture of paper, wood, yard wastes, food wastes, plastics, leather, rubber, and other combustibles, and noncombustible materials such as glass and rock.

(c) "Day" means 24 hours.

(d) "Particulate matter" means any finely divided liquid or solid material, other than uncombined water, as measured by Method 5.

§60.52 Standard for particulate matter.

On and after the date on which the performance test required to be conducted by §60.8 is initiated, no owner or operator subject to the provisions of this part shall discharge or cause the discharge into the atmosphere of particulate matter which is in excess of 0.08 gr./s.c.f. (0.18 g./NM^3) corrected to 12 percent CO_2, maximum 2-hour average.

§60.53 Monitoring of operations.

The owner or operator of any incinerator subject to the provisions of this part shall maintain a file of daily burning rates and hours of operation and any particulate emission measurements. The burning rates and hours of operation shall be summarized monthly. The record(s) and summary shall be retained for at least 2 years following the date of such records and summaries.

§60.54 Test methods and procedures.

(a) The provisions of this section are applicable to performance tests for determining emissions of particulate matter from incinerators.

(b) All performance tests shall be conducted while the affected facility is operating at or above the maximum refuse charging rate at which such facility will be operated and the solid waste burned shall be representative of normal operation and under such other relevant conditions as the Administrator shall specify based on representative performance of the affected facility.

(c) Test methods set forth in the appendix to this part or equivalent methods approved by the Administrator shall be used as follows:

(1) For each repetition, the average concentration of particulate matter shall be determined by using Method 5. Traversing during sampling by Method 5 shall be according to Method 1. The minimum sampling time shall be 2 hours and the minimum sampling volume shall be 60 $ft.^3$ corrected to standard conditions on a dry basis.

(2) Gas analysis shall be performed using the integrated sample technique of Method 3, and moisture content shall be determined by the condenser technique of Method 5. If a wet scrubber is used, the gas analysis sample shall reflect flue gas conditions after the scrubber, allowing for the effect of carbon dioxide absorption.

(d) For each repetition particulate matter emissions, expressed in gr./s.c.f., shall be determined in accordance with paragraph (c)(1) of this section corrected to 12 percent CO_3, dry basis.

SUBPART F--STANDARDS OF PERFORMANCE FOR PORTLAND CEMENT PLANTS

§60.60 Applicability and designation of affected facility.

The provisions of the subpart are applicable to the following affected facilities in portland cement plants: kiln, clinker cooler, raw mill system, finish mill system, raw mill dryer, raw material storage, clinker storage, finished product storage, conveyor transfer points, bagging and bulk loading and unloading systems.

§60.61 Definitions.

As used in this subpart, all terms not defined herein shall have the meaning given them in the Act and in Subpart A of this part.

(a) "Portland cement plant" means any facility manufacturing portland cement by either the wet or dry process.

(b) "Particulate matter" means any finely divided liquid or solid material, other than uncombined water, as measured by Method 5.

§60.62 Standard for particulate matter.

(a) On and after the date on which the performance test required to be conducted by §60.8 is initiated no owner or operator subject to the provisions of this part shall discharge or cause the discharge into the atmosphere of particulate matter from the kiln which is:

(1) In excess of 0.30 lb. per ton of feed to the kiln (0.15 Kg. per metric ton), maximum 2-hour average.

(2) Greater than 10 percent opacity, except that where the presence of uncombined water is the only reason for failure to meet the requirements for this subparagraph, such failure shall not be a violation of this section.

(b) On and after the date on which the performance test required to be conducted by §60.8 is initiated no owner or operator subject to the provisions of this part shall discharge or cause the discharge into the atmosphere of particulate matter from the clinker cooler which is:

(1) In excess of 0.10 lb. per ton of feed to the kiln (0.050 Kg. per metric ton) maximum 2 hour average.

(2) 10 percent opacity or greater.

(c) On and after the date on which the performance test required to be conducted by §60.8 is initiated no owner or operator subject to the provisions of this part shall discharge or cause the discharge into the atmosphere of particulate matter from any affected facility other than the kiln and clinker cooler which is 10 percent opacity or greater.

§60.63 Monitoring of operations.

The owner or operator of any portland cement plant subject to the provisions of this part shall maintain a file of daily production rates and kiln feed rates and any particulate emission measurements. The production and feed rates shall be summarized monthly. The record(s) and summary shall be retained for at least 2 years following the date of such records and summaries.

§60.64 Test methods and procedures.

(a) The provisions of this section are applicable to performance tests for determining emissions of particulate matter from portland cement plant kilns and clinker coolers.

(b) All performance tests shall be conducted while the affected facility is operating at or above the maximum production rate at which such facility will be operated and under such other relevant conditions as the Administrator shall specify based on representative performance of the affected facility.

(c) Test methods set forth in the appendix to this part or equivalent methods approved by the Administrator shall be used as follows:

(1) For each repetition, the average concentration of particulate matter shall be determined by using Method 5. Traversing during sampling by Method 5 shall be according to Method 1. The minimum sampling time shall be 2 hours and the minimum sampling volume shall be 60 $ft.^3$ corrected to standard conditions on a dry basis.

(2) The volumetric flow rate of the total effluent shall be determined by using Method 2 and traversing according to Method 1. Gas analysis shall be performed using the integrated sample technique of Method 3, and moisture content shall be determined by the condenser technique of Method 5.

(d) Total kiln feed (except fuels), expressed in tons per hour on a dry basis, shall be determined during each 2-hour testing period by suitable flow meters and shall be confirmed by a material balance over the production system.

(e) For each repetition, particulate matter emissions, expressed in lb./ton of kiln feed shall be determined by dividing the emission rate in lb./hr. by the kiln feed. The emission rate shall be determined by the equation, lb./hr. = $Q_s X c$, where Q_s = volumetric flow rate of the total effluent in $ft.^3$/hr. at standard conditions, dry basis, as determined in accordance with paragraph (c)(2) of this section, and, c = particulate concentration in lb./$ft.^3$, as determined in accordance with paragraph (c)(1) of this section, corrected to standard conditions, dry basis.

SUBPART G--STANDARDS OF PERFORMANCE FOR NITRIC ACID PLANTS

§60.70 Applicability and designation of affected facility.

The provisions of this subpart are applicable to each nitric acid production unit, which is the affected facility.

§60.71 Definitions.

As used in this subpart, all terms not defined herein shall have the meaning given them in the Act and in Subpart A of this part.

(a) "Nitric acid production unit" means any facility producing weak nitric acid by either the pressure or atmospheric pressure process.

(b) "Weak nitric acid" means acid which is 30 to 70 percent in strength.

§60.72 Standard for nitrogen oxides.

On and after the date on which the performance test required to be conducted by §60.8 is initiated no owner or operator subject to the provisions of this part shall discharge or cause the discharge into the atmosphere of nitrogen oxides which are:

(a) In excess of 3 lbs. per ton of acid produced (1.5 kg. permetric ton), maximum 2-hour average, expressed as NO_2.

(b) 10 percent opacity or greater.

§60.73 Emission monitoring.

(a) There shall be installed, calibrated, maintained, and operated, in any nitric acid production unit subject to the provisions of this subpart, an instrument for continuously monitoring and recording emissions of nitrogen oxides.

(b) The instrument and sampling system installed and used pursuant to this section shall be capable of monitoring emission levels within $\pm$20 percent with a confidence level of 95 percent and shall be calibrated in accordance with the method(s) prescribed by the manufacturer(s) of such instrument, the instrument shall be subjected to manufacturers recommended zero adjustment and calibration procedures at least once per 24-hour operating period unless the manufacturer(s) specifies or recommends calibration at shorter intervals, in which case such specifications or recommendations shall be followed. The applicable method specified in the appendix of this part shall be the reference method.

(c) Production rate and hours of operation shall be recorded daily.

(d) The owner or operator of any nitric acid production unit subject to the provisions of this part shall maintain a file of all measurements required by this subpart. Appropriate measurements shall be reduced to the units of the standard daily and summarized monthly. The record of any such measurement and summary shall be retained for at least 2 years following the date of such measurements and summaries.

§60.74 Test methods and procedures.

(a) The provisions of this section are applicable to performance tests for determining emissions of nitrogen oxides from nitric acid production units.

(b) All performance tests shall be conducted while the affected facility is operating at or above the maximum acid production rate at which such facility will be operated and under such other relevant conditions as the Administrator shall specify based on representative performance of the affected facility.

(c) Test methods set forth in the appendix to this part or equivalent methods as approved by the Administrator shall be used as follows:

(1) For each repetition the NO_x concentration shall be determined by using Method 7. The sampling site shall be selected according to Method 1 and the sampling point shall be the centroid of the stack or duct. The sampling time shall be 2 hours and four samples shall be taken at 30-minute intervals.

(2) The volumetric flow rate of the total effluent shall be determined by using Method 2 and traversing according to Method 1. Gas analysis shall be performed by using the integrated sample technique of Method 3, and moisture content shall be determined by Method 4.

(d) Acid produced, expressed in tons per hour of 100 percent nitric acid, shall be determined during each 2-hour testing period by suitable flow meters and shall be confirmed by a material balance over the production system.

(e) For each repetition, nitrogen oxides emissions, expressed in lb./ton of 100 percent nitric acid, shall be determined by dividing the emission rate in lb./hr. by the acid produced. The emission rate shall be determined by the equation, lg./hr. = $Q_s Xc$, where Q_s = volumetric flow rate of the effluent in $ft.^3/hr.$ at standard conditions, dry basis, as determined in accordance with paragraph (c)(2) of this section, and c = NO_x concentration in $lb./ft.^3$ as determined in accordance with paragraph (c)(1) of this section, corrected to standard conditions, dry basis.

SUBPART H--STANDARDS OF PERFORMANCE FOR SULFURIC ACID PLANTS

§60.80 Applicability and designation of affected facility.

The provisions of this subpart are applicable to each sulfuric acid production unit, which is the affected facility.

§60.81 Definitions.

As used in this subpart, all terms not defined herein shall have the meaning given them in the Act and in Subpart A of this part.

(a) "Sulfuric acid production unit" means any facility producing sulfuric acid by the contact process by burning elemental sulfur, alkylation acid, hydrogen sulfide, organic sulfides and mercaptans, or acid sludge, but does not include facilities where conversion to sulfuric acid is utilized primarily as a means of preventing emissions to the atmosphere of sulfur dioxide or other sulfur compounds.

(b) "Acid mist" means sulfuric acid mist, as measured by test methods set forth in this part.

§60.82 Standard for sulfur dioxide.

On and after the date on which the performance test required to be conducted by §60.8 is initiated no owner or operator subject to the provisions of this part shall discharge or cause the discharge into the atmosphere of sulfur dioxide in excess of 4 lbs. per ton of acid produced (2 kg. per metric ton), maximum 2-hour average.

§60.83 Standard for acid mist.

On and after the date on which the performance test required to be conducted by §60.8 is initiated no owner or operator subject to the provisions of this part shall discharge or cause the discharge into the atmosphere of acid mist which is:

(a) In excess of 0.15 lb. per ton of acid produced (0.075 kg. per metric ton), maximum 2-hour average, expressed as H_2SO_4.

(b) 10 percent opacity or greater.

§60.84 Emission monitoring.

(a) There shall be installed, calibrated, maintained, and operated, in any sulfuric acid production unit subject to the provisions of this subpart, an instrument for continuously monitoring and recording emissions of sulfur dioxide.

(b) The instrument and sampling system installed and used pursuant to this section shall be capable of monitoring emission levels within ±20 percent with a confidence level of 95 percent and shall be calibrated in accordance with the method(s) prescribed by the manufacturer(s) of such instrument, the instrument shall be subject to manufacturers recommended zero adjustment calibration procedures at least once per 24-hour operating period unless the manufacturer(s) specifies or recommends calibration at shorter intervals, in which case such specifications or recommendations shall be followed. The applicable method specified in the appendix of this part shall be the reference method.

(c) Production rate and hours of operation shall be recorded daily.

(d) The owner or operator of any sulfuric acid production unit subject to the provisions of this subpart shall maintain a file of all measurements required by this subpart. Appropriate measurements shall be reduced to the units of the applicable standard daily and summarized monthly. The record of any such measurement and summary shall be retained for at least 2 years following the date of such measurements and summaries.

§60.85 Test methods and procedures.

(a) The provisions of this section are applicable to performance tests for determining emissions of acid mist and sulfur dioxide from sulfuric acid production units.

(b) All performance tests shall be conducted while the affected facility is operating at or above the maximum acid production rate at which such facility will be operated and under such other relevant conditions as the Administrator shall specify based on representative performance of the affected facility.

(c) Test methods set forth in the appendix to this part or equivalent methods as approved by the Administrator shall be used as follows:

(1) For each repetition the acid mist and SO_2 concentrations shall be determined by using Method 8 and traversing according to Method 1. The minimum sampling time shall be 2 hours, and minimum sampling volume shall be 40 $ft.^3$ corrected to standard conditions.

(2) The volumetric flow rate of the total effluent shall be determined by using Method 2 and traversing according to Method 1. Gas analysis shall be performed by using the integrated sample technique of Method 3. Moisture content can be considered to be zero.

(d) Acid produced, expressed in tons per hour of 100 percent sulfuric acid shall be determined during each 2-hour testing period by suitable flow meters and shall be confirmed by a material balance over the production system.

(e) For each repetition acid mist and sulfur dioxide emissions, expressed in lb./ton of 100 percent sulfuric acid shall be determined by dividing the emission rate in lb./hr. by the acid produced. The emission rate shall be determined by the equation, lb./hr. = $Q_s X c$, where Q_s = volumetric flow rate of the effluent in ft.^{3}/hr. at standard conditions, dry basis as determined in accordance with paragraph (c)(2) of this section, and c = acid mist and SO_2 concentrations in lb./ft.^{3} as determined in accordance with paragraph (c)(1) of this section, corrected to standard conditions, dry basis.

APPENDIX B

SOURCE TEST GUIDELINES AND PROCEDURES, STATE OF CONNECTICUT, DEPARTMENT OF ENVIRONMENTAL PROTECTION, HARTFORD, CONNECTICUT

The material in this appendix provides an example.[1] It indicates how in the future control agencies will most likely monitor the progress of source tests. The authors feel that this is a step in the right direction. It clarifies what is required and provides standardization. The intent is to allow the state to monitor and have input to the test program without actually doing the routine work. Tables B.1 and B.2 show flow charts which are kept on file for each source. This insures that the proper documentation is made of all phases of the test program. In addition progress information on each source is keypunched and thus a status report can be obtained by calling for a printout from the computer.

SOURCE TEST GUIDELINES AND PROCEDURES

Emission tests performed in the State of Connecticut, for the purpose of complying with Connecticut air pollution regulations or of proving compliance with specific regulations, must follow Department of Environmental Protection emission test guidelines.

The guidelines are designed to ensure standardization of test requirements, a minimum standard regarding test equipment, and competence of persons intending to perform the source tests.

Test and analysis methods are specified in the April 30, 1971, and December 23, 1971, issues of the *Federal Register*. Copies of the *Federal Register* can be obtained from the Superintendent of Documents, U.S. Government Printing Office, Washington, D.C. 20402. The test methods and equipment requirements stated must be strictly complied with, unless otherwise specified by this department.

The testing procedures that follow are presented in the order in which the associated tasks are to be performed and are designed to allow monitoring of all aspects of the test program.

The appropriate forms can be obtained from the Department of Environmental Protection, Air Compliance Section, or offices of local air pollution control agencies.

SOURCE TEST PROCEDURES

A. The "Intent to Test" form, Source Test Form #1, must be completed by the tester and received by the Department no later than thirty (30) calendar days prior to the proposed test date.

B. After evaluating the completed "Intent to Test" form and, if necessary, inspecting the test site the department may require any or all of the following additional conditions:
   1. Additional tests due to adverse test conditions.
   2. Reasonable modifications of the stack or duct to obtain optimum test conditions.
   3. The process types and rates within the range of normal plant operating conditions, to be used during sampling. Production schedule conflicts can be resolved by rescheduling the test dates.
   4. This department may, at its option, supply the pollutant collecting media. Upon notification of this requirement, the tester must complete a "Collecting Media Requisition Form," Source Test Form #2. This form must be received by this department no later than ten calendar days prior to the test date.

C. If modifications to the test procedures and/or operational parameters are required, the source and testing firm will be contacted by telephone (followed by a confirming letter) in no less than fifteen (15) calendar days prior to the proposed test date. The source and testing firm may assume that no test modifications will be necessary if not otherwise advised. However, the firm to be tested and/or tester must notify this department of any modification to their test procedures as defined in the "Intent to Test" form previously submitted.

D. A department representative may observe the field test procedures and techniques. Prior to testing, calibration results of the various sampling train components must be available for inspection. This calibration information must include calibration dates, methods, data, and results. The range of calibration values must include the actual test values encountered. All components requiring calibration must be calibrated within sixty (60) calendar days prior to the actual test date. The components requiring calibration are listed as follows:
   1. Velocity measuring equipment
   2. Gas volume metering equipment
   3. Gas flow rate metering equipment
   4. Gas temperature measuring equipment

   This list may be altered at the department's discretion.

E. At the conclusion of the field test, the department's representative may, at his discretion, collect all test data sheets and sample collecting media. In those cases where the tester has supplied his own sample collecting media, the tester will be required to submit all pertinent information regarding the sample collecting media within five (5) working days after the conclusion of the field tests.

F. Laboratory Analysis

1. DEP. The laboratory analysis methods are dictated by the particular source test performed. After all specified laboratory analyses have been completed, the department will report the results to the tester.
2. Tester. The laboratory analysis methods specified for each test procedure must be strictly followed. Any changes or modifications to these procedures must have the department's prior approval.

G. Reporting of Test Results. The test results must be submitted to this Department in the form of a test report. The report shall include the following information:

1. Test results in tabular form (the units of measurement shall be consistent with units in the applicable State Regulation).
2. A copy of all field data sheets used during testing.
3. Information required by Items X, XIII of the ITT form shall be included in the report and be representative of actual conditions encountered during the field tests.
4. A sample of all formulas used in calculating results.

Date ________
Number
Assigned ________

INTENT TO TEST

Source Test Form #1

I. Source Information

Name ________________ Person or Persons Responsible for tests
Address ______________ Name ________________
________________ Telephone ______________

II. Testing Firm Information

Name ________________ Person or Persons Responsible for tests
Address ______________ Name ________________
________________ Telephone ______________

III. Premis # ________ (To be assigned by D.E.P.)
Registration # ______

Signature ______________________
Person responsible for tests

IV. Gas Stream Sampling Information. Identify all gas stream components to be sampled.

Component	Sampling Duration minutes per point	Sampling Duration total test time	Number of tests (Minimum of 3)	Expected Concentrations	Method Employed to Determine Concentration i.e., material balance, emission factor, reference (specify) etc.
1.					
2.					
3.					
4.					
5.					

V. Stack Information

1. Approx. gas temp. ________ °F
2. Approx. gas flow ________ A.C.F.M.
3. Approx. percent moisture ____ % (percent by volume)
4. Approx. gas density ________ lb/ft (at stack conditions)

VI. Purpose of Test

VII. Identification of Documents Requested on Attached Sheets

Please attach this form to the information required on Page 2.

VIII ________
IX ________
X ________
XI ________
XII ________
XIII ________
XIV ________

Please submit the following information:

VIII. Sampling Train Information

A description of each sampling train to be used. The description should include the following items:

1. A schematic diagram of each sampling train. The name, model number and date of purchase of commercially manufactured trains should be included with the diagram.
2. The type or types of media to be used to determine each gas stream component.
3. Sample tube type, *i.e.*, glass, teflon, stainless steel, etc.
4. Probe cleaning method and solvent to be used (if test procedure requires probe cleaning).

IX. Data Sheets

A sample of all field data sheets to be used in the test or tests.

X. Description of Operations

A written description of any operation, process or activity that could vent exhaust gases to the test stack. This description should include:

1. The composition of all materials capable of producing pollutant emissions used in each separate operation.
2. The feed rate and any operating parameter used to control or influence the operations.
3. The rated capacity of equipment with manufacturer's specification.

The process registration numbers, if applicable, should also be included in the description.

XI. Sampling Area Description

A cross-sectional sketch showing dimensions of the following items:

1. Stack configuration
2. Sampling port locations
3. Sampling point positions for each port.

XII. Stack and Vent Descriptions

A sketch or sketches showing the plan and elevation view of the entire ducting and stack arrangement. The sketch should include the relative position of all processes or operations venting to the stack or vent to be tested. It should also include the position of the sampling ports relative to the nearest upstream and downstream gas flow directional or duct dimensional change. The sketches should include the relative position, type, and manufacturer's claimed efficiency of all gas cleaning equipment.

XIII. Operational Parameters

A statement indicating the types of processes, operations, and operating conditions anticipated during the test or tests, on all equipment venting to the stack.

XIV. Registration

If the equipment or stack to be sampled has not been registered with the Department of Environmental Protection, the applicable registration form or forms must be completed and submitted with the Intent to Test forms.

Source Test Form #2

COLLECTING MEDIA REQUISITION FORM

Firm to be tested   Name ____________________

Address ____________________

Testing Firm   Name ____________________

Address ____________________

Person(s) Responsible
for Conducting Test(s) ____________________

Telephone ____________________

Proposed Test Date ____________________

Stack Identification Number ________ (Number Assigned on D.E.P. Registration Form)

A. Particulate Collecting Media

1. Filter type ____________________
2. Filter Diameter ____________________
3. Number of Filters Required __________

B. Gaseous Collecting Media

Pollutant(s) to be sampled	Collecting Media Description	Quantity Required

Table B.1

Status of Intent to Test Application

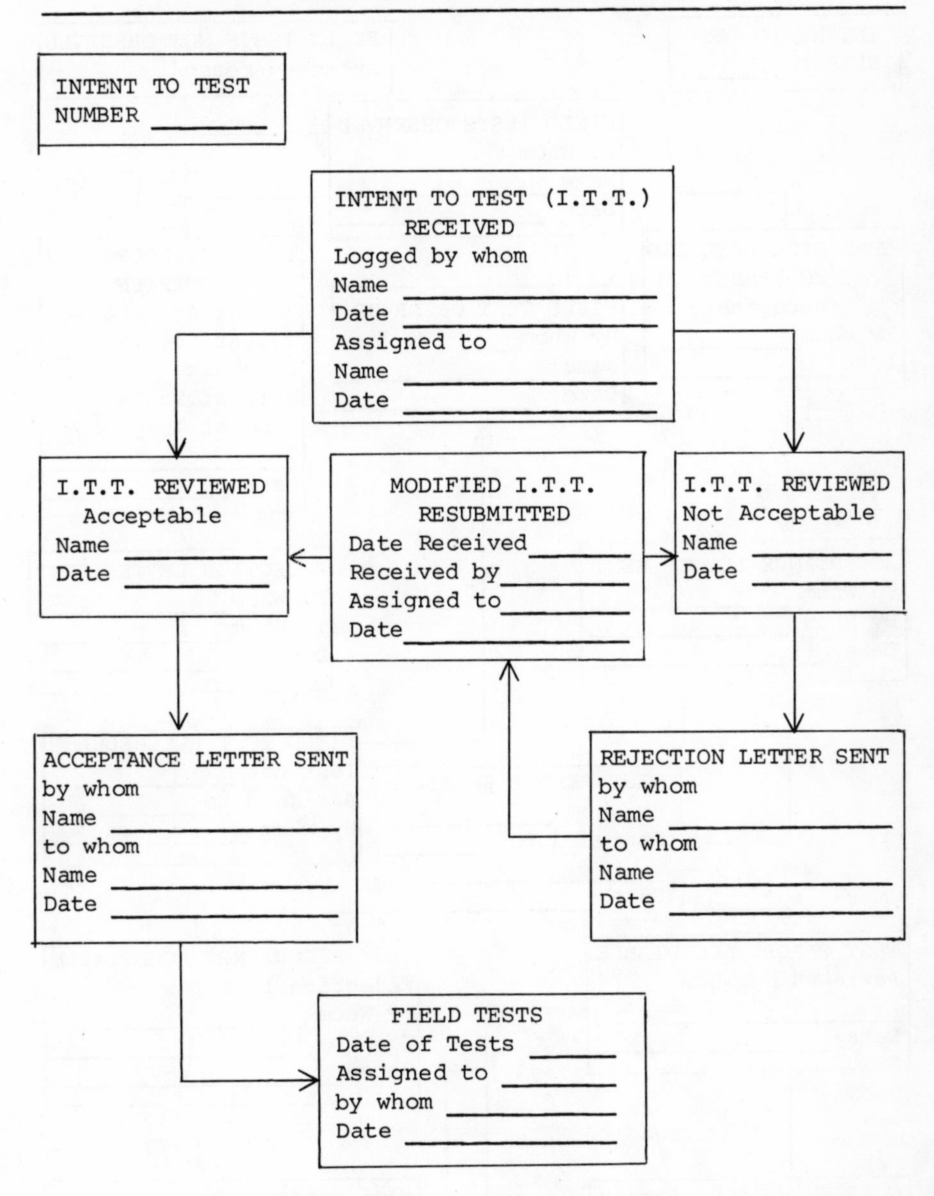

INTENT TO TEST
NUMBER ________

INTENT TO TEST (I.T.T.)
RECEIVED
Logged by whom
Name ________
Date ________
Assigned to
Name ________
Date ________

I.T.T. REVIEWED
Acceptable
Name ________
Date ________

MODIFIED I.T.T.
RESUBMITTED
Date Received ________
Received by ________
Assigned to ________
Date ________

I.T.T. REVIEWED
Not Acceptable
Name ________
Date ________

ACCEPTANCE LETTER SENT
by whom
Name ________
to whom
Name ________
Date ________

REJECTION LETTER SENT
by whom
Name ________
to whom
Name ________
Date ________

FIELD TESTS
Date of Tests ________
Assigned to ________
by whom ________
Date ________

CORRESPONDENCE: Attach Names, Dates, and Information on this form. Include all telephone calls.

Table B.2

Status of Source Test

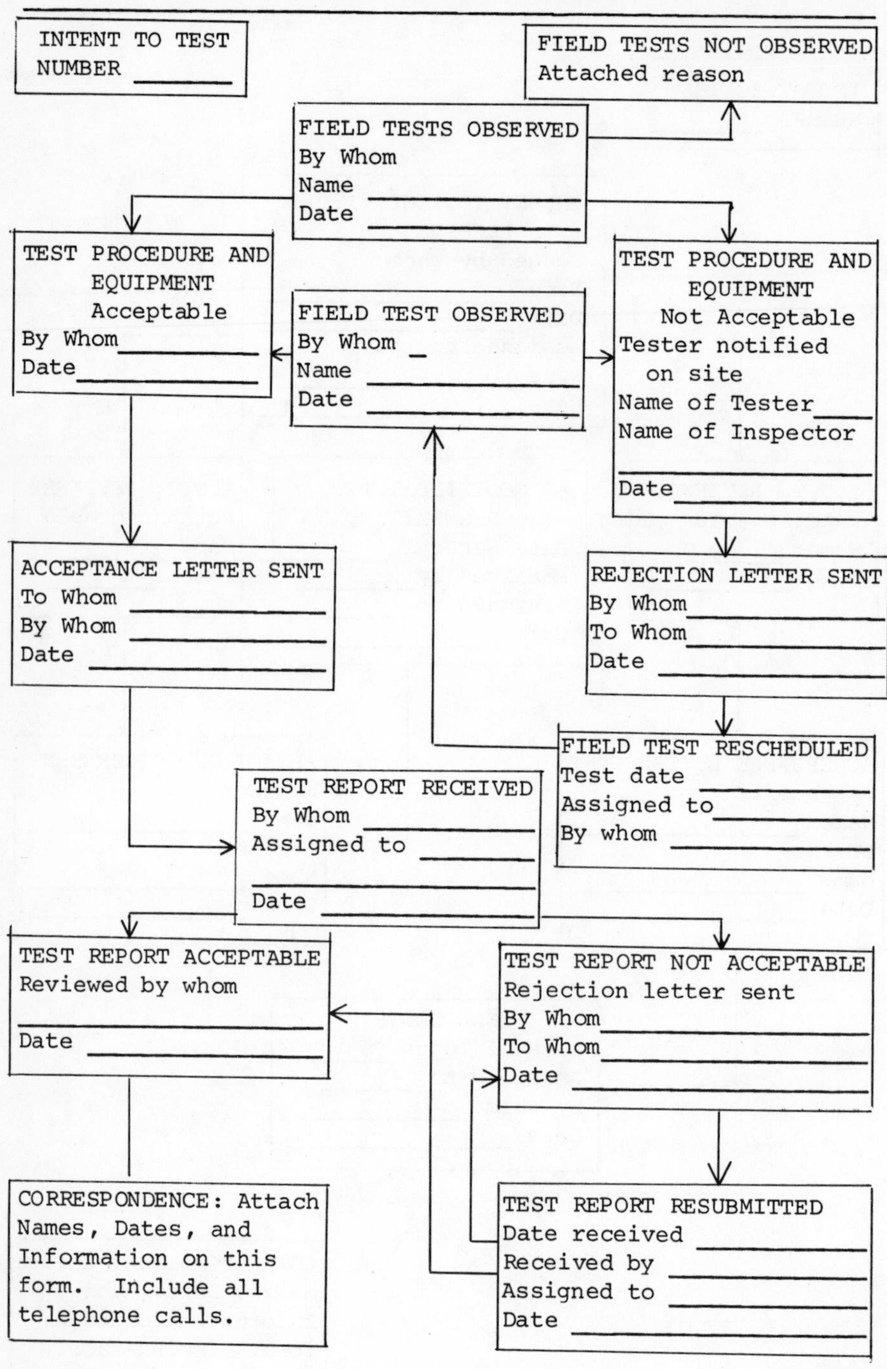

REFERENCES

1. Materials courtesy of Mr. Carl Dodge, State of Connecticut, Department of Environmental Protection, Hartford, Connecticut.

INDEX

INDEX

accuracy 146
Alundum thimbles 84
attenuation 353
beta attenuation 393
beta gauging 359
bias 147
calibration 207
chemiluminescence 355,393,416
composite sample 13
computational methods 273
conducting a source test 103
confidence intervals 153
continuous monitoring 347
cyclone 204
Dalton's Law 120
dry gas meters 88
electrochemical transducers 393,422
emission factors 1,29
emission inventory 2
EPA Method 1 128,441
EPA Method 2 128,441
EPA Method 3 441
EPA Method 4 141,441
EPA Method 5 199,308,441
EPA Method 6 225,309,337,441
EPA Method 7 243,310,337,441
EPA Method 8 235,309,441
EPA Method 9 251,441
equipment preparation 194
error propagation 156
errors in source testing 145
fiberglass filters 84
filter 203
flame photometry 356,393,432
fossil-fuel steam generators 34
gas conditioning 93
gas movers 95
grab sample 18,167
Ideal Gas Law 120
impingers 82,204
incinerators 45
instrument characteristics 360
integrated sample 18
isokinetic sampling 159,200
  field calculations 298
logistics 111
maintenance 207
manometers 71
mass and energy balances 27
maximum error 157,166
mean 152
meter box 206
mistakes 178
moisture content determination 138
molecular weight 119
  determination 132
monitoring aerosols 357
monitoring gases 348
nitric acid production 53
nomographs 123
  for isokinetic sampling 301
nozzle 79,202
orifice meters 86
Orsat analyzer 132
particle sizing 262
particulate emission calculations 285
particulate matter 16

piezoelectric crystal 393
piezoelectric
  microbalance 400
piezoelectricity 359,400
pitot tube 203
planning a source test 104
plume opacity 251
pollutant mass
  emission rate 9
portland cement
  production 49
precision 146
preparation of
  test plan 190
pressure measurement 70
pre-test plant survey 183
pre-test survey 109
priority sources 33
probes 81,202
process information 110
proportional sampling 158
purpose of sampling 1
random errors 146,150,157
report 115
representative sample 5,9
Ringelmann numbers 251
rotameter 84
S-type pitot tube 77
sample calculations 278
sample collectors 82
sampling and analysis 114
sampling box 205
sampling strategies 9,11
sampling trains 18
  for carrier gas
    constituents 136
  for moisture 141
significant digits 273
site selection 110
source monitoring
  equipment 393
source testing
  equipment 393
  procedures 9
stack gas volume
  flow rate 9,127
standard deviation 152
standard pitot tube 76
student's
  t-distribution 152
sulfuric acid production 57
systematic
  errors 146,147,172
temperature
  equivalent graph 66
  measurement 65
  scales 65
thermocouples 67
traverse points 129
umbilical cord 206
variance 152
velocity
  measurements 73,128
  of a gas stream 128

Lithoprinted by Braun-Brumfield, Inc.
Bound by Bindcrafters
Ann Arbor, Michigan